Luminescence Spectroscopy

Luminescence Spectroscopy

Edited by

Michael D. Lumb

Department of Physics, University of Manchester Institute of Science and Technology, Manchester, England

1978

Academic Press

London   New York   San Francisco

A Subsidiary of Harcourt Brace Jovanovich, Publishers

ACADEMIC PRESS INC. (LONDON) LTD
24/28 Oval Road
London NW1 7DX

United States Edition published by
ACADEMIC PRESS INC.
111 Fifth Avenue
New York, New York 10003

Library of Congress Catalog Card Number: 77-85107
ISBN: 0-12-459550-2

PRINTED IN GREAT BRITAIN BY
JOHN WRIGHT & SONS LIMITED, BRISTOL

Contributors

B. C. Cavenett, *Department of Physics, The University, Hull HU6 7RX*

Nicholas E. Geacintov, *Chemistry Department and Radiation and Solid State Laboratory, New York University, New York, New York 10003*

T. D. S. Hamilton, *The Schuster Laboratory, University of Manchester, Manchester M13 9PL, England*

G. F. Imbush, *Department of Physics, University College, Galway, Ireland*

Michael D. Lumb, *Department of Physics, University of Manchester Institute of Science and Technology, Manchester M60 1QD, England*

I. H. Munro, *Daresbury Laboratory, Daresbury, Cheshire, England*

Charles E. Swenburg, *Chemistry Department and Radiation and Solid State Laboratory, New York University, New York, New York 10003*

G. Walker, *Department of Physics, University of Manchester Institute of Science and Technology, Manchester M60 1QD, England*

For Anne, Suzy, Briony, and Nicola

Preface

"These blazes, giving more light then heat." This statement was made by Polonius to his daughter Ophelia in Hamlet (Act 1, Scene 3) and taken out of context describes the luminescence process which involves the emission of light at ambient temperatures in contrast with the light emitted by in-candescent bodies.

The phenomenon of luminescence goes back into antiquity with stories of precious gems glowing in the dark and the observation of glow-worms and the phosphorescence of the sea. The chemiluminescence of phosphorus as it oxidizes gave that element its name in the sixteenth century from the Greek word for "light bearer". The well-known mineral fluorspar gave its name in the nineteenth century to the phenomenon of fluorescence because of its photoluminescence properties. However, it was only in the middle nineteenth century that any scientific treatment of the luminescence phenomenon was founded. Becquerel developed his famous phosphoroscope and Stokes introduced his law stating that the luminescence emission always occurs at longer wavelengths than the absorption. Apart from a few curious observers the subject developed slowly until the 1930s when Jablonski contributed his ideas on energy level diagrams and meta-stable states to explain the differ-ences between the fluorescence and phosphorescence processes. In this period Vavilov and Pringsheim made their contributions to the subject. Pringsheim's book published in 1948 showed the increased activity and interest in luminescence in the 1940s which together with advances in instrumentation and purification techniques produced the surge forward in the understanding of the subject in the 1950s. In this period, not only aca-demic scientists but also industrial organizations were becoming increasingly aware of the applications of luminescence. Prior to this time, luminescent screens had been used for the detection of cathode rays, X-rays, and other ionizing radiation for academic research but the commercial use of phosphors in fluorescent tubes, television tubes, and as scintillators in the nuclear physics field was increasing. The applications in high sensitivity chemical analysis and the understanding of photochemical processes involved the applications of luminescence. The modern development of laser technology and light-emitting diodes required an understanding of the excited states of atomic or molecular systems and produced further interest in the luminescence phenomenon.

The subject of luminescence spectroscopy has in recent years been divided into the two specific areas of organic luminescence and inorganic luminescence. This monograph attempts to merge these two areas within the common ground of experimental physics.

A certain amount of necessary theory has been introduced but it is not intended that this should constitute a formal theoretical treatment as this material can be found in detail elsewhere. Also no real attempt is being made to introduce the reader to the entire field of luminescence spectroscopy which would be a task of gargantuan proportions. The aim of this monograph is to provide sufficient theoretical and experimental groundwork for experimental physicists, chemists, or biologists entering this field for the first time to grasp the essentials of luminescence spectroscopy. It is hoped that it will be a particularly useful book for first- and second-year postgraduate students.

Chapter 1 reviews the luminescence process in inorganic materials with particular reference to the field of semiconducting materials which have many industrial and commercial applications. The reader is introduced to the subject by consideration of the luminescence processes in neon, ruby and gallium arsenide and is gradually exposed to the more complex processes which give rise to luminescence in insulators and semiconductors. Chapter 2 introduces the reader to the field of organic luminescence which is associated with the molecular properties of the hydrocarbon molecules in contrast to the atomic properties of atoms or ions and their interactions with the surrounding lattice which is the realm of the inorganic materials. The luminescence properties of the organic molecule in the gas, liquid and solid phases are described in a fairly elementary manner and take the reader to a point where he should be able to consult the more specialized advanced texts on this subject.

Many of the recent advances in the luminescence field have been a product of the technological advances made in the optical and electronic field by the advent of the laser and sophisticated signal recovery techniques. Chapter 3 deals with these new advances in luminescence instrumentation and includes the use of electron–synchrotron radiation sources. However, considerable space is also devoted to the conventional equipment using classical light sources, monochromators, and detectors. An extensive bibliography is also included for the reader wishing to extend his knowledge of luminescence instrumentation.

The relatively new and interesting area of magneto-optical effects in organic systems is reviewed in Chapter 4 and the new technique of optically detected magnetic resonance (ODMR) in semiconductors is fully described in Chapter 5.

Finally, I am greatly indebted to Mrs E. Midgley for patiently typing and on occasions retyping parts of the manuscript and for assisting with the editing process. I acknowledge with gratitude the permission of many publishers, authors and colleagues from which material and ideas have been used in compiling this monograph. My thanks are conveyed to the co-authors of this work for their help and their patience and to some of the authors I extend my apologies for the times I was so rudely impatient.

Manchester, 1978   M. D. LUMB

Contents

1

Inorganic Luminescence

G. F. IMBUSCH

Department of Physics, University College,
Galway, Ireland

1.1. INTRODUCTION

In this chapter we shall be concerned with the luminescence from inorganic solid materials, that is, in the emission of electromagnetic radiation, usually in the visible or near infrared, in excess of thermal radiation from the solid. Luminescence emission involves radiative transitions between electronic energy levels of the material, and the emission is characteristic of the material. The transition originates on some excited electronic level, and after the emission of a photon a lower electronic level is occupied. In our study of luminescence we are first of all concerned with discovering the electronic levels between which the radiative emission process occurs and in understanding the nature of the radiative emission process. We are also concerned with understanding how the material can be excited, that is, how the excited electronic levels can be populated—since the intensity of the luminescence depends on the number of excited states which are populated. If the material is irradiated with electromagnetic radiation, the resultant emission is termed photoluminescence. Luminescence excited by bombardment by energetic electrons is termed cathodoluminescence. X-ray luminescence follows irradiation by X-rays. Chemiluminescence follows excitation by chemical

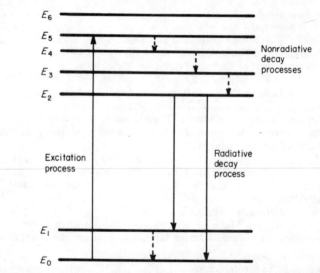

Fig. 1.1. The process of excitation and luminescence in a hypothetical material having the electronic energy level scheme seen above. At low temperatures and in the absence of any exciting mechanism level 0 alone is occupied; after excitation level 5 is occupied. Because levels 2 to 5 are close together, the excitation drops to level 2 by a non-radiative cascade process. If the gap between level 2 and the next lower level is large enough, the excitation in level 2 cannot be dissipated non-radiatively. With such a large gap the transition from level 2 to level 1 or 0 may occur radiatively by emission of a photon of electromagnetic radiation, that is, by the emission of luminescence.

reaction. Electroluminescence is luminescence excited by applying a voltage to the material.

The fundamental excitation and emission process is illustrated in Fig. 1.1 for a hypothetical material having the energy level scheme seen in the figure. The ground state has energy E_0, and E_1 to E_5 represent the energies of excited states. By one of the excitation mechanisms previously listed, the material is raised from E_0 to E_5. The energy gaps among adjacent levels between E_5 to E_2 are small, while that between E_2 and E_1 is large. Now it is found that if the gap between an excited level and the adjacent lower level is small, the material in that excited state tends to decay non-radiatively by phonon emission—releasing the energy as heat to the material. Radiative decay, that is, the transition from a higher to a lower electronic level which is accompanied by the emission of a photon, only occurs when the gap to the adjacent lower level is above a critical value. Consequently when the material of Fig. 1.1 is raised to the E_5 level, it loses energy as heat by cascading from level 5 to level 4, thence to level 3, ultimately ending up on level 2. The gap between levels 2 and 1 is above the critical value, so the material decays radiatively from level 2 emitting a photon, and ending on level 1 or 0. If the material decays radiatively to state 1, it then decays non-radiatively through the small gap to the ground state. Two possible luminescence transitions can occur with frequencies given by ν_0 and ν_1, and with angular velocities given by ω_0 and ω_1, where

$$\left.\begin{aligned} h\nu_1 &= \hbar\omega_1 = E_2 - E_1 \\ h\nu_0 &= \hbar\omega_0 = E_2 - E_0 \end{aligned}\right\} \tag{1.1}$$

(It is common nowadays to refer to ω as the "frequency" of the transition.) Clearly it is important for us to know not only the position and nature of the energy levels involved in the radiative process, but also the position and nature of the other levels which may be involved in the excitation process, and in the non-radiative process whereby the excitation arrives at the radiative level. The simplest way to investigate the energy levels of the material is by absorption spectroscopy, and in our study of luminescence we shall be concerned with the process of absorption of radiation as well as with the process of emission of radiation.

There is a wide variety of materials which exhibit luminescence, and the type of luminescence from a given material may depend on the method of preparation of the material and on its temperature. Altogether we are confronted with a bewildering assortment of luminescence phenomena. Our purpose will be to classify the various types of luminescent materials and to try to understand the distinctive features of the luminescence from each type.

This survey of inorganic luminescence will involve us in a number of difficult fields of study—the structure of solids, group theoretical classification of states, the interaction of radiation and matter, lattice vibrations, magnetism, semiconductor physics, etc. Since this is an introductory article and not a review article, references will not be included in the body of the text. At the end of the chapter, however, we shall list source material for each section, give some references, and indicate textbooks and papers which cover in more satisfactory detail the various aspects of luminescence touched upon here.

Single isolated atoms constitute the simplest electronic system capable of emitting light. We meet such systems when we study low pressure gas discharges where the atoms of the gas are raised to excited levels by the electrical discharge, after which the atoms may lose their energy by emission of photons. Because of the low pressure, the individual atoms can be considered as acting independently of each other. Hence in order to understand the luminescence from such a gas discharge, we need only consider one representative emitting atom, and then apply appropriate statistical considerations. An emitting solid is a much more complicated electronic system; it consists of a very large number of atoms (or ions), each one interacting with many of its neighbours. The energy levels of the solid are characteristic not only of the constituent atoms or ions, but also of the way in which the atoms or ions combine together in the material and, as we shall see, impurities and defects may have a strong effect on the luminescence properties.

As the discussion on the hypothetical material of Fig. 1.1 showed, a material must have a reasonably large energy gap between adjacent energy levels if it is to emit luminescence. Consequently it is found that inorganic luminescence solids are either large band gap semiconductors or insulators, since both of these are characterized by a filled electron band separated by a sufficiently large gap from an adjacent unfilled electron band. In the case of an insulator, the band gap corresponds to the energy of an ultraviolet photon, and visible luminescence is not expected from the pure material. Impurity atoms and defects, however, are always present, and these may possess electronic levels separated by a gap which corresponds to a photon of visible light. *The luminescence from insulators is almost always associated with such impurity atoms and defects.*

Many semiconductor materials possess a fairly large band gap, one which corresponds to a visible or near infrared photon. Irradiating such materials with visible light will raise electrons from the valence to the conduction band and will result in an increased electrical conductivity. These materials are said to be *photoconductive*. The decay of the electron from the conduction band to the valence band might be expected to result in the emission of a photon whose energy corresponds to the band gap. Once again, however, *defects and impurities play a very significant role and most luminescence is associated with them.*

At this point it is interesting to compare the luminescence behaviour of the following three distinct electronic systems:

(a) a single atom, e.g. Ne,

(b) a doped insulator, e.g. $Al_2O_3 : Cr^{3+}$,

(c) a semiconducting photoconductive material, e.g. GaAs,

as these exhibit quite distinct luminescence properties, and together they cover many of the luminescence features normally encountered in inorganic solids. The emission of light from isolated Ne atoms, such as we find in a low pressure Ne gas discharge, is not normally termed luminescence, but it is instructive for us to spend some time in analyzing it, as the single atom is a reasonably simple well-understood electronic system. We can then see in what way the luminescence from solids differs from the light emission from this simplest and most fundamental electronic system.

1.1.1. Spectrum of Ne Discharge

When an electrical discharge occurs in a gas of neon atoms at low pressure, some of the neon atoms are bombarded by energetic electrons and ions, and the atoms are raised to excited states from which they decay with the emission of light. The bright neon discharge is commonly used in commercial displays and is the active medium for the important He–Ne laser. A simplified energy level diagram for the Ne atom is shown in Fig. 1.2. The ground state has the electronic configuration $1s^2 2s^2 2p^6$, all the electronic shells are either fully occupied or empty, and this is a very inert state. It requires a relatively large amount of energy to remove an electron from the tightly bound 2p shell and raise it to the 3s state. As the diagram shows, there is a very large gap of 16 eV between the $2p^6$ ground state and the lowest excited state which has the $2p^5 3s$ outer electron configuration. The other excited states are within a few electron volts of the $2p^5 3s$ state.

The relationship between the energy of a photon (E) in electron volts (eV), the energy in wavenumbers (cm^{-1}), and the wavelength (λ) in microns (μ) is given by

$$\left. \begin{array}{c} \lambda\ (\mu) = \dfrac{1\cdot 24}{E\ (\text{eV})} \\[2mm] E\ (\text{cm}^{-1}) = 8060E\ (\text{eV}) \end{array} \right\} \tag{1.2}$$

and 1 micron equals 10,000 Å (ångström units). From this we see that a gap of 16 eV corresponds to an ultraviolet photon. Photons in the visible region (0·4–0·7 μ) correspond to energy gaps from 3 eV to 1·75 eV.

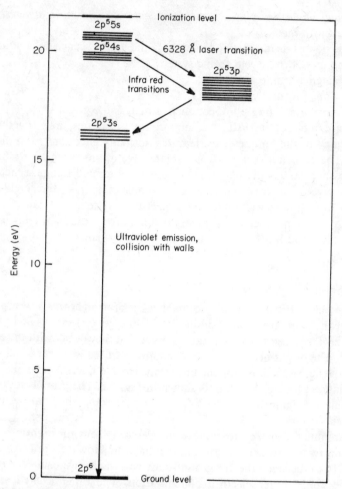

Fig. 1.2. Some of the electronic levels and transitions in the Ne atom. The fine structure is not drawn to scale. All excited levels can be reached by electron or ion bombardment, such as occurs in a low pressure gas discharge.

Electron bombardment in the discharge raises Ne atoms to excited states. Some of the optical transitions which occur in the decay of these states are shown. The yellow-red visible light is due to transitions between $2p^5 5s$ and $2p^5 3p$. The near infrared transitions occur between $2p^5 4s$ and $2p^5 3p$. Having reached the $2p^5 3s$ state, the Ne atoms decay to the ground state by emission of ultraviolet radiation or by collisions with the walls of the gas container.

Just like the energy levels of other free atoms, the electronic levels of the Ne atoms are sharp and easily distinguishable from each other. Consequently the emission spectrum consists of sharp lines. Once a free atom is raised to an excited state, it generally decays by the emission of a photon. Such photon emission is an electric dipole process with a decay time of around 10^{-8} s. In the rare cases where photon emission from an excited state is forbidden, the excited atoms can lose their excitation by collision with other atoms or with the walls of the container vessel.

It should be noted that electron (and ion) bombardment is a very convenient way to excite free atoms and to cause photon emission.

It is interesting to think of the process of excitation of the Ne atom in terms of electrons and holes. In the ground state, the 1s, 2s, and 2p shells are filled with electrons, and the shells of higher energy are empty. When the atom is excited, an electron is taken from the 2p shell and goes into one of the higher empty shells leaving a "hole" in the 2p shell. The excitation of a Ne atom can be regarded as the generation of an electron–hole pair on the Ne atom. Because the Ne atoms are essentially isolated from each other, the electron–hole pair stays in the same atom—in contrast to the case of semiconductor materials where, because of the proximity of other atoms, the electron and hole may drift apart. The emission of radiation by the Ne atom and the return of the atom to its ground state can be regarded as a recombination of the electron and hole pair, where the excess energy is emitted as a photon (or photons) of radiation.

Before leaving this brief discussion on the Ne spectrum we should allude once again to the large 16 eV gap between the ground state and the lowest excited state, and should try to get some general rules about the size of the gap between levels and to understand how this depends on the nature of the electronic states. Consider the case where all the electrons on an atom or ion are in filled shells. There is only one energy level associated with this electronic configuration; this is the ground level of the atom or ion. The next highest energy level belongs to a different configuration, one where the outer shell is no longer filled; one of its electrons has been raised to a previously empty higher shell. The energy gap between the ground state, with its filled shells, and the next highest state—which belongs to a different electronic configuration in which some of the shells are partially filled—is usually very high, much too high to be bridged by a photon of light. An electron configuration in which some of the shells are partially filled has a number of different energy levels available to it, and the gaps between these energy levels are not very large. These gaps correspond to photons of visible or infrared radiation.

The energy level diagram for Ne displays these facts very clearly. The one level available to the filled shell $1s^2 2s^2 2p^6$ configuration is the ground state, and a large gap of 16 eV occurs between it and the next level belonging to a different configuration. This next highest configuration ($1s^2 2s^2 2p^5 3s$) has partially filled shells, and consequently has a number of closely spaced energy levels associated with it. The other excited configurations have similar splittings. Further, the energy gap between these electronic configurations with partially filled shells need not be large. In the case of Ne, the gap between them corresponds to photons of visible, near ultraviolet, or near infrared radiation.

In Fig. 1.3 we compare the energy levels of Ne, Na, and Pr^{3+}. The ground state of Ne, as we saw, has a filled shell electronic configuration, hence a large gap exists between this single level and the next highest levels. The ground state of Na has the electronic configuration $1s^2 2s^2 2p^6 3s$ with a partially filled shell, and it is a spin doublet whose two energy levels coincide in the absence of a magnetic field. The next highest levels corresponding to the $1s^2 2s^2 2p^6 3p$ configuration are about 2 eV above the ground state, and this gap corresponds to a photon of yellow light (the strong absorption and luminescence transition of sodium vapour). The ground state of Pr^{3+} has an outer electronic configuration $4f^2$, and a large number of different levels exist for this unfilled shell configuration. Some are sketched in Fig. 3. The gaps between them correspond to infrared and visible photons.

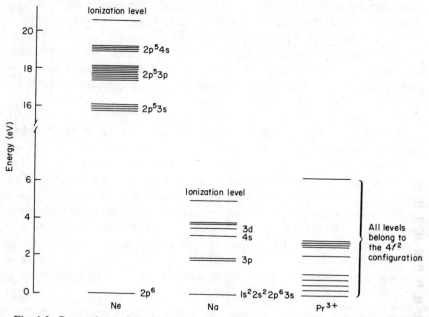

Fig. 1.3. Comparison of the low-lying electronic energy levels of single atoms of Ne and Na, and of the single ion Pr^{3+}. Na and Pr^{3+} each has a partially filled shell of electrons and as a result has a number of excited electronic levels adjacent to the ground state. Ne, which has only filled shells of electrons in the ground state, has a large energy gap to the nearest excited state.

From these observations we can state the general rule that if an atom or an ion with a filled shell electronic configuration does not have nearby electronic levels, visible light cannot be absorbed by such an atom or ion, and consequently it is optically inert. On the other hand, atoms or ions whose ground states have partially filled shells possess nearby electronic levels separated by gaps corresponding to visible or infrared photons, and are, in general, optically active.

The notion that filled shells of electrons are tightly bound is very helpful in understanding which ions or atoms are optically active. The s *sub*shell, however, is not always a tightly bound entity, and an ion with such an outer filled subshell can be optically active, e.g. Tl^+ which has the $6s^2$ outer configuration in its ground state.

1.1.2. Luminescence from Al_2O_3:Cr^{3+}

Al_2O_3:Cr^{3+} (ruby) is an example of an insulating luminescent material. Pink ruby consists of Al_2O_3 (sapphire) in which about 0.1% of the Al^{3+} ions are replaced by Cr^{3+} ions. The sapphire is called the *host material* and the Cr^{3+} ions are the *dopants*. If sapphire is doped with about 1% Cr^{3+}, a deep red ruby is obtained. The red colour of ruby is caused by the dopant ions.

Let us consider the host material on its own. Sapphire can be regarded as made up of Al^{3+} and O^{2-} ions bound electrostatically to each other. This is a very strong binding, and sapphire is one of the hardest materials known. Both Al^{3+} and O^{2-} have closed shell electronic configurations in their ground states; consequently neither can absorb visible radiation, and so sapphire is transparent. The higher electronic states of sapphire are separated from the ground state by a large gap, and the material has absorptions in the ultraviolet.

Each of the Cr^{3+} ions, on the other hand, has a partially filled shell electronic configuration, and so it has low-lying excited states. These ions can absorb photons in the visible regions of the spectrum and the absorbed energy is released as luminescence. The absorption spectrum of ruby in the infrared, visible, and ultraviolet is represented in Fig. 1.4. The absorptions in the visible are caused by transitions involving the Cr^{3+} ions. The strong band at around 7 eV is a *charge-transfer band*. In this transition one of the electrons of the tightly bound O^{2-} electronic configuration is removed from the oxygen ion and transferred to the Cr^{3+} ion. This transition can be represented by the formula

$$O^{2-}(2p^6) + Cr^{3+}(3d^3) \longrightarrow O^-(2p^5) + Cr^{2+}(3d^4)$$

and the energy is supplied by the 7 eV photon. The very strong absorption from about 9 eV upwards is caused by absorption in the host material itself.

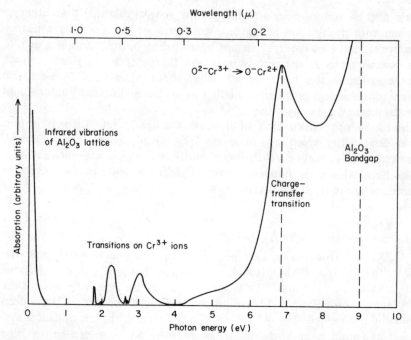

Fig. 1.4. The absorption spectrum of ruby in the infrared, visible, and ultraviolet. This trace has not been obtained with a single ruby crystal, but is a composite trace made by joining together separate infrared, visible, and ultraviolet absorption spectra.

This intrinsic absorption causes charge transfer from O^{2-} to Al^{3+} as represented by

$$O^{2-}(2p^6) + Al^{3+}(2p^6) \longrightarrow O^-(2p^5) + Al^{2+}(2p^6 3s).$$

This 9 eV energy can be regarded as the band gap of the sapphire.

Let us now turn our attention to the transitions which occur only in the Cr^{3+} ions and which appear in the visible region of the spectrum. In pink ruby the Cr^{3+} ions are sufficiently far away from each other that we can regard each Cr^{3+} ion as acting on its own, surrounded by the inert host material. So in order to understand the optical properties of ruby we need only confine our attention to one representative Cr^{3+} ion and determine how it interacts with visible light. The optical behaviour of the material in this case is not unlike the gas of non-interacting optically active Ne atoms in the first example. Here we have, as it were, a condensed gas of optically active Cr^{3+} ions isolated from each other by being imbedded in the inert host material.

It is interesting to compare the calculated energy levels of a Cr^{3+} ion with the visible absorption spectrum shown in Fig. 1.4. The calculated energy levels are plotted in Fig. 1.5, where the absorption spectrum and the luminescence spectrum (both taken at 77 K) are also shown. The labels on the levels are the quantum labels which characterize the electronic states of the ion in the solid, and they are analogous to the familiar ^{2}S, ^{4}P, labels for the electronic levels of free atoms. The labels A_2, E, T_1, T_2 refer to the symmetry properties of the orbitals of the ions in the solid. The superscripts 2 and 4 again refer to spin multiplicities; they are used to describe the spin states of the ion in the solid as well as the spin states of the free atoms.

We first of all notice that the strength of the absorption transitions from the $^4A_2$ ground state to the $^4T_2$ and $^4T_1$ excited states are much stronger than the others. This happens because these two transitions are spin-allowed while the absorption transitions to the ^{2}E, $^2T_1$, and $^2T_2$ states are spin-forbidden. So besides knowing where the excited states are positioned, one must also take into account the *selection rules* of the transitions. Next we must refer to the fact that whereas the electronic states of the Cr^{3+} ion have sharp energy levels, some of the transitions appear as broad bands. This comes about because of the interaction of the Cr^{3+} ion with the neighbouring ions of the host material. We shall have to learn how to take that interaction into account later when we attempt to put the process of luminescence (and absorption) on a firmer theoretical footing.

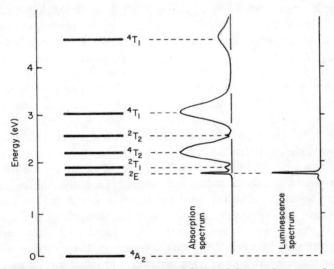

Fig. 1.5. The low-lying electronic levels of Cr^{3+} in Al_2O_3 are shown on the left. The fine structure in the levels is omitted. In the centre is the observed absorption spectrum of ruby plotted vertically for easy comparison with the Cr^{3+} energy levels. On the right is a similar plot of the luminescence spectrum. Luminescence is emitted only from the lowest excited level—as this is the only excited level with a large energy gap to the next lower level.

The two strong absorption bands of ruby fall in the yellow-green and in the blue. Consequently, under ordinary white light the yellow-green and blue are strongly absorbed and the crystal transmits only red light, with some lesser transmission in the violet and blue. This gives ruby its rich red colour. If, by the absorption of a photon of blue light, a Cr^{3+} ion is raised to, say, the $^4T_1$ excited state, it decays non-radiatively by phonon emission through the various excited states until it ends on the lowest excited ^2E state. There is a large gap of 14,000 cm^{-1} between this and the ground state, and with such a large gap a radiative decay process from ^2E to $^4A_2$ is more probable than a non-radiative process. Consequently, the material emits luminescence at a wavelength which corresponds to the energy gap between ^2E and $^4A_2$. This is seen in Fig. 1.5. Just as in the case of Ne, we can consider the excitation and de-excitation of the Cr^{3+} ion in terms of the generation and recombination of an electron–hole pair on the isolated Cr^{3+} ion.

There are a number of ways in which the process of excitation and luminescence of the Cr^{3+} ion in ruby differs from the excitation and luminescence process of the Ne atoms in the low pressure gas discharge. Excitation by electron bombardment is not so suitable for ruby, as it excites only Cr^{3+} ions near the surface. Irradiation with visible light—particularly in the yellow-green and blue—is more effective, since the exciting light can penetrate some distance into the bulk of the material. This method of excitation is often termed *optical pumping*. Next we notice that whereas the excited states of the free atom could decay only by photon emission, the excited states of the ion in the solid may decay by phonon emission (non-radiatively) as well as by photon emission (radiatively). In developing a theoretical basis for the luminescence process we shall have to understand how these two competing processes compare. This has an important bearing on the efficiency of the luminescence process. We say that the quantum efficiency is unity if, for every photon of pumping light which is absorbed, a photon of luminescence is emitted. Ruby has a quantum efficiency under optical pumping of 80–100%.

The strength of the luminescence from ruby depends on the number of Cr^{3+} ions raised to excited states, and on the probability that an ion raised to the luminescent level will decay by emission of a photon. The number of ions raised to excited states depends upon the strength of the absorption transitions. As we saw in discussing the decay processes in ruby, no matter what excited state is reached by optical pumping, the Cr^{3+} ion ultimately ends up in the ^2E excited state. Thus the strong broad absorption transitions in the yellow-green and blue serve to populate the ^2E state. The $^2E \rightarrow {}^4A_2$ radiative process is a weak spin-forbidden transition, but it does, however, have a greater probability of occurring than the non-radiative process between ^2E and $^4A_2$. Thus the ions which are in the ^2E state tend to decay radiatively. The precise strength of the radiative emission process is not as important as the fact that the radiative process is more probable than the competing non-radiative process. The intensity of the $^2E \rightarrow {^4A_2}$ luminescence in ruby, then, is a measure of the strength of the $^4A_2 \rightarrow {^4T_2}$ and $^4A_2 \rightarrow {^4T_1}$ absorption transitions.

1.1.3. Luminescence from GaAs

Gallium arsenide is an example of a large gap ($E_g = 1·52$ eV) semiconductor material which exhibits photoconductivity and emits luminescence in the near infrared. In contrast to ruby, where the main host material (sapphire) is unaffected by visible light and where the absorption and luminescence are due to small amounts of substitutional atoms, in GaAs the material itself is sensitive to light. Some of the electronic energy levels in GaAs are sketched in a very simple fashion in Fig. 1.6. First of all we note that we have bands of electronic levels rather than the discrete electronic levels seen in the other two examples. We shall see that the energy within the band is related to the mobility of the electrons which do not stay localized on a specific atom or ion in this material. At the lowest temperatures the valence band is filled, the conduction band is empty, and no conductivity occurs. If light whose photons have energy greater than the band gap falls on the material, the light may be absorbed, and for each photon absorbed

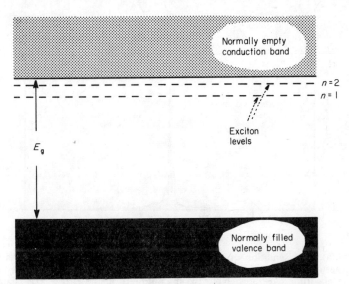

Fig. 1.6. Simplified energy level diagram of GaAs. The electronic levels occur in bands separated by an energy gap of size E_g. The broken lines indicate exciton levels, which are discussed at a later point in the chapter.

an electron is raised from the valence to the conduction band, leaving a hole behind. The hole and the electron may drift away from each other, and both may contribute to conductivity. This is termed photoconductivity. As long as the energy of the photon is greater than the band gap, the photon may be absorbed, so the absorption spectrum should be in the form of a broad band from the band gap (E_g) to higher energies. The absorption spectrum of very pure GaAs at low temperatures has the shape seen in Fig. 1.7, generally in agreement with our expectations. The fine structure in the absorption just below the gap is due to *excitons*—which we shall have to consider in more detail later.

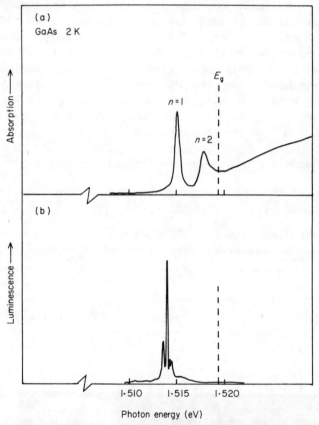

Fig. 1.7.(a) Representation of the absorption spectrum of very pure GaAs at low temperatures in the vicinity of the band gap—drawn from the published data to show the strong absorption features. Note the discontinuity in the energy axis (data from Sell, *Phys. Rev.* **6B**, 3750 (1972)). (b) Photoluminescence spectrum of very pure GaAs at low temperatures. All the luminescence occurs near the band gap. The spectrum is drawn from the data of Sell (*Phys. Rev.* **7B**, 4568 (1972)).

When an electron in the conduction band recombines with a hole in the valence band, a photon of light may be emitted. Usually the conduction electrons occupy the lowest states of the conduction band and the holes occur in the highest states of the valence band. So the recombination of electrons and holes should result in radiation at an energy just above the energy gap. The actual luminescence observed in very pure GaAs is shown in Fig. 1.7 where it is seen to occur at an energy lower than E_g. Some of the luminescence appears at the same energy as the excitons seen in absorption, but the major luminescence features occur at slightly lower energies. This is due to the recombination of an electron–hole pair at an impurity centre in the GaAs crystals. No matter how pure the GaAs material, there are always trace amounts of impurities and defects present. The reasons why even these trace amounts of defects dominate the luminescence has to do with the high mobility of the electrons and holes. The conduction electrons and the holes can combine best at the defects. Because of their high mobility, these electrons and holes move individually about the material until they meet at a defect, where recombination takes place. Consequently the luminescence in "pure" GaAs occurs at defects, where the energy of recombination is slightly below the energy gap.

The behaviour of GaAs is typical of most large band gap semiconductors. The luminescence of "pure" materials is complicated by the fact that the electrons and holes move easily through the material, tending to seek out the unintentional trace amounts of impurities, and the recombination radiation is characteristic of the band gap and of the nature of the impurities. Doping the materials purposely with other substitutional atoms can perturb the band shapes and alter the recombination process. Sometimes doping with optically active atoms can lead to luminescence more characteristic of the dopant than of the semiconductor material itself. There is a wide variety of luminescence spectra which the same semiconductor can exhibit, depending on the defects and substitutional centres in the material. We shall attempt to categorize these later.

If a large gap semiconductor, such as GaAs, is grown with substitutional donor atoms, electrons are placed in the conduction band, and the conductivity is improved. We now have n-type GaAs. These conduction electrons, however, cannot drop to the valence band and emit light, since the valence band is essentially filled—the luminescence process is the coming together of an electron and a hole. Some additional excitation process must generate holes for luminescence to occur. Similarly, p-type GaAs with holes in the valence band will not automatically lead to luminescence; we also need electrons in the conduction band. But if a p–n junction is made of GaAs, and if a sufficiently large forward bias is applied across the junction, electrons in the conduction band of the n-type may be injected into the conduction band of the p-type, and these conduction electrons may re-combine with the holes and emit luminescence. The luminescence occurs at the p–n junction only. This process, which is illustrated in Fig. 1.8, is called *injection luminescence* and is the basic process occurring in light-emitting diodes (LEDs) and in injection lasers.

The coupling of optical and semiconducting processes in the same materials is termed *opto-electronics*. It is clear that opto-electronic devices are going to play an increasingly important role in many aspects of the technology of communications.

The two materials so far looked at, namely ruby and gallium arsenide, have quite different luminescence features, and each is representative of a large class of luminescence materials. We can effectively assign any inorganic luminescent solid into one or other of these two classes. Ruby luminescence is characteristic of insulators doped with optically active dopants or defects,

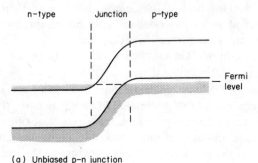

(a) Unbiased p–n junction

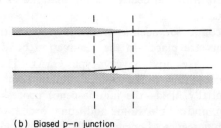

(b) Biased p–n junction

Fig. 1.8.(a) Filled electron levels (dark shading) in an unbiased p–n junction. (b) When the junction is forward biased the recombination of excited electrons and holes may lead to the emission of radiation. The radiation is at an energy corresponding approximately to the band gap energy, but the actual process of recombination may be quite complicated.

while gallium arsenide luminescence is characteristic of large band gap semiconductors. Both materials have band gaps. Insulators have large band gaps, of an energy corresponding to an ultraviolet photon, and in equilibrium there are essentially no electrons in states above the gap. In semiconductors the band gap must be smaller to allow for a non-negligible population of electrons in the states above the gap. In addition, the material must have the property that when an electron–hole pair is formed, the electron and the hole can drift apart, and each can contribute to the conductivity. For the semiconductor to be able to emit visible or near infrared luminescence, the band gap must correspond to a photon of visible or near infrared radiation.

Because they differ so much in their luminescence properties we shall consider each of these two classes separately, and we shall attempt to develop for each class an appropriate understanding of its luminescence behaviour.

1.2. INORGANIC INSULATING MATERIALS

1.2.1. Host Materials

Insulators which act as hosts for optically active dopant ions or defects include:

(i) alkali halides, e.g. NaCl,
(ii) alkali earth halides, e.g. CaF_2,
(iii) some II–VI compounds, such as ZnS,
(iv) oxides, e.g. Al_2O_3, MgO, $Y_3Al_5O_{12}$ ("YAG"),
(v) tungstates, silicates, and molybdates, e.g. $CaWO_4$,
(vi) glass,
(vii) diamond.

These range from the mainly ionically bonded crystals, e.g. NaCl, to covalently bonded crystals, e.g. diamond, and all are characterized by a large energy gap between the highest filled electron band and the lowest empty electron band. This band gap corresponds to a photon of ultraviolet radiation. These materials appear as transparent crystals, since they are unaffected by visible radiation. We can regard them as optically inert.

Let us consider why these materials are optically inert. Consider NaCl. The neutral atom of sodium has eleven electrons in the ground state electronic configuration $1s^2 2s^2 2p^6 3s^1$. Ten of these are in closed shells, but the eleventh is not in a closed shell. We call it an unpaired electron. This outer electron is loosely bound to the rest of the atom, and it can be raised to the next highest 3p state by the absorption of a photon of yellow light. So the sodium atom on its own is optically active since it can be affected by visible light. Similarly Cl atoms are optically active. But when Na and Cl atoms come together to form NaCl, they combine as Na^+ and Cl^- ions.

The electronic configurations of these Na^+ and Cl^- ions are, respectively, $1s^2 2s^2 2p^6$ and $1s^2 2s^2 2p^6 3s^2 3p^6$; they both contain only filled shells of electrons. Now electrons in filled shells, as we saw, are very tightly bound and cannot be raised to higher states by the absorption of visible light. The Na^+ and Cl^- ions, then, are optically inert, and since the material NaCl is formed mainly by the electrostatic coupling between these optically inert ions, NaCl itself is optically inert. We can generalize this result and say that although most neutral atoms are optically active, in the formation of ionic inorganic insulators atoms come together as ions all of whose electrons are in closed shells. They are, then, optically inert. We must point out that we exclude transition metal atoms from this general rule, as well as post-transition metal atom, rare earth atoms, and actinides. It will be clear later why these atoms are not expected to obey this general rule. The situation with covalently bonded crystals is more complex. Here, however, the electrons go into filled bands characteristic more of the compound than of the isolated atoms. If the covalently bonded crystal is an electrical insulator, there is a large gap to the next higher vacant band which cannot be bridged by a photon of visible light, and so the material is optically inert and may make an excellent host material for the optically active dopant atoms or ions.

1.2.2. Optically Active Dopant Ions and Defects

Let us start by considering the inclusion of optically active transition metal ions in oxide host materials. The pure oxide materials are optically inert; they form hard colourless transparent crystals. When they are doped with trace amounts of transition metal ions, they become brightly coloured. For example, when about 1% of the Al^{3+} ions in Al_2O_3 is replaced by Ti^{3+} ions, the material is a deep blue colour. The substitution of similar amounts of Cr^{3+} for Al^{3+} results, as we saw, in a deep red colour, and the material is called ruby. Ni^{3+} substitution colours the material green. The colour in each case comes about because of the transition metal ions which are optically active.

The optical activity of the transition metal ions can be understood by considering their ground state electronic configurations which are listed in Table I. Since the 3d shell requires 10 electrons to fill it, we see that each of these ions has a partially filled 3d shell. As a result they possess electronic levels close to the ground state, and the ions are optically active. The low-lying energy levels of Cr^{3+} in ruby were shown in Fig. 1.5, and the absorption and luminescence transitions were discussed. Some of the low-lying energy levels of Mn^{2+} are seen in Fig. 1.9. The Mn^{2+} free ion levels are split up when the ion is placed in a host crystal, and visible as well as ultraviolet absorptions occur on the dopant Mn^{2+} ion. When the Mn^{2+} ion is raised to one of the higher excited states, it decays non-radiatively to the adjacent lower excited

TABLE I

Electronic Configuration of the Transition Metal Ions

Ti^{3+}	$1s^2\,2s^2\,2p^6\,3s^2\,3p^6\,3d$
V^{3+}	$1s^2\,2s^2\,2p^6\,3s^2\,3p^6\,3d^2$
$V^{2+},\,Cr^{3+}$	$1s^2\,2s^2\,2p^6\,3s^2\,3p^6\,3d^3$
Mn^{3+}	$1s^2\,2s^2\,2p^6\,3s^2\,3p^6\,3d^4$
$Mn^{2+},\,Fe^{3+}$	$1s^2\,2s^2\,2p^6\,3s^2\,3p^6\,3d^5$
$Fe^{2+},\,Co^{3+}$	$1s^2\,2s^2\,2p^6\,3s^2\,3p^6\,3d^6$
Co^{2+}	$1s^2\,2s^2\,2p^6\,3s^2\,3p^6\,3d^7$
Ni^{2+}	$1s^2\,2s^2\,2p^6\,3s^2\,3p^6\,3d^8$
Cu^{2+}	$1s^2\,2s^2\,2p^6\,3s^2\,3p^6\,3d^9$

states until it reaches the lowest excited $^4T_1$ state. The 2·2 eV gap between this and the ground state is too large to be bridged by non-radiative decay, so the $^4T_1$ state decays radiatively by the emission of a yellow-orange photon. Manganese is a very important active element in commercial phosphors.

Let us now consider the inclusion of optically active rare earth ions in host materials, specifically in $LaCl_3$. This is a mainly ionic insulator containing La^{3+} and Cl^- ions, both of which have filled shell electronic configurations, so the $LaCl_3$ crystal is optically inert. The rare earth ions can enter substitutionally for La^{3+} as triply charged ions, and they have the general electronic configuration

$$1s^2\,2s^2\,2p^6\,3s^2\,3p^6\,3d^{10}\,4s^2\,4p^6\,4d^{10}\,5s^2\,5p^6\,4f^n$$

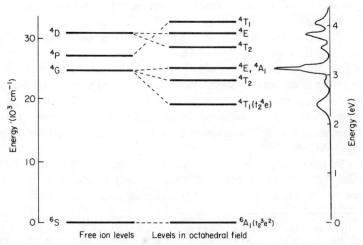

Fig. 1.9. Energy levels of the free $Mn^{2+}(3d^5)$ ion, and of Mn^{2+} in a typical host material where the manganese ion is in a site of octahedral symmetry. The absorption spectrum of Mn^{2+} ions in a solid host material is shown on the right.

TABLE II
Number of Electrons (n) in the 4f Shell of Triply Charged Rare Earth Ions

Ce^{3+}	1
Pr^{3+}	2
Nd^{3+}	3
Pm^{3+}	4
Sm^{3+}	5
Eu^{3+}	6
Gd^{3+}	7
Tb^{3+}	8
Dy^{3+}	9
Ho^{3+}	10
Er^{3+}	11
Tm^{3+}	12
Yb^{3+}	13

where n varies from 1 to 13. Since it takes 14 electrons to fill the 4f shell, these ions contain a partially filled shell; consequently they have low-lying excited states and are optically active.

In Table II the number of electrons in the unfilled 4f shell is listed for each triply charged rare earth ion.

The observed energy levels of the trivalent rare earth ions in $LaCl_3$ are shown in Fig. 1.10. A pendant semicircle attached to a level indicates that luminescence occurs from this level of the ion when it is included in $LaCl_3$. A very wide range of luminescence transitions in the visible and near infrared is made available to us by these trivalent rare earth dopant ions.

Divalent rare earth ions can also be incorporated in a number of host materials, and they are also optically active.

Sometimes two different rare earth ions are introduced as substitutional dopants in the same host crystal. A rare earth ion and a transition metal ion may both be incorporated in the same host. For example, $Y_3Al_5O_{12}$ (yttrium aluminium garnet, known colloquially as YAG) can have some rare earth ions substituting for Y^{3+} and some transition metal ions substituting for Al^{3+}, e.g. YAG:Nd,Cr.

So far we have been discussing dopant ions where the electrons which are responsible for the optical properties are strongly localized on the dopant ions and where the effect of the environment is mainly to introduce an

---

Fig. 1.10. Energy levels of the trivalent rare earth ions in $LaCl_3$. The width of a level represents the total crystal field splitting in $LaCl_3$. A pendant semicircle indicates that luminescence occurs from this level in $LaCl_3$. (This figure summarizes work done by the late Professor Dieke's group at Johns Hopkins University and is reproduced by courtesy of Professor H. M. Crosswhite.)

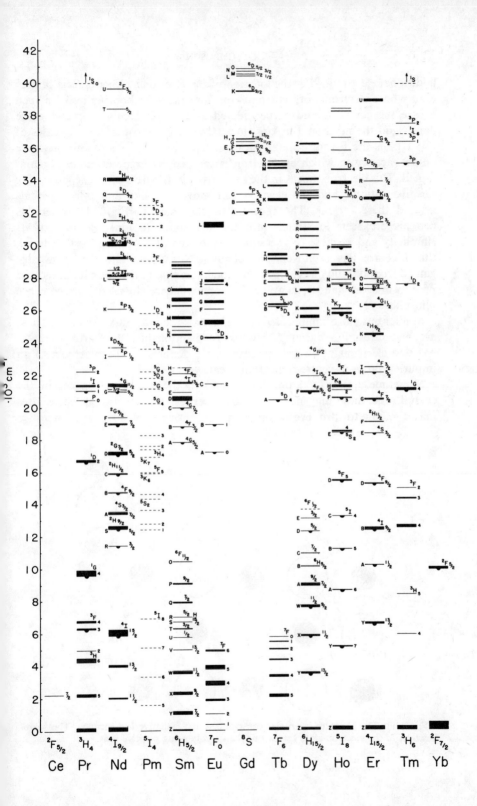

additional electric field at the location of the electrons. Other defects centres exist whose electronic orbits stretch out into the neighbouring ions. In such a case the centre should be considered as a larger complex involving the defect and the neighbouring host ions. One of the most interesting cases of this type is the F centre found in alkali halides. This consists of a negative anion vacancy in which an electron is trapped. Charge neutrality is maintained. This electron moves in the attractive electric field of the neighbouring positive cations, and the electron can move in s-like and p-like orbitals around these cations. The F-centre is illustrated in Fig. 1.11, where its orbitals are seen to stretch into the neighbouring ions. Because of its simplicity and because of the simplicity of the alkali halide host material, this F-centre has been very thoroughly investigated, and the knowledge gained from its study has helped our understanding of the basic luminescence process in alkali halides, as well as in more commercially useful luminescence phosphors.

Similar but more complicated centres are also found in alkali halides. These are F'-centres, F_A-centres, V-centres, to mention just three. An F'-centre has two electrons at an anion vacancy, F_A denotes an F-centre next to an impurity ion. A V-centre is a cation vacancy.

Sometimes impurity atoms are found in the interstitial spaces in the crystal, and these may have optical properties. Finally, two defect centres may combine to form even more complex centres.

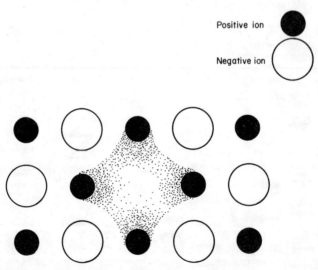

Positive ion

Negative ion

Fig. 1.11. The F-centre is an electron trapped at a negative ion vacancy. The wavefunction of the electron state spreads to the neighbouring positive ions.

Since the transition metal ions and the rare earth ions constitute the most common optically active centres found in luminescent insulating materials, we shall concentrate on these two groups of ions from now on. We will, however, occasionally mention some other centres.

1.2.3. Interaction of Radiation with Optically Active Centres

So far we have considered the types of optically active centres which are found in insulators. In this section we will set down the basic formulae governing the optical transitions which occur at these centres.

Electromagnetic radiation can interact with the centre through the electric field of the radiation (an electric dipole process) or through the magnetic field of the radiation (a magnetic dipole process). Because of this interaction the centre may change from an initial state i to a final state f with the absorption or emission of a photon of light, where the energy difference between states i and f equals the energy of the photon.

The probability of a transition from state i to state f, in which a photon of energy $\hbar\omega$ is absorbed, is given by

$$P_{if} = (2\pi/\hbar)\,|V_{fi}|^2\,\delta(E_f - E_i - \hbar\omega) \qquad (1.3)$$

where V_{fi} is the matrix element, $\langle f|V|i \rangle$, of the transition, and V is the operator denoting the energy of interaction of the centre with the radiation. When the transition is an electric dipole process, the interaction term is $V = \mathbf{p} . \mathbf{E}$, where $\mathbf{p}$ is the electric dipole moment operator and $\mathbf{E}$ is the electric field intensity of the radiation field. The electric dipole moment is given by $\mathbf{p} = \sum_i e\mathbf{r}_i$, where the summation is over all the electrons of the centre. In evaluating the matrix element of $\mathbf{p} = \sum_i e\mathbf{r}_i$, we consider that only one electron changes state in the optical transition, so that we must only evaluate the matrix element for a single electron, $e\langle f|\mathbf{r}|i \rangle$. In the case where the matrix element of $\mathbf{p} . \mathbf{E}$ between states i and f is zero, the electric dipole process is forbidden. The transition may still proceed by a magnetic dipole process because of the interaction of the centre with the magnetic component of the radiation field. In this case the interaction term is $V = \mu . \mathbf{B}$, where μ is the magnetic dipole operator and $\mathbf{B}$ is the strength of the magnetic field of the radiation. The magnetic dipole process is a weaker process, since the maximum possible values of the matrix element of $\mu . \mathbf{B}$ are much less than the maximum possible values of the matrix element of $\mathbf{p} . \mathbf{E}$. Nevertheless, when a radiative transition is forbidden by an electric dipole process, it may occur by the weaker magnetic dipole process.

An examination of the matrix element of $\mathbf{r}$ and μ between states i and f will tell us whether or not an optical transition can occur, and so the selection rules can be obtained. Matrix elements of $\mathbf{r}$ and of μ between states of different total spin (S) are zero. This gives us the *spin selection rule*, which says that $\Delta S = 0$ for an optical transition. This rule only holds rigorously if total spin is a good quantum number. Taking spin–orbit coupling into account relaxes this selection rule somewhat, but transitions between different total spin states are, in general, orders of magnitude weaker than transitions between states of the same total spin.

The operator, $\mathbf{r}$, for electric dipole processes has odd parity. Consequently, if the wavefunctions for states i and f have the same parity, e.g. both have even parity or both have odd parity, then $\langle f|\mathbf{r}|i\rangle = 0$. This is the basis for the rigorous *parity selection rule* which says that for an allowed electric dipole transition, the initial and final states must have opposite parity. The magnetic dipole operator, $\boldsymbol{\mu}$, has even parity, so we also have a selection rule which says that for an allowed magnetic dipole transition, the initial and final states must have the same parity.

From the $\mathbf{E}$ and $\mathbf{B}$ terms which occur in V_{fi} it is clear that the probability of the transition increases linearly with the intensity of the light. In the case where we are dealing with an electric dipole process in which light with its $\mathbf{E}$ vector along, say, the $\hat{x}$ direction is involved, the appropriate operator is $\mathbf{p}.\mathbf{E} = p_x E$. Hence the matrix element of the $\hat{x}$ component of $\mathbf{p}$ must be evaluated between states i and f, that is, $e\langle f|x|i\rangle$. Because of the nature of the electronic states, the matrix element $\langle f|x|i\rangle$ may be zero while $\langle f|y|i\rangle$ may be non-zero. This means that $\hat{x}$ polarized light will not be emitted or absorbed in an optical transition between states i and f, while $\hat{y}$ polarized light will be emitted or absorbed. These selection rules are all contained in the matrix elements. One can calculate the intensity ratios among the different polarized transitions from these matrix elements. The absolute value of the emission or absorption strength is more difficult to estimate, as it involves full evaluation of the complete P_{ij} formula and its summation over all allowed modes of radiation.

A quantity of interest in luminescence is the probability per second that an atom or ion in an excited state i will spontaneously decay by emission of a photon to a lower state f. This is the *Einstein spontaneous transition probability*, A_{fi}. A calculation of A_{fi} can be carried out by perturbation theory as outlined above, a second quantized description of the radiation field is usually employed, and a summation over all photon modes and polarizations must be made. We will assume that the i and f states have degeneracies g_i and g_f, respectively. In the case where we are dealing with an electric dipole process we obtain

$$A_{fi} = \frac{1}{g_i} \sum_{i,f} \frac{1}{4\pi\varepsilon_0} \frac{4e^2}{3\hbar c^3} \omega^3 \left[\left(\frac{E_{\mathrm{eff}}}{E_0} \right)^2 n \right] |\langle f|\mathbf{r}|i\rangle|^2 \qquad (1.4)$$

In this calculation we have summed over final states and averaged over initial states. The velocity of light in vacuum (c) enters through the density of photon modes. n is the refractive index of the material in which the centre is imbedded. E_{eff}/E_0 is the "local field correction" and makes allowance for the fact that the electric field at the site of the centre may be different from the average field in the medium. For ions one can use the formula $E_{\text{eff}}/E_0 = (n^2+2)/3$ for this correction factor.

The Einstein spontaneous transition probability from states i to f, A_{fi}, is related to an experimentally measured quantity, namely, the radiative *decay time* from the state i. If there are N_0 centres in an excited state i at time $t = 0$, and if the centres can decay by spontaneous emission of a photon and end on state f, then the number of excited centres will decrease as time increases, as described by the formula

$$N(t) = N_0 \exp(-t/\tau) \tag{1.5}$$

where τ is the lifetime of the transition. Since the intensity of luminescence from the state i varies as the number of centres in the state i, the intensity decreases as

$$I(t) = I_0 \exp(-t/\tau) \tag{1.6}$$

A_{fi} and τ are related by the formula

$$1/\tau = A_{fi} \tag{1.7}$$

In the case where there are a number of final states, f, we have

$$1/\tau = \sum_f A_{fi} \tag{1.8}$$

Another quantity of interest is the probability that a centre in the ground state i will absorb a photon of light from a beam of radiation and be raised to an excited state f. For an electric dipole process this probability will depend on $|\langle f|x|i\rangle|^2$, where $\hat{x}$ is the direction of the polarization of the beam of light, and will depend on the intensity of the light beam and on its velocity. Before giving the formula for this quantity we shall relate it to the *absorption coefficient* of the material, which is an experimentally measured quantity.

The intensity of a beam of light of frequency ω is reduced on travelling through a material. The absorption coefficient, $k(\omega)$, is defined by

$$I_\omega(d) = I_\omega(0) \exp[-k(\omega)d] \tag{1.9}$$

where $I_\omega(d)$ is the intensity of the light of frequency ω after traversing a thickness d of material as shown in Fig. 1.12. $k(\omega)$ is the absorption coefficient. If the radiation which is being absorbed is polarized in the $\hat{x}$ direction, we label the absorption coefficient by $k^x(\omega)$. $\int k^x(\omega)\,d\omega$ is a measure of the fractional amount of radiation absorbed by the material of unit thickness

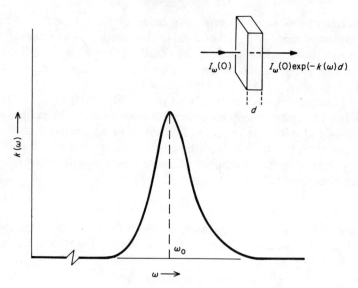

Fig. 1.12. Definition of absorption coefficient, and typical shape of $k(\omega)$ function.

when the incident radiation is polarized in the $\hat{x}$ direction. The absorption per centre is given by

$$\int \sigma^x(\omega)\, d\omega = \frac{1}{N} \int k^x(\omega)\, d\omega \qquad (1.10)$$

where N is the number of centres per unit volume. This quantity may be calculated by perturbation theory, and if the process is an electric dipole transition one gets

$$\int \sigma^x(\omega)\, d\omega = \frac{1}{4\pi\varepsilon_0} \frac{1}{g_i} \sum_{i,f} \frac{4\pi^2 e^2}{\hbar c}\, \omega_0 \left[\left(\frac{E_{\mathrm{eff}}}{E_0}\right)^2 \frac{1}{n} \right] |\langle f|x|i \rangle|^2 \qquad (1.11)$$

where ω_0 is the frequency at the peak of the absorption (Fig. 1.12).

Long before the quantum mechanical analysis of the interaction of matter and radiation was developed, there was a classical theory due to Lorentz and others. One of the parameters of that theory was the *oscillator strength*, which was the number of electric dipole oscillators on the atom (or ion) which could be stimulated by the radiation field. This oscillator strength is given the symbol f. It is a dimensionless quantity, and for strongly allowed optical transitions it has a value close to unity. This is a useful concept of convenient size, and it is still retained in the quantum mechanical treatment. It is, however, redefined in quantum mechanical terms. There is some ambiguity in the formula for oscillator strength in the case of a centre in a material—sometimes refractive index terms are included. We shall define it in a manner appropriate for free atoms, without any refractive index terms. The oscillator strength of a transition between initial and final states i and f for an electric dipole transition with $\hat{x}$ polarized light is

$$f_{if}{}^{x}(\text{ed}) = \frac{1}{g_i} \sum_{i,f} \frac{2m}{\hbar} \omega_0 |\langle f|\hat{x}|i\rangle|^2 \qquad (1.12)$$

and a total oscillator strength can be defined by

$$f_{if}(\text{ed}) = \tfrac{1}{3}[f_{if}{}^{x}(\text{ed}) + f_{if}{}^{y}(\text{ed}) + f_{if}{}^{z}(\text{ed})] = \frac{1}{g_i} \sum_{if} \frac{2m}{3\hbar} \omega_0 |\langle f|\mathbf{r}|i\rangle|^2 \quad (1.13)$$

For an allowed electric dipole transition, where we might expect the matrix element of x to be about 0·5 Å, the value of f is of the order of unity. Using these definitions we can relate the measured absorption to the oscillator strength. We obtain

$$\int \sigma^x(\omega)\,d\omega = \frac{1}{4\pi\varepsilon_0} \frac{2\pi^2 e^2}{mc} \left[\left(\frac{E_{\text{eff}}}{E_0}\right)^2 \frac{1}{n}\right] f_{if}{}^{x}(\text{ed}) \qquad (1.14)$$

We can also relate the Einstein spontaneous transition probability, A_{fi}, which equals $1/\tau$, to the total oscillator strength, and we get

$$\frac{1}{\tau} = \frac{1}{4\pi\varepsilon_0} \frac{2\omega^2 e^2}{mc^3} \left[\left(\frac{E_{\text{eff}}}{E_0}\right)^2 n\right] f_{if}(\text{ed}) \qquad (1.15)$$

By inserting numerical values for the physical constants, and by expressing the local correction in terms of the refractive index, this equation can be written (SI units)

$$f(\text{ed})\,\tau = 1\cdot5 \times 10^4 \frac{\lambda_0^2}{[\tfrac{1}{3}(n^2+2)]^2\, n} \qquad (1.16)$$

where λ_0 is the wavelength in vacuum and n is the refractive index. For $\lambda_0 = 5000$ Å and $n = 1$, the quantity on the right has a value of around 4×10^{-9} s. This means that for an allowed electric dipole transition ($f = 1$) on a free atom or ion, the decay time is $\tau = 4 \times 10^{-9}$ s. Lifetimes of around 10^{-8} s are experimentally observed for radiative decay times in gases. For an ion in a solid we might typically have $n = 1\cdot7$, giving a lifetime of around 10^{-9} s. The observed decay times of dopant ions in solids are usually much longer—in the range 10^{-4} s to 1 s. Oscillator strengths of dopant ions in solids are much less than unity. For example, the broad band absorption transitions on the Cr^{3+} ion in ruby have oscillator strengths of around 10^{-4}, much smaller than absorption transitions in free atoms. But one must bear in mind that in a 0·1% ruby there are about 10^{21} Cr^{3+} ions per cm³, in contrast to about 10^{15} atoms per cm³ in a low pressure gas discharge. So the ions make up in number for their weak absorption oscillator strengths, and efficient absorption of yellow-green and blue light occurs in ruby. Once the ions are excited, they decay non-radiatively to the ^2E level. The ^2E state decays by spontaneous photon emission to the ground state—giving rise to the ruby luminescence transition. With an oscillator strength of around 4×10^{-7} this luminescence transition has a small transition probability, but it is still greater than the transition probability of a non-radiative process between ^2E and the ground state. It is important to realize that all the optical transitions in dopant transition metal ions have oscillator strengths which are orders of magnitude smaller than the radiative transitions normally observed in free atoms. The optical transitions in dopant trivalent rare earth ions are similarly very small. With such low oscillator strengths one must consider the possibility that magnetic dipole processes are effective in optical transitions in dopant ions. One can develop similar formulae for oscillator strength, absorption strength, and radiative decay time for the case where the transition is due to the $\mu.B$ term in the interaction between the ion and the radiation field. We first of all define a magnetic dipole oscillator strength, $f(\text{md})$, which is a dimensionless number:

$$f_{ij}^x(\text{md}) = \frac{1}{g_i} \sum_{i,f} \frac{2m}{\hbar} \frac{\omega}{e^2 c^2} |\langle f|\mu_x|i\rangle|^2 \tag{1.17}$$

and

$$f_{ij}(\text{md}) = \frac{1}{g_i} \sum_{i,f} \frac{2m}{3\hbar} \frac{\omega}{e^2 c^2} |\langle f|\mu|i\rangle|^2 \tag{1.18}$$

In comparing $f(\text{md})$ with $f(\text{ed})$ we notice that they are identical if the r matrix element of the electric dipole formula is replaced by the matrix element of μ/ec. For an allowed magnetic dipole transition such that the matrix element of μ is about one Bohr magneton, we find $f(\text{md}) \simeq 10^{-6}$. This is six orders of magnitude less than the electric dipole value, yet it is typical of oscillator strengths of dopant ions in solids.

Next we relate absorption strength and decay time to $f(\text{md})$. The relevant formulae are:

$$\int \sigma^x(\omega)\, d\omega = \frac{1}{4\pi\varepsilon_0} \frac{2\pi^2 e^2}{mc} n f_{ij}^x(\text{md})$$

$$= \frac{\mu_0}{4\pi} \frac{2\pi^2 e^2 c}{m} n f_{ij}^x(\text{md}) \tag{1.19}$$

and

$$\frac{1}{\tau} = \frac{1}{4\pi\varepsilon_0} \frac{2\omega^2 e^2}{mc^3} n^3 f_{ij}(\text{md})$$

$$= \frac{\mu_0}{4\pi} \frac{2\omega^2 e^2}{mc} n^3 f_{ij}(\text{md}) \tag{1.20}$$

Finally, we can arrange the last equation to read (SI units)

$$f(\text{md})\,\tau = 1\cdot5 \times 10^4(\lambda_0{}^2/n^3) \tag{1.21}$$

when n is the refractive index of the material. There is no local field correction here, as we assume that we are dealing with non-magnetic materials. It is noticeable, also, that the refractive index enters to different powers in the electric dipole and magnetic dipole formulae. This comes about because the ratio of **B** to **E** of the radiation field depends on the dielectric constant (which is n^2). Also it should be remembered that the $\hat{x}$ polarization referred to in the $f^x(\text{md})$ and σ^x-formulae is the direction of the magnetic field of the radiation, and is at right angles to what is normally termed the polarization of the light.

With their weak oscillator strengths, optical processes on dopant ions in solids are usually either very weakly allowed electric dipole processes or magnetic dipole processes. To see why this is so we must take into account the nature of the low-lying states of the dopant transition metal ions and rare earth ions in solids.

1.2.4. Energy Levels of Dopant Transition Metal Ions

The transition metal ions enter substitutionally for positively charged ions in a variety of hosts. Oxide crystals are particularly suitable as hosts. In the oxides the transition metal ions are in sites which are surrounded by negatively charged oxygen ions. One common arrangement has six oxygen ions which are approximately equidistant from the transition metal ion in the $\pm x$, $\pm y$, and $\pm z$ directions. This situation is shown in Fig. 1.13. When the distances to the six oxygen ions are exactly the same, the transition metal ion is in a site of perfect octahedral symmetry. In solving for the energy levels and wavefunctions of the transition metal ion in this site, one must take into account the effect of the electrostatic *crystal field* of the neighbouring ions on the 3d electrons.

In many oxide hosts the arrangement of the six oxygen ions is distorted from the perfect octahedral symmetry. If the distortion is in the form of an elongation or compression along the [111] direction (Fig. 1.13), the resultant crystal field has trigonal symmetry; a distortion along the [100] direction results in a tetragonal symmetry. More complicated distortions lead to other symmetries. In most cases, however, the distortion from perfect octahedral symmetry is small, and the main part of the electrostatic crystal field has octahedral symmetry. The energy levels of all the transition metal ions in fields of octahedral symmetry have been calculated, and, to a first approximation, the predicted energy levels agree with the observed spectra. The

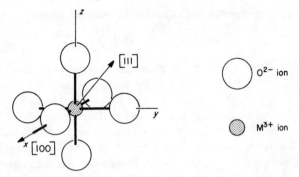

Fig. 1.13. A situation often encountered in oxide crystals—the central cation (M) is surrounded by six oxygen ions. If all the oxygen ions are equidistant from the cation and along orthogonal axes, the environment of the cation has octahedral symmetry. In most cases the oxygen ions are slightly displaced from a situation of perfect octahedral symmetry.

departures from perfect octahedral symmetry cause a small shift and splitting of these energy levels. We will now discuss the task of estimating the energy levels of the ion in an octahedral field.

Let us start by considering the isolated free ion, where there is no crystal field effect. For the transition metal ions all the electron shells up to the 3d shell are filled; the 3d shell is partially filled. The inner filled electron shells are unaffected by the optical transitions, so we disregard them. The outer electrons have the configuration $(3d)^n$. The wavefunctions for these 3d electrons are found by first of all neglecting the interaction between the 3d electrons, then solving for each 3d electron state in the central field of the nucleus plus the inner filled shells of electrons. Next the interactions among the 3d electrons are considered, and the many-electron states are labelled by their values of total spin (S) and total orbital angular momentum (L)— as ^{2S+1}L. The separation in energy of the various ^{2S+1}L states is expressed in terms of Racah parameters (A, B, C), which describe the strength of the electrostatic interaction between the electrons. Finally, the weak spin–orbit coupling term splits the ^{2S+1}L levels into sublevels, with different values of total angular momentum (J) and the J value is given as a subscript—$^{2S+1}L_J$, e.g. $^4P_{\frac{5}{2}}$. We can label $\mathscr{H}_{FI}$ the Hamiltonian which describes these free ion interactions. The $(3d)^n$ configuration, then, splits into a number of $^{2S+1}L_J$ states in the case of the free ion. These levels stretch over an energy range of around 4 eV, and the gaps between these levels can correspond to photons of visible and near infrared radiation. All these free ion states are formed from 3d single electron states. The 3d states have even parity, hence all the free ion states formed from the $(3d)^n$ configuration have even parity and electric dipole transitions are forbidden between them. The $(3d^{n-1}4p)$

configuration is some 6 eV higher, and it also splits up into a numʋ. $^{2S+1}L_J$ states. These have odd parity.

Let us now consider how one would try to solve for the energy levels and wavefunctions of the outer electrons of the transition metal ion when the ion is in a solid. The interaction of the 3d electrons with the crystal field is denoted by V_c, and we need to solve for eigenstates of $\mathscr{H}_{\rm FI}+V_c$. Although the ions in a solid are constantly vibrating, we shall initially assume, for simplicity, that the dopant ion and the neighbouring ions are in their time-average positions. In other words, we are going to assume a static crystal field. This static assumption is a good approximation to make when one is calculating energy levels. The time-varying part of the crystal field is of importance in the analysis of the shape and intensities of the optical transitions.

For the transition metal ions the effect of the static crystal field on the 3d electrons is of the same order of magnitude as the energy of the electrostatic interaction between electrons, and both are much greater than the spin–orbit interaction energy. Two approaches are now possible to the solution of the problem. Each starts off by neglecting spin–orbit coupling. In the first approach (the intermediate field scheme) one starts with the free ion ^{2S+1}L states and diagonalizes the matrix containing the V_c terms. In the other approach (the strong field scheme) the effect of the crystal field on the individual 3d electron states is first taken into account and using these as basis functions the electrostatic interaction among the electrons is next taken into account. Finally, spin–orbit coupling is considered for both schemes.

We take Ti^{3+} as the first sample. It is particularly simple as it contains only one 3d electron and we merely have to take into account the effect of the crystal field on this 3d electron. In the absence of a crystal field the

$$d_{3z^2-r^2} = \sqrt{\left(\frac{4\pi}{5}\right)}\,Y_2^0$$

$$d_{x^2-y^2} = \sqrt{\left(\frac{2\pi}{5}\right)}(Y_2^2+Y_2^{-2})$$

$$d_{xy} = -i\sqrt{\left(\frac{2\pi}{5}\right)}(Y_2^2-Y_2^{-2})$$

$$d_{zx} = (-)\sqrt{\left(\frac{2\pi}{5}\right)}(Y_2^1-Y_2^{-1})$$

$$d_{yz} = i\sqrt{\left(\frac{2\pi}{5}\right)}(Y_2^1+Y_2^{-1})$$

$$(1.22)$$

Hamiltonian of the outer electron has spherical symmetry, and the angular eigenfunctions are the spherical harmonics, Y_l^m. The 3d state has $l = 2$, and so it is fivefold degenerate—all five Y_2^m functions having the same energy. Any five independent linear combinations of the Y_2^m functions are also valid orbitals for describing these degenerate orbital states. Equation 1.22 defines particularly useful 3d orbitals for discussing the effect of the crystal field. The shapes of these five orbitals are shown in Fig. 1.14. Consider now the situation when the Ti^{3+} ions finds itself in a crystal field of octahedral symmetry, such as we had in Fig. 1.13, where the ion is surrounded by six oxygen ions. In Fig. 1.14 the six oxygen ions are shown as well as the above five orbitals. Three of the orbitals—d_{xy}, d_{yz}, d_{zx}—are affected in a similar way by the oxygen ions; they have the same energy. The lobes of the orbitals point between the oxygen ions. The $d_{3z^2-r^2}$ orbital, on the other hand, has lobes which point towards the oxygen ions which lie along the $\pm z$ axes.

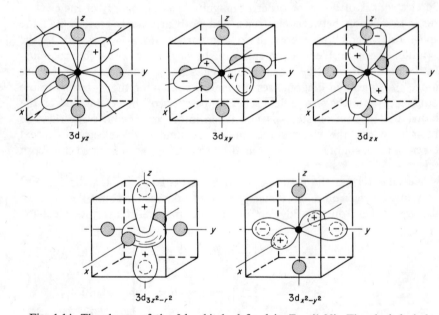

Fig. 1.14. The shapes of the 3d orbitals defined in Eq. (1.22). The shaded circles represent the six oxygen ions in an arrangement of octahedral symmetry at the site of the central ion.

Since the electron in this orbital gets closer to O^{2-} ions, this orbital will have a larger energy than the three previous orbitals. It turns out that the electron in the $d_{x^2-y^2}$ orbital has exactly the same extra energy as the $d_{3z^2-r^2}$ orbital. So the five 3d orbitals are no longer degenerate in energy. The octahedral crystal field splits this 3d state in two, into a threefold degenerate state (with orbitals d_{xy}, d_{yz}, d_{zx}) and a higher energy twofold degenerate state (with orbitals $d_{3z^2-r^2}, d_{x^2-y^2}$). The difference in energy between the two states is expressed as $10Dq$, and Dq is a measure of the strength of the octahedral field. The threefold degenerate orbitals are called t_{2g} orbitals, while the doubly degenerate orbitals are called e_g orbitals. The single 3d electron of Ti^{3+} in an octahedral field has two different energy levels available to it. These are seen in Fig. 1.15. In $Al_2O_3 : Ti^{3+}$ the crystal field at the Ti^{3+} site is slightly distorted from octahedral symmetry. Neglecting the small deviation

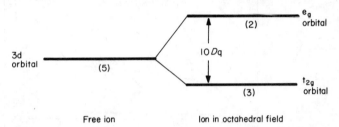

Fig. 1.15. Separation in energy of the five d orbitals when the d electron experiences an electrostatic field of octahedral symmetry.

from octahedral symmetry, the 3d state is split in two, with the separation $10Dq$ of around 20,000 cm^{-1}. This corresponds to a photon of green light. A band of light centred at 5000 Å is absorbed by the crystal. It transmits in the blue and red, and the dominant colour is blue.

The situation is more complicated when the ion has more than one 3d electron. In that case the electrostatic interactions among the 3d electrons as well as the effect of the crystal field must be taken into account. Consider the case of a $(3d)^3$ ion. If we neglect the crystal field, the three 3d electrons can combine to form ^{2S+1}L states of total S and L; these are ^4F, ^4P, ^2P, ^2D, ^2F, ^2G, ^2H. From Hund's rule the ^4F lies lowest in energy. These states are arranged according to increasing energy on the left-hand side of Fig. 1.16. On the other hand, if we neglect the electrostatic interaction between the electrons and concentrate on the effect of the crystal field only, we would have products of three single electron orbitals where each single electron orbital is either a t_{2g} or an e_g orbital. Four different combinations are possible: t_{2g}^3, $(t_{2g}^2 e_g)$, $(t_{2g} e_g^2)$, and e_g^3, and of these t_{2g}^3 has the lowest energy. When the situation is intermediate between these two extremes (zero crystal field, and zero electrostatic interactions among the 3d electrons) the situation is more complex and the levels are shown in Fig. 1.16. This shows how the free ion levels split up as the ratio of the crystal field to inter-electron interaction (Dq/B) increases. The energy levels are calculated for typical values of B and C. The broken vertical line (a) corresponds to a value of Dq/B which is suitable for V^{2+} in $KMgF_3$, and the broken line (b) corresponds to the

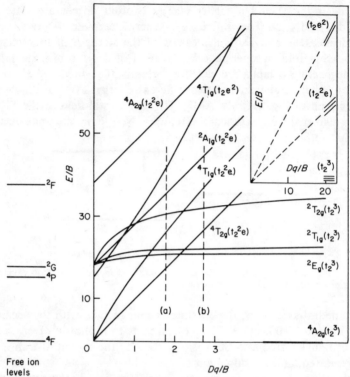

Fig. 1.16. Energy levels of the $(3d)^3$ electronic configuration. The free ion levels are shown on the left, and their splittings in an increasingly large octahedral crystal field are shown on the right. The dependence of E/B on Dq/B will vary slightly for different values of the Racah parameters B and C, and the values of B and C vary from ion to ion, and they vary on the same ion from host to host. The broken vertical lines (a) and (b) are drawn at values of Dq/B which correspond to V^{2+} in $KMgF_3$ and Cr^{3+} in Al_2O_3, respectively. The intersects of the vertical lines and the energy curves give the approximate energy values in these two cases. The insert shows the situation in the high field limit where the splitting is determined mainly by the crystal field.

case of Cr^{3+} in Al_2O_3. We see that the value of the ratio Dq/B determines the order in increasing energy of some of the levels.

The labels A_2, E, T_1, T_2 are group theoretical labels which relate to the symmetry of the orbital wavefunction in the octahedral crystal field. The use of group theory to classify crystal field orbital states is now a well-established practice, and is very elegantly presented in a number of textbooks.

We must also mention the g subscript used with these orbital labels. As we saw, free ion orbitals have either even or odd parity. Similarly, since the octahedral crystal field due to the arrangement of neighbouring ions in Fig. 1.13 has inversion symmetry, the electron orbitals in this crystal field have either even or odd parity. Since all the low-lying states in Fig. 1.16 are formed from 3d one-electron functions, which have even parity, it follows that the orbital states of the $(3d)^n$ configuration in the octahedral field have even parity, and this is symbolized by the g subscript. Crystal field states belonging to the $3d^{n-1}4p$ configuration have odd parity, and are labelled by the u subscript. They are too far above the ground state in Fig. 1.16 to be seen.

Since we have not included the effect of spin–orbit coupling, all these states are eigenstates of total spin (S) and the $2S+1$ multiplicity is used as a superscript in the labels describing the states.

In the energy level diagram shown in Fig. 1.16 we see that as the crystal field energy begins to dominate (towards the right-hand side of the figure), the energy levels adopt one of three distinct slopes. $^4A_2$, ^2E, $^2T_1$, and $^2T_2$ have the same (zero) slope. $^4T_2$, $^4T_1$, $^2A_1$ have identical slopes which are non-zero, while $^4T_1$ (among other higher states not shown here) has a slope which is twice as great. The explanation for this is that the $^4A_2$, ^2E, $^2T_1$, and $^2T_2$ states are all formed from t_{2g}^3 orbitals—in which each of these 3d electrons are in t_{2g} orbitals. The next three states—$^4T_2$, $^4T_1$, $^4A_1$—are all formed from $t_{2g}^2 e_g$ orbitals in which two of the three 3d electrons are in t_{2g} orbitals and the third is in an e_g orbital. The highest $^4T_1$ state seen in the figure is formed from $t_{2g} e_g^2$ orbitals. In the limit where $Dq \gg B$ (the strong field limit) the energy levels split up into three groups. The levels whose states are formed from t_{2g}^3 orbitals constitute the lowest group, the levels whose states are formed from $t_{2g}^2 e_g$ orbitals constitute the second group and they are $10Dq$ above the lowest group, and the group of levels whose states are formed from $t_{2g} e_g^2$ orbitals and are a further $10Dq$ higher in energy. This limiting behaviour is shown in the insert of Fig. 1.16.

By comparing the optical spectrum of a $(3d)^3$ ion with the calculated energies one can determine experimental parameters for Dq, B, and C. The value of Dq can vary considerably from host to host for the same dopant ion, and this can have a significant bearing on the optical properties of the dopant ion. For example, the absorption bands and the energy levels to which they are attributed, for Cr^{3+} in Al_2O_3 (ruby), for Cr^{3+} in $Be_3Al_2(SiO_3)_6$ (emerald), and for Cr^{3+} in $LiNbO_3$, are shown in Fig. 1.17. Since we are dealing with three different host materials, the crystal field energy is different in the three cases. In each case the energy separation between $^4A_2$ and ^2E is approximately the same, which is to be expected since the two states in question are formed from similar (t_{2g}^3) orbitals. The energy separation between $^4A_2$ and $^4T_2$ states is quite different in the three cases, which is a reflection of the fact that the $^4T_2$ state is formed from $(t_{2g}^2 e_g)$ orbitals and as a result its energy separation from the $^4A_2$ (t_{2g}^3) state is very sensitive to the value of the crystal field (Dq). Dq has the largest value in the case of ruby, its value is smaller in emerald, and smaller still in $LiNbO_3 : Cr^{3+}$. The $^4A_2 \rightarrow {}^4T_2$ and

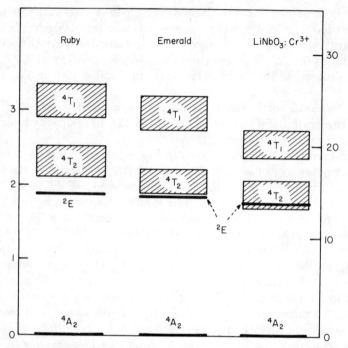

Fig. 1.17. Experimentally observed absorption levels of Cr^{3+} in Al_2O_3 (ruby), Cr^{3+} in $Be_3Al_2(SiO_6)_3$ (emerald), and Cr^{3+} in $LiNbO_3$. The strength of the crystal field is strongest in ruby and weakest in $LiNbO_3:Cr^{3+}$.

$^4A_2 \rightarrow {}^4T_1$ transitions give rise to the strong absorptions in all three cases and they occur at different wavelengths. In ruby the absorption transitions are in the yellow-green and blue, hence the red colour of the crystal. In emerald, on the other hand, they occur in the orange-yellow and blue, hence the green colour of emerald. In both of these cases the lowest excited state is ^2E, and sharp-line luminescence occurs from the ^2E state for ruby and emerald. For $LiNbO_3:Cr^{3+}$, the $^4T_2$ absorption band stretches below the sharper ^2E level, and in this material luminescence occurs from $^4T_2$, from the part of the $^4T_2$ band which is below ^2E. This luminescence is in the form of a broad band and is in the near infrared.

The splittings of the Mn^{2+} levels in an octahedral crystal field are shown in Fig. 1.9 for a typical value of Dq. The octahedral labels are also given. All of these orbitals are of even parity—since they belong to the $3d^5$ electronic configuration. It is not uncommon to omit the g subscript, as is done in Fig. 1.9.

Let us consider this matter of parity a little further. The energy of the electrons of the ion in an octahedral field is described by the Hamiltonian

$\mathcal{H} = \mathcal{H}_{FI} + V_c$. Both $\mathcal{H}_{FI}$ and V_c are invariant under inversion of co-ordinates, hence the eigenstates have either even or odd parity. Sometimes the distortion from octahedral symmetry introduces some odd parity crystal field terms and the Hamiltonian is no longer invariant under inversion. The eigenstates of this Hamiltonian no longer have even or odd parity—their parity is mixed. Since the distortion from octahedral symmetry is small, this odd parity crystal field term is small and the eigenstates "almost" have a definite parity. But although the mixing of parity is very small, it can have a profound effect on the optical transitions between the states.

1.2.5. Radiative Transitions in Dopant Transition Metal Ions

The low-lying states of the transition metal ions in sites of octahedral symmetry are formed from $(3d)^n$ orbitals, all have even parity, and consequently electric dipole transitions are not allowed between them. That optical transitions are seen for transition metal ions in solids must be explained by one or more of the following reasons: (i) these are magnetic dipole transitions; (ii) the ions are in sites where the crystal field lacks inversion symmetry, their eigenstates are of mixed parity, and as a result electric dipole transitions can occur between them; (iii) although the time-average crystal field has inversion symmetry, lattice vibrations of odd parity can momentarily destroy the inversion symmetry and allow electric dipole transitions to occur. As a simple example of a site which lacks inversion symmetry, we can take the arrangement of ions shown in Fig. 1.13 but imagine that the M^{3+} ion is displaced from its central position.

Situation (ii) is what obtains in the case of ruby. The environment of the Cr^{3+} ion in ruby lacks inversion symmetry, and so there is a small odd parity crystal field term, V_u, in the Hamiltonian describing the system. We can regard V_u as a perturbation on the mainly octahedral functions. Let us use perturbation theory to examine the effect of V_u on the optical transition in ruby. We start with the octahedral basis functions. The low-lying octahedral states have even parity, as their g subscripts indicate, while the high energy states belonging to the $(3d)^2 4p$ configuration have odd parity, as their u subscripts indicate, and are some 6 eV or more above the ground state. These are represented by Γ_u and they can be spin doublets ($^2\Gamma_u$) or spin quartets ($^4\Gamma_u$). The effect of V_u is to mix these g and u states together. For example, the ground state, which we label by $\Psi(^4A_2)$, is given by

$$\Psi(^4A_2) = |^4A_{2g}\rangle + \sum_{^4\Gamma_u} \frac{\langle ^4\Gamma_u | V_u | ^4A_{2g}\rangle}{E(^4A_{2g}) - E(^4\Gamma_u)} |^4\Gamma_u\rangle \qquad (1.23)$$

where $E(\Gamma)$ is the energy of the $|\Gamma\rangle$ state and where the summation is over all possible quartet odd parity functions. $|^4A_{2g}\rangle$ is the ground state; we use the octahedral field labels for these levels. The energy denominator in

Eq. (1.22) is large, and so the admixture of $^4\Gamma_u$ into the ground state is very small. Similarly, the excited $^4T_2$ state, which we label by $\Psi(^4T_2)$, is given by

$$\Psi(^4T_2) = |^4T_{2g}\rangle + \sum_{^4\Gamma_u'} \frac{\langle ^4\Gamma_u' | V_u | ^4T_{2g} \rangle}{E(^4T_{2g}) - E(^4\Gamma_u')} |^4\Gamma_u' \rangle \qquad (1.24)$$

Once again the summation is over all odd parity states, and since the energy denominator is large, the amount of admixture of odd parity states into the $^4T_2$ state is small.

Consider now the radiative transitions which can occur between $\Psi(^4A_2)$ and $\Psi(^4T_2)$. The magnetic dipole operator can connect $|^4A_{2g}\rangle$ to $|^4T_{2g}\rangle$, as well as $|^4\Gamma_u\rangle$ to $|^4\Gamma_u'\rangle$, but this latter contribution to the magnetic dipole transition between $\Psi(^4A_2)$ and $\Psi(^4T_2)$ is very small. So a magnetic dipole transition can occur between these states. The electric dipole operator connects $|^4A_{2g}\rangle$ and $|^4\Gamma_u'\rangle$—when the $^4\Gamma_u'$ state has the $^4T_{2u}$ form—and also connects $|^4\Gamma_u\rangle$ and $|^4T_{2g}\rangle$, so electric dipole transitions can also occur. The relative strength of the magnetic dipole and electric dipole transitions goes as the ratio of the squares of the appropriate matrix elements. Taking just one of the electric dipole matrix elements we see that this ratio goes approximately as

$$\frac{|\langle ^4T_{2g} | \boldsymbol{\mu} | ^4A_{2g} \rangle|^2}{e^2 c^2} \Bigg/ \frac{|\langle ^4\Gamma_u' | \mathbf{r} | ^4A_{2g} \rangle|^2 \langle ^4\Gamma_u' | V_u | ^4T_{2g} \rangle|^2}{[E(^4\Gamma_u') - E(^4T_{2g})]^2} \qquad (1.25)$$

The matrix element squared of $\mathbf{r}$ is about six orders of magnitude greater than the matrix element square of $\boldsymbol{\mu}/ec$. Thus, although the V_u term is small, and the energy denominator is large—both of which reduce the effectiveness of the electric dipole process—the significantly greater matrix element of $\mathbf{r}$ ensures that the electric dipole process is more effective than the magnetic dipole process. The oscillator strength of the $^4A_2 \rightarrow ^4T_2$ transition in ruby is about 10^{-4}; if it were a magnetic dipole process, it would be around 10^{-6}.

Let us now look at the $^2E \rightarrow ^4A_2$ transition in ruby which gives rise to the strong luminescence from this material. This transition is spin-forbidden. But we have neglected to take into account the spin–orbit coupling interaction (V_{SO}) in our description of the electronic states. Non-zero matrix elements of $\langle ^4T_{2g} | V_{SO} | ^2E_g \rangle$ exist, so the ^2E state contains small amounts of $^4T_{2g}$; and because of the V_u term, electric dipole matrix elements connect the $^4T_2$ and $^4A_2$ states: hence an electric dipole transition occurs between $^4A_2$ and ^2E. Its oscillator strength is weaker by

$$\left| \frac{\langle ^4T_{2g} | V_{SO} | ^2E_g \rangle}{E(^4T_{2g}) - E(^2E_g)} \right|^2 \qquad (1.26)$$

than the $^4A_2 \rightarrow ^4T_2$ oscillator strength.

When the Cr^{3+} ion substitutes for Mg^{2+} in MgO it can enter a site of perfect octahedral symmetry. The sharp $^2E \to {}^4A_2$ luminescence line from this ion has been shown to be a magnetic dipole transition. It has an oscillator strength of around 10^{-8}. This very small value is a result of the fact that not only is it a magnetic dipole transition, but it is almost spin forbidden as well.

The energy level scheme for Mn^{2+} in an octahedral crystal field is shown in Fig. 1.9. All the absorption transitions are both spin- and parity-forbidden, and so are very weak. The $^4T_1 \to {}^6A_1$ luminescence transition has a small oscillator strength, but it is important in many commercial phosphors. Because of the weakness of the absorption transitions, direct optical pumping of the Mn^{2+} ions is not an effective excitation mechanism. Other methods of indirectly exciting Mn^{2+} ions will be mentioned later.

In this section, as well as in the last, we have concentrated mainly on the Cr^{3+} and Mn^{2+} ions, as they are the major luminescent transition metal ions. Energy levels and radiative transitions in other transition metal ions can be similarly analyzed. Details of the analyses are given in the books referred to at the end of the chapter.

1.2.6. Energy Levels of Dopant Rare-earth Ions

These ions have a partially filled 4f shell of electrons. In contrast to the case of the transition metal ion, where the electrons in the partially filled 3d shell occupy orbits on the outside of the ion, 4f electrons in the rare-earth ion occupy inner orbits, since the filled 5s, 5p shells are outside the 4f shell. The 4f electrons, then, are shielded from the electric field of the neighbouring ions, and the effect of the crystal field on the 4f electrons is small. We are in the weak field limit. The procedure for calculating wavefunctions and energy levels of the rare-earth ions in a solid is first of all to calculate the free ion $^{2S+1}L_J$ states, and then to consider the effect of the crystal field using perturbation theory. One consequence of the weakness of the crystal field is that the energy levels of the rare-earth ions do not vary greatly from one host material to another.

The observed energy levels of the trivalent rare-earth ions when substituted for the La^{3+} ion in $LaCl_3$ are shown in Fig. 1.10. In this host material the crystal field at the La^{3+} site lacks inversion symmetry. The states are labelled by their $^{2S+1}L_J$ free ion values. The effect of the crystal field is to cause a fine splitting of these levels, but this splitting is so small that it can only be represented on the energy level scheme as a broadening of the levels. A semicircle drawn under a level indicates that luminescence is observed from this level in $LaCl_3$. In other hosts the energy levels of the trivalent ions are essentially in the same positions, but the splittings of levels will be different, reflecting the different strengths and symmetries of the different crystal fields. Also, the strengths of the radiative transitions and the question as to whether or not luminescence is observed from a given level will depend in part on the symmetry of the crystal field. It should be remembered that, for each trivalent ion, the states with energies up to around 6 eV belong to the same $4f^n$ configuration, and so these states have the same parity. States with the $4f^{n-1}5d$ configuration are over 6 eV above the ground state, and these have the opposite parity to the ground state.

Divalent rare-earth ions can be incorporated in many hosts. We might hope to be able to predict energy levels for divalent ions from the trivalent energy level diagram by looking up the appropriate $4f^n$ levels. For example, we might hope that the Pr^{2+} ($4f^3$ configuration) would have energy levels almost concident with those of Nd^{3+} which has a similar $4f^3$ configuration. Such, however, is not always the case, and the level assignment for divalent ions is not nearly as well understood as it is for the trivalent ions. One of the complicating factors is the fact that for the divalent ions, the levels belonging to states with the $4f^{n-1}5d$ configuration are often within 3 eV of the ground state, and such levels give rise to broad absorption transitions.

1.2.7. Radiative Transitions in Dopant Rare-earth Ions

The energy levels of any one of the rare earth ions shown in Fig. 1.10 all belong to the same $4f^n$ configuration, and so all these states have the same parity. Transitions between these states can occur either as magnetic dipole transitions, for which the free ion selection rules are $\Delta L = 0$, $\Delta S = 0$, $\Delta J = 0$, ± 1, or as weakly allowed electric dipole transitions, where the parity forbidden selection rule is partially lifted if the rare-earth ion is in a site which lacks inversion symmetry. The crystal field term, in this case, contains an odd component V_u. This odd component of crystal field mixes some $4f^{n-1}5d$ states into the $4f^n$ states. However, as the amount of mixing is small, the low-lying states are predominantly $4f^n$ and so are predominantly of the same parity. Since electric dipole transitions occur because of mixing of opposite parity states, and as this mixing is difficult to estimate, it is not easy to derive very exact electric dipole selection rules.

The sharpness of the levels and the weakness of the visible transitions mean that optical pumping with broad band sources is not a very efficient way of exciting luminescence in these materials.

For divalent ions the low lying $4f^{n-1}5d$ states permit strong absorption transitions in the visible from the ground $4f^n$ state. Further, the 5d orbital is much more sensitive to the environment than the 4f orbital, and as a result, the $4f \rightarrow 5d$ transitions are broad. The presence of broad strong absorption bands causes a deep colour in these divalent crystals. They also allow efficient optical pumping of the divalent ion, which can result in strong luminescence from these ions.

1.2.8. Energy Levels and Radiative Transitions in an F-centre

The F-centre is a defect centre which can be induced in alkali halide crystals, and it consists of a negative anion vacancy at which an electron is trapped. This electron moves in the attractive electric field of the neighbouring positive ions. The orbitals of the electron overlap the neighbouring ions, and so the electronic states are very strongly coupled to the lattice—in contrast to the $4f^n$ orbital states of the rare-earth ion. We can think of the electron in the F-centre as resembling a particle in a box—the box in question being the cage formed by the neighbouring ions. Excited states of the F-centre occur within the energy of visible photons from the ground state, and a strong optical absorption occurs from the ground to the first excited state. The oscillator strength is close to unity, indicating an allowed electric dipole transition. Strong luminescence occurs also from this excited state. Figure 1.18 shows the absorption and luminescence shapes for a typical F-centre in an alkali halide. A number of points are worth noting. Even though the

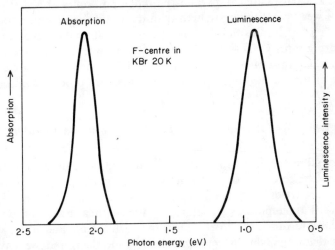

Fig. 1.18. Absorption and luminescence transitions between the *same* two electronic levels in the F-centre in KBr. The luminescence occurs at a lower energy than the absorption (Stokes' shift).

absorption and emission transitions take place between the same two states—the ground state and the first excited state—the luminescence occurs at a lower energy than the absorption. We also note the large width of the transitions—hardly what we expect between simple one-electron states; and, finally, the oscillator strength in emission can be significantly smaller than the oscillator strength in absorption.

The difference between the absorption and the luminescence transitions has to do with the strong coupling between the F-centre and the lattice. In dealing with the rare earth ions it is a good approximation to regard the effect of the environment as a small perturbation on the free ion energy levels. Hence the levels are reasonably sharp—as one would expect from free ion spectra. In the transition metal ions, the interaction of the optically active 3d electrons and the neighbouring ions of the host lattice is stronger. Consequently, we sometimes find broad band transitions on the transition metal ions, and broad bands are an indication of a strong interaction with the lattice. In the F-centre the approximation of a weak interaction with the lattice is quite untenable. Because of the strong interaction with the lattice we can no longer regard the trapped electron on its own (in some electrostatic box) as the optically active centre. The electron and the neighbouring ions, all strongly coupled together, must be regarded as the optically active centre. When the electron is raised to its excited state, the neighbouring ions find themselves in a different electrostatic environment, and they relax to a different equilibrium arrangement of lower energy. Luminescence takes place from this relaxed excited state, and the radiation occurs at a lower energy and has a weaker oscillator strength than the original absorption transition. The width of the transitions comes from the vibrational motion of the neighbouring ions. Clearly, we will need to take into account the dynamic lattice in which the optically active ions and defects occur if we are to gain a complete picture of the absorption and luminescence processes in solids.

1.2.9. Interaction of the Optically Active Centre with the Vibrating Lattice: A General View

In this and in the next two sections we shall consider how the absorption and luminescence of the optically active centre is affected by the fact that the centre is part of a crystal lattice which can undergo static and dynamic distortions. For simplicity we shall regard the centre as a dopant ion. Up to now we have regarded the environment in which the ion finds itself as doing no more than supplying a constant electrostatic field, and we have been concerned with determining the electronic levels of the ion in the presence of this static crystal field. In reality the environment is not static. The neighbouring ions vibrate about some average positions, so there is a dynamic as well as a static crystal field, and this dynamic crystal field affects the electronic states of the dopant ion. In addition, the environment may be affected by changes in the electronic state of the dopant ion. For example, when the dopant ion changes its electronic state, the neighbouring ions may take up new average positions and the nature of their vibrations about their new average positions may not be the same as before. It is clear that we should take into account the total ion-plus-lattice state when describing optical transitions.

The state of the lattice is taken into account as follows. The ions are connected to each other and we assume that they are maintained in their average positions by harmonic restoring forces. The ions can vibrate about these average positions and the normal modes are collective modes of vibration called *lattice vibrations*. If there are N ions in the crystal, there are $3N$ different allowed modes of lattice vibration, and these cover a range of frequencies from zero to around $\omega = 10^{14}$ s^{-1}. We have a full description of the state of the lattice if we know (a) the average arrangement of the lattice ions and (b) the vibrational state of the lattice ions. If the average arrangement did not change then we would only have to concern ourselves with the vibrational state of the lattice.

We adopt the following quantum mechanical description of the vibrating lattice. The energy of vibration at frequency, ω, is quantized in units of energy $\hbar\omega$. Each such quantum of energy is called a phonon. We can describe the state of the vibrating lattice by specifying the number of phonons of each frequency which are present. This description is given by the state function $|n_1, n_2, n_3, n_4, \ldots\rangle$ where n_i is the number of phonons in the ith vibrational mode. For simplicity this state function is often written as $|n\rangle$. The $|n\rangle$ functions form an orthonormal set. Sometimes when we are specifically concerned with the number of ith mode phonons and are not so concerned with the other modes, we describe the phonon state simply by $|n_i\rangle$. Or we might be concerned only with i and j mode phonons, in which case we would describe the lattice state by $|n_i, n_j\rangle$.

The full ion-plus-lattice system is described by the Hamiltonian $\mathscr{H} = \mathscr{H}_{\mathrm{FI}} + V_c + \mathscr{H}_L$ where $\mathscr{H}_{\mathrm{FI}}$ is the free ion Hamiltonian of the dopant ion, $\mathscr{H}_L$ is the Hamiltonian describing the lattice, and V_c is the interaction energy of the ion in the crystal field of the lattice. This crystal field term consists of two parts: a static term V_c^0 which is the interaction energy when all the ions are in their average positions, and a dynamic term V_c^v, which is the interaction energy of the dopant ions in the time-varying crystal field. The full Hamiltonian is $\mathscr{H} = \mathscr{H}_{\mathrm{FI}} + V_c^0 + V_c^v + \mathscr{H}_L$.

In the previous sections we concerned ourselves with determining the states of the dopant ion in the average crystal field, hence we considered only the $\mathscr{H}_{\mathrm{FI}} + V_c^0$ terms and expressed the eigenstates as $|i\rangle$. We shall now consider the other terms in the Hamiltonian. The way in which these are handled depends on the strength of the coupling between the dopant ion and the neighbouring ions. We shall consider three cases: vanishingly small coupling, weak coupling, and strong coupling.

Let us first consider the case of a vanishingly small coupling. In that case the dopant ion is unaffected by the environment and so $V_c = 0$. Further, the lattice is unaffected by electronic changes in the dopant ion so the $\mathscr{H}_{\mathrm{FI}}$ and $\mathscr{H}_L$ terms are independent of each other. $\mathscr{H}_L$ is the dynamic lattice

Hamiltonian whose eigenstates are $|n\rangle$. The eigenstates of the full Hamiltonian, $\mathcal{H}_{FI} + \mathcal{H}_L$, are the product states $|i\rangle|n\rangle$ which we can also write as $|i,n\rangle$. The $|i\rangle$ states are the free ion wavefunctions.

At first sight it might seem that being a part of the vibrating lattice has no effect on the dopant ion, since the electrostatic crystal field of the neighbours does not affect the dopant ion. Participating in the vibrational motion of the lattice, however, does affect the optical properties of the ion, since it causes the absorption or emission to be Doppler-modulated. The situation is quite analogous to the absorption or emission of gamma rays by the nucleus of an ion which is part of a vibrating lattice (Mössbauer effect). Just as in the case of the Mössbauer effect, we shall find that the optical spectrum of the vibrating ion consists of a sharp line and a sideband. We shall use a simple semi-classical model to show this effect.

In this model we regard the optically active dopant ion as a classical oscillator which is emitting (or absorbing) continuously at the optical frequency ω_0. (In quantum terms $\hbar\omega_0$ is the energy separation of the initial and final states of the ion.) The classical oscillator is also participating in the vibrational motion of the lattice. Let us assume that it is participating in the lattice vibrational mode of vibrational frequency ω_L. If due to this vibrational motion the ion is moving in the $\pm x$ direction, we can describe its displacement from equilibrium by $x = x_0 \sin \omega_L t$ and its velocity in the x direction is given by $v(t) = \omega_L x_0 \cos \omega_L t$. If the observer detects the radiation from the oscillator, when the oscillator is moving towards him with velocity $v(t)$, then the observed frequency is Doppler shifted, and the frequency has value

$$\omega_0\left(1 + \frac{v(t)}{c}\right) = \omega_0\left(1 + \frac{\omega_L x_0}{c} \cos \omega_L t\right) \tag{1.27}$$

The output amplitude function of the vibrating classical oscillator can thus be written as $\sin \phi(t)$, and

$$\frac{d\phi}{dt} = \omega = \omega_0\left(1 + \frac{\omega_L x_0}{c} \cos \omega_L t\right) \tag{1.28}$$

From this we obtain by integration $\phi(t) = \omega_0 t + (\omega_0 x_0/c) \sin \omega_L t$. We can expand this, and we obtain for $\sin \phi(t)$

$$\sin \omega_0 t \cos\left(\frac{\omega_0 x_0}{c} \sin \omega_L t\right) + \cos \omega_0 t \sin\left(\frac{\omega_0 x_0}{c} \sin \omega_L t\right) \tag{1.29}$$

Let us now assume that $\omega_0 x_0/c \ll 1$, which is valid in the case of optical frequencies, so the formula for the amplitude function can be simplified to read

$$\sin \phi(t) = \sin \omega_0 t + \cos \omega_0 t \left(\frac{\omega_0 x_0}{c} \sin \omega_L t \right)$$

$$= \sin \omega_0 t + \frac{\omega_0 x_0}{2c} (\sin (\omega_0 + \omega_L) t - \sin (\omega_0 - \omega_L) t). \quad (1.30)$$

As Eq. (1.30) shows, the output consists of a component at the "carrier frequency" ω_0, as well as two sidebands a distance ω_L on either side of ω_0. This is the familiar result of classical frequency modulation (FM) theory, and is illustrated in Fig. 1.19(a).

If we had carried out this FM analysis without making approximations, we would have found that the total intensity (the sum of the intensity in the carrier wave and the intensity in the sidebands) would not change with the degree of modulation. The greater x_0, the greater is the degree of modulation, and the greater the relative intensity in the sidebands, but the total intensity is unchanged.

Returning to our approximate formulae, we see from Eq. (1.30) that if the intensity at the carrier wave frequency, ω_0, is unity, the intensity of the two sidebands goes as $\omega_0^2 x_0^2 / 2c^2$. Hence the fraction of the intensity at ω_0 is $(1 + \omega_0^2 x_0^2 / 2c^2)^{-1}$. Since the intensity in the sidebands is small, this expression for the fraction of the intensity at ω_0 can be approximated by $\exp(-\omega_0^2 x_0^2 / 2c^2)$. When there are a large number of lattice vibrations of different amplitudes x_0^i and frequencies ω_L^i, the expression becomes

$$\exp\left(-\omega_0^2 \sum_i (x_0^i)^2 \Big/ 2c^2 \right) \quad (1.31)$$

and there is a large number of FM sidebands, each of intensity $\omega_0^2 (x_0^i)^2 / 4c^2$ and positioned a distance ω_L^i on either side of the sharp line at ω_0. This is shown in Fig. 1.19(b). Although this output spectrum has been derived for an emitting ion which is vibrating, the absorption spectrum of such a vibrating ion would be identical on this semi-classical model.

A quantum mechanical model would predict a different shape for the spectrum at very low temperatures—one in which absorption and emission are mirror images of each other about the sharp line. The fractional intensity at ω_0, however, is the same as that given in expression (1.31) but expressed in a quantum mechanical language:

$$\exp\left(-\frac{\omega_0^2 \langle u^2 \rangle}{2c^2} \right)$$

where $\langle u^2 \rangle$ is the mean square displacement due to lattice vibrations. This is the Debye–Waller factor which is familiar to us from Mössbauer theory.

On the quantum mechanical model the sideband transition at $\omega_0 + \omega_L$ would correspond to the absorption of a photon of energy $\hbar(\omega_0 + \omega_L)$. Of this energy, an amount $\hbar\omega_0$ is used to excite the ion, and $\hbar\omega_L$ is used to generate a phonon of frequency ω_L. For each phonon mode, a sideband

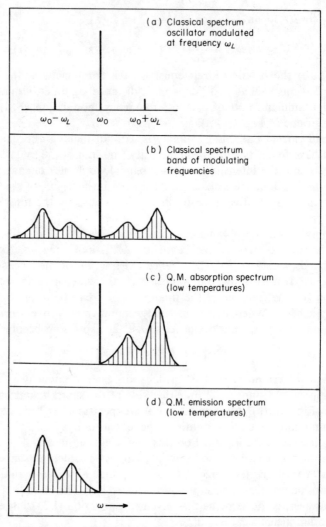

Fig. 1.19.(a) Result of classical FM analysis when an oscillator of natural frequency ω_0 is frequency modulated (weakly) at a single modulation frequency ω_L. (b) Result of classical FM analysis when the oscillator is being frequency modulated at a band of frequencies. (c) Result of quantum mechanical analysis showing the frequency distribution in the low temperature absorption spectrum when an ion of natural frequency ω_0 is participating in the vibrational motion of the host lattice in which it is embedded. The spectrum of vibrational frequencies is the same as in (b), and the strength of the modulation is also the same as in (b). (d) Result of a similar quantum mechanical analysis of the low temperature luminescence spectrum.

transition is allowed. The sum total of all the individual sideband transitions produces the absorption sideband spectrum seen in Fig. 1.19(c), where the shape of the sideband spectrum is determined by the density of phonon states. The sideband transition in emission at frequency $\omega_0 - \omega_L$ corresponds to the case where the ion loses energy $\hbar\omega_0$, of which $\hbar\omega_L$ is used to excite a phonon and the remainder is emitted as a photon of frequency $\omega_0 - \omega_L$. Such a spectrum is shown in Fig. 1.19(d). It should be noted that these are one-phonon sidebands in which just one phonon is absorbed or emitted during the optical process. The sharp line is called the *no-phonon line* on this quantum mechanical model.

Returning to the Debye–Waller factor which gives the fractional intensity in the sharp line, we notice that whereas for nuclear gamma rays ($\omega_0 \simeq 10^{20} \text{s}^{-1}$) the fraction in the sharp line can vary from a few per cent to over 90%, in the case of visible radiative ($\omega_0 \simeq 3 \times 10^{15} \text{ s}^{-1}$) essentially all the intensity is in the sharp line, and there is essentially none in the sidebands.

The preceeding analysis shows that, if the optically active electrons on the dopant ion were shielded from the neighbouring ions, the optical absorption and luminescence spectrum would consist of exceedingly sharp Mössbauer-type lines. Such a situation is never realized in practice because the optically active electrons are never fully shielded from the neighbouring ions. For rare earth dopant ions where the shielding is most effective, the spectra consist mainly of sharp lines, but the lines are not exceedingly sharp and some bands are also observed.

It is interesting to use the same semi-classical model to analyze the effect of the vibrating lattice on the optically active ion in the case of a weak crystal field. The positions of the electronic levels of the ion depend on the value of the crystal field, as does the frequency of the radiation emitted or absorbed by the ion. Because of lattice vibrations the crystal field is being modulated, and this modulates the frequency of the radiation emitted or absorbed. The resultant crystal-field-modulated FM sidebands, rather than the Doppler-modulated FM sidebands associated with the Mössbauer effect, are what are observed in the optical spectra of ions in solids. When the crystal field modulation is weak—as in the case of rare earth ions—these sidebands are weak and the spectra are characterized mainly by sharp lines. When the modulation is stronger—as in the transition metal ions—most of the transitions occur in sidebands. In the case of very strong modulation—as in F-centres—the sideband is very broad and the sharp line is reduced to zero. Examples of such spectra are shown in Fig. 1.20. In the example shown in Fig. 1.20(a), (b), where the crystal field modulation of the transitions is not very strong, the sideband is due to transitions each involving a photon along with a single phonon. In the examples shown in Fig. 1.20(c), (d), the crystal field modulation is strong, and the sideband is due to transitions each

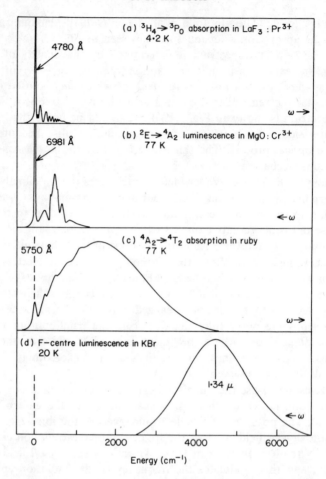

Fig. 1.20. Four examples of low temperatures sideband transitions found in optical spectra. The positions of the no-phonon lines are drawn directly above each other, and the sidebands are drawn to the right, irrespective of whether it is an absorption or a luminescence spectrum. In examples (a) and (b) the crystal field modulation is weak, the sidebands are one-phonon sidebands, and they can be explained by the simple crystal field FM analysis presented. (c) and (d) are examples of strong crystal field modulation, the sidebands are multiphonon transitions, and they cannot be explained by the same simple analysis. In (c) the no-phonon line is weak. In (d) the modulation is so strong that the no-phonon line does not appear. The dashed line shows where it would appear if it were allowed.

of which involves a photon along with a number of phonons. Hence its much larger width.

According to the crystal field-modulation analysis presented above, the frequency of the no-phonon line, ω_0, should be determined by the static crystal field of the ions in their average positions, and it should be exceedingly sharp. In practice it is found that the lines have a width much larger than predicted by the above analysis. The reason for this is that strains and defects in the structure of the crystal cause the average separation between dopant ion and neighbouring ion to vary from place to place. This causes a range of values of ω_0 from dopant ions in different parts of the crystal. When we observe the light from a macroscopic volume of crystal, the no-phonon line is a composite of exceedingly sharp no-phonon lines from different microscopic sections of the crystal, each with a different frequency. This inhomogeneous broadening causes the finite width of the observed no-phonon line. Reducing strains in the crystal by careful crystal growth can reduce the inhomogeneous broadening. In addition, optical pumping of a microscopic section of the crystal by a focused laser beam can lead to exceedingly narrow no-phonon lines.

At high temperatures additional broadening mechanisms become effective, and in general it is found that all features of the spectrum broaden with increasing temperature.

Although the semi-classical model is useful for gaining an understanding of the kind of spectrum expected from a solid, we must return to a quantum mechanical treatment if we are to obtain an adequate explanation for the interaction of the dopant ion with the lattice.

1.2.10. Quantum Mechanical Analysis of the No-phonon Sharp Line and One-phonon Sideband from Dopant Ions

We look for eigenstates of the Hamiltonian $\mathscr{H}_{FI} + V_c^0 + V_c^v + \mathscr{H}_L$ in the case where $V_c^0 + V_c^v$ is small. In particular, we assume that $\mathscr{H}_L$ is independent of the electronic state of the dopant ion. In that case $\mathscr{H}_L$ is the Hamiltonian for the vibrations of the perfect lattice whose eigenstates are $|n\rangle$. $\mathscr{H}_{FI} + V_c^0$ is the Hamiltonian describing the dopant ion in the time average crystal field, and the electronic eigenstates for this Hamiltonian are written $|i\rangle$. The product states $|i\rangle|n\rangle$ (or $|i,n\rangle$) are the eigenstates of $\mathscr{H}_{FI} + V_c^0 + \mathscr{H}_L$. We regard V_c^v, the ion-vibrating lattice interaction, as a small perturbation which can be handled by perturbation theory, and we are interested in its effect upon the energy levels and radiative transitions.

The probability of an optical transition from $|i,n\rangle$ to $|l,m\rangle$ depends on the square of the matrix element $\langle l,m|\mathbf{d}|i,n\rangle$, where $\mathbf{d}$ is the appropriate dipole operator ($\mathbf{r}$ or $\boldsymbol{\mu}$). This operator does not act on phonon states, so the matrix element is a product of an electronic and a vibrational matrix element, $\langle l|\mathbf{d}|i\rangle\langle m|n\rangle$.

Since the phonon states are orthonormal, $\langle m|n\rangle = 0$ unless $m = n$. Hence the only optical transition allowed is that in which there is no change in the phonon state—only sharp no-phonon transitions are allowed. We now consider the ion-vibrating lattice interaction, V_c^v, as a perturbation on these basis functions and see how this affects the optical transitions.

We expect that V_c^v will depend on the strain, ε, that is, on the distortion of the near-neighbour separation from its equilibrium value. We express V_c^v as a power series in the strain:

$$V_c^v = V_1\varepsilon + V_2\varepsilon^2 + \ldots \tag{1.32}$$

V_1 and V_2 are functions of the electronic coordinates of the dopant ion. Each lattice vibration of frequency ω introduces a strain ε_ω, so the full strain ε is given by the sum of the ε_ω's. Using a second quantized form for the strains we get

$$\varepsilon = \sum_\omega \varepsilon_\omega = \sum_\omega \sqrt{\left(\frac{\hbar\omega}{2Mv_\omega^2}\right)}(a_\omega + a_\omega^+) \tag{1.33}$$

where M is the mass of the crystal, v_ω is the velocity of sound at frequency ω, and a_ω and a_ω^+ are, respectively, annihilation and creation operators for phonons of frequency ω. These operators have the following properties:

$$\left. \begin{aligned} a_\omega|n_\omega\rangle &= \sqrt{n_\omega}|(n-1)_\omega\rangle \\ a_\omega^+|n_\omega\rangle &= \sqrt{(n+1)_\omega}|(n+1)_\omega\rangle \end{aligned} \right\} \tag{1.34}$$

The ion-plus-lattice state which was originally described by the basis function $|i, n\rangle$ is now described in first-order perturbation theory by $\psi(i, n)$, where

$$\psi(i, n) \simeq |i, n\rangle + \sum_{j,\omega} \alpha_j^\omega |j, (n+1)_\omega\rangle + \sum_{j,\omega} \beta_j^\omega |j, (n-1)_\omega\rangle \tag{1.35}$$

and $\alpha_j^\omega, \beta_j^\omega$ are given by

$$\left. \begin{aligned} \alpha_j^\omega &= \sqrt{\left(\frac{\hbar\omega}{2Mv_\omega^2}\right)} \frac{\langle j|V_1|i\rangle\langle(n+1)_\omega|a_\omega^+|n_\omega\rangle}{E_i - E_j - \hbar\omega}, \\ \beta_j^\omega &= \sqrt{\left(\frac{\hbar\omega}{2Mv_\omega^2}\right)} \frac{\langle j|V_1|i\rangle\langle(n-1)_\omega|a_\omega|n_\omega\rangle}{E_i - E_j + \hbar\omega} \end{aligned} \right\} \tag{1.36}$$

E_i and E_j are the energies of the electronic states $|i\rangle$ and $|j\rangle$, and $\hbar\omega$ is the energy of the phonon created or destroyed.

The state $|j, n+1\rangle$ differs from $|i, n\rangle$ in that the ion is now in electronic state j while the lattice state has one more phonon in some mode ω. States of one phonon more or one phonon less are mixed into the original state because of the creation and annihilation operations in the ion-vibrating lattice interaction.

Consider now a no-phonon transition between $\psi(i, n)$ and $\psi(l, n)$. Since the dipole operator, **d**, connects only ion-plus-lattice states where the phonon occupancy is the same, the parts of $\psi(i, n)$ and $\psi(l, n)$ between which non-zero matrix elements of **d** exist are indicated in Fig. 1.21(a). The value of matrix element I (in Fig. 1.21) is $\langle l|\mathbf{d}|i\rangle$, while matrix elements II and III have values $\alpha_j^\omega(\alpha_{j'}^\omega)^*\langle j'|\mathbf{d}|j\rangle$ and $\beta_j^\omega(\beta_{j'}^\omega)^*\langle j'|\mathbf{d}|j\rangle$, respectively. Since the α and β factors are exceedingly small when V_c^v is small, matrix elements II and III are essentially negligible.

Next we consider a *one-phonon transition* between $\psi(i,n)$ and $\psi(l,n+1)$. The parts of the wavefunctions between which non-zero matrix elements of **d** exist are shown on Fig. 1.21(b). Matrix elements of **d** exist only between states whose phonon occupancies are the same. We see that for each phonon mode ω, a one-phonon transition is allowed because of the application of first-order perturbation theory. We would have to go to second-order perturbation theory in $V_1\varepsilon$, or consider $V_2\varepsilon^2$ in first-order perturbation theory, in order to see how a two-phonon transition could occur. We assume that such processes are unlikely when V_c^v is small.

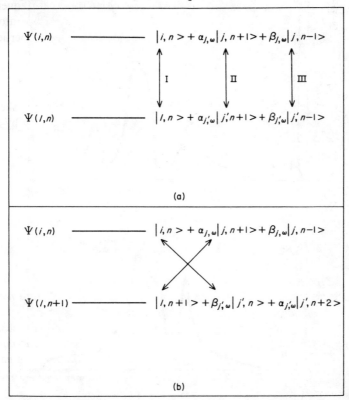

Fig. 1.21.(a) Arrows indicate states between which matrix elements of **d** exist, and these matrix elements lead to a no-phonon transition between $\Psi(i,n)$ and $\Psi(l,n)$. (b) Arrows indicate states between which matrix elements of **d** exist, and these matrix elements lead to a one-phonon sideband transition between $\Psi(i,n)$ and $\Psi(l,n+1)$.

Although it is rather cumbersome, we can write down a formula for the intensity in the sideband at phonon frequency, ω, $I(\omega)$. This is

$$I(\omega) = I_0 \left| \sum_j \alpha_j{}^\omega \langle l | \mathbf{d} | j \rangle + \sum_j \beta_j{}^\omega \langle j | \mathbf{d} | i \rangle \right|^2 \rho(\omega) \Big/ |\langle l | \mathbf{d} | i \rangle|^2 \quad (1.37)$$

where I_0 is the intensity in the no-phonon line, and $\rho(\omega)$ is the density of lattice modes at frequency ω. In this expression the dipole operator, $\mathbf{d}$, has two forms; it is either the magnetic dipole operator, μ/ec, or the electric dipole operator $\mathbf{r}$. The specific form of $\mathbf{d}$ need not be the same for all the matrix elements in the numerator, and the form in the denominator is not dependent on the form used in the numerator.

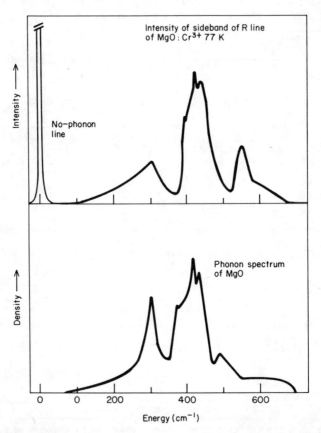

Fig. 1.22. Comparison of the shape of the sideband of the R line of MgO:Cr^{3+} at 77 K and of the density of phonon states for MgO (data from Peckham, *Proc. Phys. Soc.* **90**, 657 (1967)).

If the α_j^ω and β_j^ω factors are constants which are independent of ω, then the intensity of the sideband at phonon frequency ω would vary only as $\rho(\omega)$, and the sideband shape would then reflect accurately the density of states. The α_j^ω and β_j^ω factors, however, are not constants. Nevertheless, the sideband shape does reflect many of the strong features of the density of phonon states. A comparison of the one-phonon luminescence sideband of MgO:Cr^{3+} with the true density of phonon states for MgO is shown in Fig. 1.22.

Let us return to the points made in relation to Eq. 1.37, that the form of $\mathbf{d}$, whether it is $\boldsymbol{\mu}/ec$ or $\mathbf{r}$, need not be the same for all the matrix elements, and that the α_j^ω and β_j^ω factors need not be constants independent of ω. For this discussion let us consider only the first term in V_c^y, that is, $V_1\varepsilon$. The amount of $|j, (n+1)_\omega\rangle$ which is mixed into $|i, n_\omega\rangle$ by $V_1\varepsilon$ is

$$\frac{\langle j|V_1|i\rangle\langle (n+1)_\omega|\varepsilon|n_\omega\rangle}{E_i - E_j - \hbar\omega} \tag{1.38}$$

Now the distortion of the crystal field due to the dynamic strain, ε, may be of odd parity. This means that V_1 in the above formula can be an odd parity function. Figure 1.23 shows how an odd parity lattice vibration can destroy the inversion symmetry of the dopant ion's environment, thereby inducing

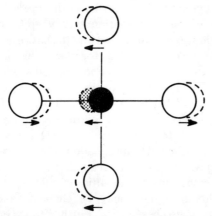

Fig. 1.23. Two-dimensional example of how an odd parity mode of vibration can destroy the inversion symmetry at the site of the central ion.

a V_1 component of odd parity in the ion-vibrating lattice interaction. This odd parity V_1 component can mix some odd parity j state with an even parity i state, and vice versa, as expression (1.38) shows. This can have a profound effect upon the strength of the sideband transition.

Let us assume that states i and l (Fig. 1.21(a), (b)) have the same *even* parity. Then only a magnetic dipole matrix element can exist between i

and l, and this is the dominant matrix element in the no-phonon line (I in Fig. 1.21(a)). Because of the odd parity component of V_1, some of the j and j' states of Fig. 1.21(a), (b) will have *odd* parity. We can then have electric dipole matrix elements connecting $|i, n\rangle$ and $|j', n\rangle$, and connecting $|l, n+1\rangle$ and $|j, n+1\rangle$ in Fig. 1.21(b), and these are the matrix elements which are responsible for the sideband transitions. Since electric dipole processes are so much stronger than magnetic dipole processes, this can intensify the sideband process relative to the no-phonon line. In Fig. 1.22 where the luminescence transition of $MgO:Cr^{3+}$ is shown, the sharp no-phonon line is a magnetic dipole process, but the sideband is almost certainly an electric dipole process caused by lattice vibrations which introduce dynamic crystal field distortions of odd symmetry at the site of the Cr^{3+} ion. Most broadband transitions on dopant ions in solids are electric dipole processes caused by dynamic lattice distortions of odd symmetry (vibronic transitions).

1.2.11. Analysis of the Broadband Transitions on Dopant Ions in Crystals

The broad bands found in the spectra of ions in crystals are due to optical transitions in which many phonons are generated in addition to an electronic transition on the dopant ion. Such multi-phonon processes are indicative of strong coupling between the ion and the environment, and they cannot be handled using the simple perturbation theory approach of the previous section. We can no longer regard the ion and the lattice as almost independent systems which are weakly coupled together. One of the very helpful simplifications in the previous section was the fact that the lattice Hamiltonian, $\mathcal{H}_L$, was independent of the electronic state of the dopant ion. This is no longer true when the coupling is strong. When the dopant ion is in the ground state, the coupling between the dopant ion and the lattice has a certain strength and this influences the average arrangement of the neighbouring ions. All the ions vibrate about their average positions in any of a number of allowed vibrational frequencies. When the dopant ion is in an excited state, the coupling to the lattice is different, the average arrangement of the neighbouring ions may be changed, and the spectrum of vibrational modes may also be changed. Further, as the environment vibrates in one of these modes, the electronic energy of the dopant ion is being strongly modulated. Clearly, the dopant ion and the lattice are part of the same tightly bound system.

In order to discuss optical transitions on this tightly bound system we shall have to adopt some approximations; an exact description is impossible. The first is the Born–Oppenheimer approximation. This says that the vibrational motion of the ion is so slow by comparison with the motion of the electrons within the ion, that the electronic state of the ion is constantly adjusting itself to the comparatively slowly varying crystal field. So we should first solve for the vibrational motion of the ions. Then we should solve for the electronic states of the ion, regarding the distances to the neighbouring ions as adjustable parameters. The wavefunction for the dopant ion would be expressible as $\Psi(r_i, R_1, R_2, ...)$ where r_i represents the electrons of the dopant ion, the R's are the distances from the dopant ion to the various neighbouring ions, and these R's are regarded as slowly varying parameters. Now because of the wide spectrum of available lattice vibrations, it is impossible to take into account all the possible arrangements of R values when solving for the electronic wavefunction of the dopant ion. So we introduce the second simplifying approximation—that we need consider only one mode of lattice distortion and that all static and vibrational distortions are of this type. This mode of vibration is the "breathing mode" in which the surrounding lattice pulsates in and out about the dopant ion. We assume that a harmonic oscillator model describes these pulsations. This model has the great advantage that we need only one parameter to describe the lattice distortion. This parameter is the distance from the dopant ion to the nearest-neighbour ions, it is called the configurational coordinate, and it is given the label Q. We also assume that the vibrations occur only at one frequency, ω. The wavefunctions of the dopant ion can be written as $\Psi(r_i, Q)$ to indicate that the crystal field is represented by one single parameter—the configurational coordinate Q. The value of this parameter oscillates about its average value at one frequency ω. The average value of Q may be different for different states of the same dopant ion, and the frequency of the breathing mode vibration may also be different. This one-parameter one-frequency harmonic oscillator model gives us a very simplified picture of the ion-plus-lattice system. It has the advantage that we can represent the electronic and lattice states on a single energy level diagram, and this is called a *configurational coordinate diagram*.

The configurational coordinate diagram showing the ground and excited states of a hypothetical ion-plus-lattice system is seen in Fig. 1.24. Energy is plotted on the vertical axis, and the value of Q is plotted on the horizontal axis. When the ion is in the ground electronic state g, the configurational coordinate has the average value Q_0. The ground state harmonic oscillator parabola, $V = \frac{1}{2}M\omega^2(Q - Q_0)^2$, is also shown. M is the equivalent ionic mass which is engaged in the vibrations, and ω is the characteristic vibrational frequency. Within the parabola we have sketched some of the quantized harmonic oscillator levels—the nth vibrational state is described by the harmonic oscillator function $\chi_n(Q)$ and it has energy $(n + \frac{1}{2})\hbar\omega$. When the ion is in the excited state e, the average value of the configurational coordinate is Q_0', which is different from the ground state value. The harmonic oscillator parabola appropriate to this state is also shown. We have drawn

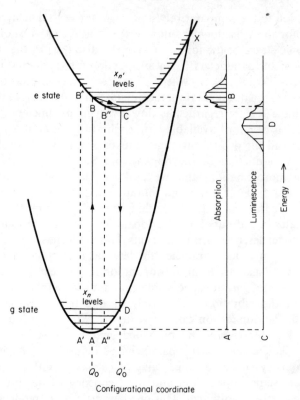

Fig. 1.24. Configurational coordinate diagram to analyze transitions between two electronic states *g* and *e*. The theoretical shapes and energies of the absorption and luminescence transitions are shown on the right.

the excited state parabola such that the vibrational frequency in the excited state, ω', is less than the vibrational frequency in the ground state. Some of the quantized harmonic oscillator levels in the excited state parabola are also shown. These vibrational states are described by the harmonic oscillator functions $\chi_{m'}(Q)$ with energy $(m'+\frac{1}{2})\hbar\omega'$. The $\chi_n(Q)$ functions form an orthonormal set. The $\chi_{m'}(Q)$ functions also form an orthonormal set, but they are distinct from the previous set, since the origin of the Q coordinate is different in the two cases and the frequencies may also be different. Note that by drawing the vibrational levels as horizontal lines, we have neglected to show the variation in the electronic energy as Q varies within the same vibrational state.

In Fig. 1.25 we show the probability amplitude squared in the zero vibrational state ($n = 0$) and in a higher ($n = 20$) vibrational state. We see that in the zero vibrational state the most likely value of Q is its value at the lowest point of the parabola, while in high vibrational states the most likely values of Q are at the turning points of the oscillation, that is, at the edges of the parabola.

Now in the spirit of the Born–Oppenheimer approximation we say that an optical transition on the dopant ion occurs so rapidly that the lattice arrangement does not change during the transition. On the configurational coordinate diagram we therefore show the optical transition by a vertical arrow, indicating that the initial and final values of Q are the same. This is called the Franck–Condon principle.

Let us consider the absorption transition between the electronic states g and e at low temperatures (Fig. 1.24). At low temperatures only the zero vibrational level of the g electronic state is occupied. Since the most likely value of the configurational coordinate is Q_0, we draw the vertical arrow upwards from the point A on the zero vibrational level. Which vibrational level of the electronic e state does the arrow end on? The matrix element for the optical transition between the zero vibrational level of the ground state, when the configurational coordinate has some value Q, and the m'th vibrational level of the excited state, when the configurational coordinates has the same value, is

$$\langle e(Q) | \mathbf{d} | g(Q) \rangle \chi_{m'}{}^*(Q) \chi_0(Q) \tag{1.39}$$

The probability of an optical transition between the zero vibrational level of the ground state and the m'th vibrational level of the excited state varies as the square of the matrix element

$$\int \langle e(Q) | \mathbf{d} | g(Q) \rangle \chi_{m'}{}^*(Q) \chi_0(Q) \, dQ \tag{1.40}$$

The range of values of Q which can occur in the ground state is shown by the width of the probability amplitude function in the zero vibrational state (Fig. 1.25). If we assume that this range of values is small enough so that the electronic states of the dopant ions, $g(Q)$ and $e(Q)$, do not vary significantly over this range of Q values, then we can replace the variable Q in the electronic matrix element by the fixed value Q_0. This allows us to separate the two integrals in expression (1.40), and the matrix element becomes

$$\langle e(Q_0) | \mathbf{d} | g(Q_0) \rangle \int \chi_{m'}{}^*(Q) \chi_0(Q) \, dQ = \langle e(Q_0) | \mathbf{d} | g(Q_0) \rangle \langle \chi_{m'} | \chi_0 \rangle \tag{1.41}$$

The electronic matrix element in the above expression is the same irrespective of the vibrational levels involved in the transition, so the relative probability that the absorption transition ends up on the m'th vibrational level varies as $|\langle \chi_{m'} | \chi_0 \rangle|^2$. This will have a maximum value when $|\chi_0|^2$ and $|\chi_{m'}|^2$ have maxima at the same value of Q. If m' is a large quantum number, the $\chi_{m'}$

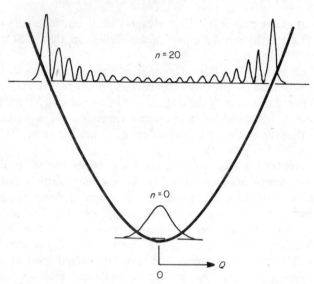

Fig. 1.25. Shape of $|\chi_n(Q)|^2$ for $n = 0$ and $n = 20$. The $E = \frac{1}{2}M\omega^2 Q^2$ parabola is also drawn, and the energy levels appropriate to the $\chi_n(Q)$ vibrational wave functions for $n = 0$ and $n = 20$ are also drawn. For large values of n, $|\chi_n(Q)|^2$ has its maximum at the values of Q where the nth energy level touches the parabola.

function has its maxima (and minima) at the values of Q corresponding to the end-points of the oscillation, that is, where the m'th energy level touches the excited state parabola. This is seen in the $n = 20$ example in Fig. 1.25. χ_0 has its maximum at Q_0. The most likely m'th vibrational state to be reached by the upward transition then is the state whose $\chi_{m'}$ function has maxima at Q_0, that is, the vibrational state whose energy level touches the excited state parabola at Q_0. Thus we represent this most probably low temperature absorption by the vertical arrow AB. Of course the transition may end up on higher or lower m' levels, but its probability will be less than that of the AB transition. We can calculate the width of the absorption transition by finding the highest and lowest m' values for which the overlap integral with χ_0 is non-zero. Since χ_0 falls to zero roughly at these values of Q where the zero vibrational level touches the ground state parabola, vertical lines drawn upwards from these edge points to the excited state parabola will give approximate values of the maximum and minimum energies—$A'B'$ and $A''B''$, respectively.

What we have represented pictorially here can be expressed in more precise mathematical terms, and values of $|\langle \chi_{m'}|\chi_n \rangle|^2$ have been derived in terms of the configurational coordinate parameters. In Fig. 1.24 where we illustrated the absorption processes on the configurational coordinate diagram, we show the individual absorption transitions to the different vibrational levels of the e state at the side. A fairly smooth envelope is drawn over the individual absorption transitions, and this curve represents the full absorption band from g to e. Some of the individual transitions project a little from this smooth curve. In particular, the transition between the zero vibrational levels of e and g, when it is allowed, appears as a sharp singularity at the side of the band. These features are seen in the absorption band shown in Fig. 1.20(c). This zero-vibrational transition—so called because there is no change in vibrational state—is analogous to the no-phonon transition discussed in the previous section.

At high temperatures the absorption transition can start from higher vibrational levels of the ground state and the absorption band becomes even broader.

At low temperatures the total strength of the absorption band varies as

$$\sum_{m'} |\langle e(Q_0)|\mathbf{d}|g(Q_0)\rangle|^2 |\langle \chi_{m'}|\chi_0 \rangle|^2 = |\langle e(Q_0)|\mathbf{d}|g(Q_0)\rangle|^2 \qquad (1.42)$$

where we use the completeness of the harmonic oscillator functions. Therefore on the configurational coordinate model the strength of the full absorption band can be estimated using the purely electronic oscillator strength formulae given previously. According to this analysis the absorption strength of the *broadband* transition between the g and e states of a dopant ion in a vibrating lattice is equal to the absorption strength of the *sharp line* transition between the same g and e states in the case where the dopant ion is in a static lattice, and where the neighbouring ions are fixed in their time-average positions. The effect of the vibrations of the lattice on the optical transition is merely to change the shape but not the strength of the transition. This satisfying theoretical result, unfortunately, rarely occurs in reality. The above model, since it considers only the totally symmetric even parity breathing mode vibrations, neglects the vibrations which introduce odd parity crystal field effects; and these latter vibrations can allow electric dipole processes to occur. Particularly if the transition represented by the $\langle e|\mathbf{d}|g\rangle$ matrix element is a magnetic dipole process or a very weakly allowed electric dipole process, the main part of the absorption band will be due to vibrationally induced electric dipole processes (vibronic transitions) and will be much stronger than the configuration model predicts. This is generally the case for transitions involving transition metal dopant ions. Although the configurational model may not predict the correct strength for the transition, it does predict quite accurately the shape of the absorption transition.

After the ion-plus-lattice system under study in Fig. 1.24 is excited by the absorption transition, it may decay by luminescence emission. The configurational coordinate model can be used to predict the shape of the luminescence transition. Again we will assume that the material is at low temperatures. After the absorption transition $(A \to B$, say), some higher vibrational levels of the excited state parabola are populated. This vibrational energy is transmitted to the rest of the crystal until the zero vibrational state of the excited state parabola is reached $(B \to C)$. The ion-plus-lattice is now said to be in the *relaxed excited state*. The probability of a radiative transition from this relaxed excited state to the nth vibrational level of the ground state can be calculated as before, and it is represented by the matrix element

$$\langle g(Q_0') | \mathbf{d} | e(Q_0') \rangle \langle \chi_n | \chi_0' \rangle \tag{1.43}$$

where Q_0' is the average configurational coordinate appropriate to the excited state. Using the same arguments as before, we can estimate the shape of the luminescence transition. The peak of the luminescence, that is, the most likely luminescence transition, will occur at an energy given by the arrow CD. The zero-vibrational transition, if it occurs, will occur at the same energy in absorption and in luminescence. The shapes of the parabolas in Fig. 1.24 are different, and this leads to the difference in the shapes of the absorption and luminescence bands. This model predicts a shift to lower energy between the peaks of the absorption and luminescence bands, and this shift which is found experimentally is called the *Stokes' shift*. Because of its very strong coupling to the lattice, the F-centre exhibits a large Stokes' shift, as seen in the case of KBr in Fig. 1.18. In this case the configurational coordinate displacement $(Q_0' - Q_0)$ is too large for a zero-vibrational transition to occur. Its analysis on a configurational coordinate diagram is shown in Fig. 1.26.

The existence of a Stokes' shift is very helpful in the broadband luminescence process, for it means that the luminescence occurs at a lower frequency than the absorption, and as a result the luminescence is not reabsorbed by the material. Reabsorption can only occur for the no-phonon line, and it can be quite severe if the line is narrow and of strong oscillator strength.

Broad transitions are so often interpreted on the configurational co-ordinate model that it is worth while to write down some formulae involving the parameters of the model. We assume, for simplicity, that all parabolas have the same shape and vibrational frequency. The lateral displacement between the two parabolas is ΔQ. A parameter, S, called the *Huang-Rhys parameter* is defined by

$$\tfrac{1}{2} M\omega^2 (\Delta Q)^2 = S\hbar\omega \tag{1.44}$$

so S is a measure of the displacement between the parabolas and is a dimensionless quantity. From Fig. 1.24 we see that the Stokes' shift, ΔE_S, is given by

$$\Delta E_S = M\omega^2 (\Delta Q)^2 - \hbar\omega = (2S - 1)\hbar\omega \tag{1.45}$$

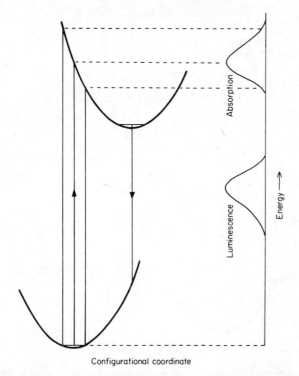

Configurational coordinate

Fig. 1.26. Configurational coordinate diagram for the case where the two parabolas have the same shape and where the displacement $Q_0' - Q_0$ between the parabolas is large. The predicted shapes of the absorption and luminescence transitions are shown on the right. In this case the transition shapes are mirror images of each other and the no-phonon line does not occur. This is appropriate for the case of the F-centre in KBr.

The square of the overlap integral of harmonic oscillator functions, which is useful in calculating band shapes, is given by

$$|\langle \chi_{n'} | \chi_m \rangle|^2 = \exp(-S) \frac{m!}{n'!} S^{n'-m} |\mathscr{L}_m{}^{n'-m}(S)|^2 \qquad (1.46)$$

where $\mathscr{L}$ is a Laguerre polynomial. When $n' = m = 0$ we get

$$|\langle \chi_0' | \chi_0 \rangle|^2 = \exp(-S) \qquad (1.47)$$

From this we see that $\exp(-S)$ is the fraction of the intensity which is contained in the sharp no-phonon line—according to the configurational coordinate model which neglects vibronic processes.

Let us apply the configurational coordinate model to analyze the absorption and luminescence transitions involving the $^4A_2$ ground state and the ^2E and $^4T_2$ excited states of Cr^{3+} in ruby. The appropriate diagram is shown in Fig. 1.27. When Cr^{3+} is in the $^4A_2$ ground state, its three outer 3d electrons

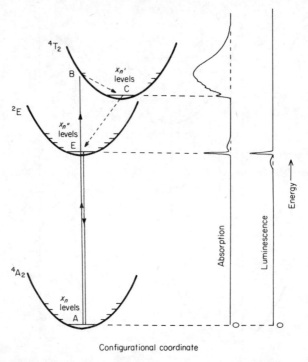

Fig. 1.27. Configurational coordinate diagram suitable for discussing the transition between the $^4A_2$, ^2E, and $^4T_2$ states of Cr^{3+} in ruby or MgO. The predicted absorption and luminescence spectra are shown on the right.

occupy t_{2g} orbitals. (We neglect the small deviation from octahedral symmetry in this analysis.) The coupling of the ion to its neighbours depends in large measure on the orbital states of these outer electrons. In the ground state the average value of the configurational coordinate is Q_0, and the appropriate harmonic oscillator parabola is drawn. When Cr^{3+} is in the ^2E state, the same three t_{2g} orbitals are occupied, and as a result the coupling of the ion to its neighbours is very similar to what it was in the ground state. So we have drawn the harmonic oscillator parabola for the ^2E state vertically above the $^4A_2$ state parabola. The shift in average Q value between them is too small to be seen in the figure. When Cr^{3+} is in the $^4T_2$ state, two of the outer 3d electrons are in t_{2g} orbitals while the third is in an e_g orbital. This changes the coupling to the neighbours, and the average Q value is different from the outer two cases. For simplicity we have assumed that the same vibrational frequency and harmonic oscillator parabola is appropriate for all three electronic states.

The $^4A_2 \rightarrow ^4T_2$ absorption spectrum at low temperatures is calculated in the manner employed in Fig. 1.24; its shape, as determined by the configurational coordinate model, is shown on the side of Fig. 1.27; and its observed shape in the case of ruby is shown in Fig. 1.20(c).

When the ^2E parabola is below $^4T_2$, as in the case of ruby, one does not observe luminescence at low temperatures from the $^4T_2$ state. The reason is that when the relaxed $^4T_2$ state is occupied by optical pumping it decays non-radiatively to the relaxed ^2E state (path CE). At low temperatures the luminescence occurs from the zero-vibrational state of ^2E. The probability of radiative emission from the χ_0'' zero-vibrational level of the ^2E state to the χ_n vibrational level of the $^4A_2$ ground state varies as

$$ |\langle ^4A_2(Q_0)|\mathbf{d}|^2E(Q_0)\rangle|^2 |\langle \chi_n(Q)|\chi_0''(Q)\rangle|^2 \qquad (1.48) $$

The electronic matrix element has been discussed previously. The shape of the transition is determined by the overlap integral of the vibrational states. If the ^2E and $^4A_2$ parabolas are directly above each other, then $\langle \chi_n(Q)|\chi_0''(Q)\rangle = 0$ unless $n = 0$, since both sets of vibrational levels will be identical to each other. This means that only the zero-vibrational transition (the no-phonon line) is allowed and no sideband should occur. This result is not unexpected since the approximations adopted to obtain expression (1.47) above include neglecting the modulation of the electronic levels by the vibrations. If we wish to take this into account, this is most easily done by using the perturbation theory approach of Section 1.2.10. As we saw in that section, the ion-vibrating lattice interaction will explain why a sideband occurs in addition to the no-phonon line. Returning to the configurational coordinate diagram, we can suppose that the ^2E and $^4A_2$ parabolas do not lie directly above each other but they are laterally displaced by a small amount. This means a small Huang–Rhys parameter S. Consequently $\exp(-S)$ is not quite unity, so some amount of the luminescence occurs in a band beside the no-phonon line. The shape of the $^2E \rightarrow ^4A_2$ luminescence transition in $MgO:Cr^{3+}$ is seen in Fig. 1.22; it shows the strong no-phonon line and weak sideband predicted by this analysis.

At this point it is worth while clarifying the relationship between the strength of the ion-lattice coupling and the width of the transitions. We shall want to qualify the normally accepted statement that broad optical transitions are indicative of strong ion-lattice coupling, while sharp optical transitions are indicative of weak ion-lattice coupling. The Cr^{3+} ion in the $^4A_2$, ^2E, and $^4T_2$ states interacts strongly with the lattice. If the interactions were identical in the two of these states, as it almost is for $^4A_2$ and ^2E, then we should only get a sharp line transition between $^4A_2$ and ^2E. It is the occurrence of a large *difference in the ion-lattice coupling* in the initial and final states which causes the optical transition between such states to be broad.

In the case of an F-centre, the lattice distortion in the excited state is so different from what it is in the ground state that the relaxed excited electronic state (from which the luminescence occurs) can be significantly different from the excited electronic state reached by the absorption transition. Because of this difference in the electronic states, the electronic matrix elements for absorption and emission can be different in absorption and in luminescence.

In this and in the previous sections we have dealt at some length with the effect of the environment—particularly the vibrating environment—on the radiative transitions, since it is central to the understanding of the shapes of the transitions. Yet it must be realized that we are using very simple models in these explanations. The fact that they sometimes accurately predict band shapes (which they are designed to do) does not mean that they are an accurate description of all that is happening. For example, the con-figurational coordinate model is a harmonic model which neglects anharmonic terms, and such anharmonic terms can be important in non-radiative relaxation processes. This model makes no allowance for odd parity vibrations which can profoundly affect the strength of the transitions; and it seems to lose sight of the effect of the vibrations on the energy levels of the electronic states.

1.2.12. Non-radiative Transitions

For both magnetic dipole and electric dipole radiative processes, we saw that the transition probability increases as ω^3, where $\hbar\omega$ is the energy of the photon involved. Consequently, when an ion is in an excited state, the probability of radiative decay to a lower state varies as the cube of the energy gap between initial and final states. Let us now consider the prob-ability of non-radiative decay from the excited state, and ask how it varies with the energy gap to the next lower level. In non-radiative decay the energy is released as phonons. We shall find that the probability of non-radiative decay *decreases* as the gap increases, in contrast with the radiative process. It is a matter of considerable technical importance to know whether we can speak of a critical gap—above which radiative processes should dominate and below which non-radiative processes should dominate. For dopant rare-earth ions it would seem that the concept of a critical gap has validity. The critical gap depends on the host material, but for a given host material it is roughly the same for all rare-earth ions. This state of affairs occurs probably as a result of the weakness of the ion–lattice interaction. For transition metal ions, on the other hand, the ion–lattice interaction is so strong, and it varies so much from state to state of the same ion, that the concept of a single critical gap for all transition metal ions is not nearly so valid.

Non-radiative decay occurs because of the interaction of the dopant ion with the vibrating lattice. The probability of non-radiative decay depends on the strength of the coupling between the ion and the vibrating lattice, V_c^v, and in the case where this coupling is weak, we shall attempt to analyze non-radiative behaviour using perturbation theory. We will assume that V_c^0, the static part of the ion-lattice coupling, has already been taken into account in calculating the electronic states i. We regard V_c^v as a perturbation. We shall also assume that the lattice vibrations are the same whatever the electronic state i. The basis functions are $|i, n\rangle$, and we are interested in calculating the non-radiative transition probability from $|i, n\rangle$ to $|f, m\rangle$ due to the V_c^v perturbation. We express V_c^v as the usual power series in the strain ε: $V_c^v = V_1 \varepsilon + V_2 \varepsilon^2 + \ldots = \sum_l V_l \varepsilon^l$ and we use the second quantized formula for ε in terms of annihilation and creation operators (Eq. 1.33).

The first term is linear in a_ω and a_ω^+, and gives rise to processes where one phonon is created. The second term, when used in first-order perturbation theory, gives rise to two-phonon processes. A two-phonon process can similarly arise due to the first term ($V_1 \varepsilon$) in second-order perturbation theory. In general the $V_n \varepsilon^n$ term in first-order perturbation theory gives rise to an n-phonon relaxation process, the $V_1 \varepsilon$ term in nth-order perturbation theory does likewise, as will intermediate terms in lower order. Clearly the calculation of an n-phonon relaxation process by this procedure is tedious.

How many phonons are involved in typical decay processes? A radiative transition in the green involves a gap of around $18,000 \text{ cm}^{-1}$, but the active phonons of highest energy typically have energies of around 500 cm^{-1}. Therefore if a non-radiative transition occurs across this $18,000 \text{ cm}^{-1}$ gap, it involves the creation of around 36 phonons! So we are dealing with a process of high order in perturbation theory. The greater the gap, the greater is the number of phonons involved, and the higher the order of perturbation theory needed. The higher the order of perturbation theory needed, the smaller is the probability. So the larger the gap the smaller is the probability of non-radiative decay.

Let us write $W(n)$ as the probability of an n-phonon relaxation process. Let us postulate that the ratio of the nth to the $(n-1)$th process is a constant, which is characteristic of the host crystal, characteristic perhaps of the ion, but essentially independent of n:

$$\frac{W(n)}{W(n-1)} = \gamma \ll 1 \tag{1.49}$$

This gives an exponential dependence of $W(n)$ on n. If the gap has energy ΔE and if $\hbar\omega$ is the energy of the dominant phonon, then $\Delta E = n\hbar\omega$, and we can write

$$W = A \exp\left(-\frac{\Delta E}{\hbar\omega}\right) \tag{1.50}$$

where W is the probability of occurrence of a non-radiative transition across the gap ΔE. For a given host material, $\hbar\omega$ will be the same irrespective of the dopant ion. The parameter A will be characteristic of the host and will depend as well upon the sensitivity to strain of the levels of the dopant ion. This formula is found to be obeyed quite well by the trivalent rare-earth dopant ions, and the value of A is approximately the same for different rare-earth dopant ion in the same host material, as Fig. 1.28 shows for the host material $YAlO_3$.

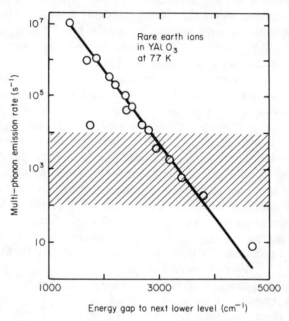

Fig. 1.28. A plot of the non-radiative decay rate against energy gap to the next lowest level in the case of a number of excited states of various rare earth ions in $YAlO_3$. The data are obtained from transitions on trivalent Nd, Eu, Ho, Er, and Tm. The shaded area indicates the range of values of radiative decay rates in the same ions (after Weber, *Phys. Rev.* **8B**, 54 (1973)).

The radiative decay times of the trivalent rare-earth ions are typically in the range 10^{-4} s to 10^{-2} s—in the region covered by the shaded area in Fig. 1.28. From a comparison of the radiative and non-radiative rates we see that the critical gap is 3000—4000 cm^{-1}. The active phonons in this material are in the 550—600 cm^{-1} range. Hence we can adopt the general rule for this material that, if the gap from the excited state of the dopant rare-earth ion to the next lowest state can be bridged by 7 phonons or less, the excited state will decay non-radiatively, while for larger gaps the state will decay radiatively. Similar behaviour is found for some other host materials—the critical gap corresponding roughly to an energy of about 7 phonons.

Dopant transition metal ions do not lend themselves to the same mathematical treatment because of their larger and more complex ion–lattice interaction. It would seem, however, that there is some validity to a rule that a multi-phonon decay process involving no more than 20—25 phonons is more efficient than a radiative process across the same gap. We see that non-radiative decay processes are more probable for transition metal ions than for rare-earth ions, as is expected from the difference in the coupling to the lattice in the two cases. The smallest energy gap across which transition metal ion luminescence occurs would seem to be found in Co^{2+} and Ni^{2+}, in both of which the first excited state is about 7000 cm^{-1} above the ground state. For Co^{2+} in $KMgF_3$, the radiative and non-radiative transition probabilities are approximately equal at low temperatures. For rare-earth ions, however, luminescence can be found at a wavelength beyond 3 μ.

In the energy level diagrams of the transition metal ions, it is noticeable that the largest gap usually occurs between the ground state and the first excited state. The higher excited states are separated by gaps of less than 7000 cm^{-1} and so are bridged by non-radiative processes. Luminescence, when it occurs in the transition metal ions, usually occurs across the large gap between the lowest excited state and the ground state. On the other hand, luminescence can occur from many excited states of the same rare-earth ion.

In the case of dopant ions and centres with a strong coupling to the lattice, we might hope that the configurational coordinate model could be of help. Indeed, the following interesting fact has been observed. Let us consider the absorption process from g to e which we analyzed with the aid of Fig. 1.24. If the absorption band populates excited vibrational levels of the e parabola which are higher than X—the point of intersection of the e and g parabolas— then the excited state will decay non-radiatively to the ground state. Even if the vibrational levels which are initially populated are below X, thermal activation may cause the levels at X to become populated, and these will decay non-radiatively. As a result the quantum efficiency will decrease with increasing temperature; the luminescence is said to become *quenched* as the temperature is raised.

When we attempt to get numerical values for the non-radiative decay rates using configurational coordinate diagrams, the results can be disappointingly poor. For example, the $^2E \rightarrow {}^4A_2$ non-radiative relaxation rate in Cr^{3+}, as calculated from a configurational coordinate model, is too small by a factor of around 10^{16} by comparison with what is believed to be the observed value! The trouble with the use of the configurational coordinate model in this case would seem to stem from its neglect of anharmonic effects. Since this large-gap transition involves about 30 phonons, anharmonic effects could be very significant in such a high-order decay process. Non-radiative relaxation rates calculated on the configurational coordinate model may be much more realistic when the gap is small.

1.2.13. Doubly Doped Materials and Energy Transfer among Ions

Consider Mn^{2+} which when excited sometimes emits yellow luminescence; its energy level diagram is seen in Fig. 1.9. We notice that all the absorption transitions are spin-forbidden, the absorption bands are therefore weak, and the Mn^{2+} ions cannot be efficiently raised to excited states by direct optical pumping. Nevertheless, Mn^{2+} is one of the most important luminescence centres in commercial phosphors. For example, the double-doped phosphor $Ca_5(PO_4)_3F:Sb^{3+},Mn^{2+}$ is used in commercial fluorescent lamps where it converts mainly ultraviolet light from a mercury discharge into visible radiation. When 2536 Å mercury radiation falls on this material, the radiation is absorbed by the Sb^{3+} ions, not by the Mn^{2+} ions. Some excited Sb^{3+} ions emit their characteristic blue luminescence, while other excited Sb^{3+} ions transfer their energy to Mn^{2+} ions, and these excited Mn^{2+} ions emit their characteristic yellow luminescence. The efficiency of transfer of ultraviolet photons through the Sb^{3+} ions to the Mn^{2+} ions can be, perhaps, as high as 80%. The strong absorption strength of the Sb^{3+} ions, along with the efficient transfer of excitation from Sb^{3+} to Mn^{2+}, are responsible for the strong manganese luminescence from this material: this is an example of *energy transfer* and it is illustrated schematically in Fig. 1.29. The ion which emits the light and which is the active element in the material is called the *activator*; the ion which helps to excite the activator and which makes the material more sensitive to pumping light is called the *sensitizer*.

Let us look into the transfer of energy between two ions, say from ion A to ion B where A and B are dissimilar ions. We are not speaking about a transfer which takes place radiatively, in which A emits a photon of light which is subsequently absorbed by B. Rather it is a non-radiative transfer of excitation from A to B. There must be some interaction between A and B for this transfer to occur. Electric and magnetic forces between the ions can cause the transfer, as can the exchange interaction. All these forces fall off rapidly as the separation between ions increases, consequently the probability of energy transfer between ions decreases rapidly as the separation between the ions increases.

In the transfer between two dissimilar ions (A and B in Fig. 1.30) the energy gaps will not be exactly equal so some of the energy is released as phonons. The process can be regarded as phonon-assisted energy transfer. In calculating the transition probability that excitation will be transferred from A to B and that the excess energy will be released as phonons, one

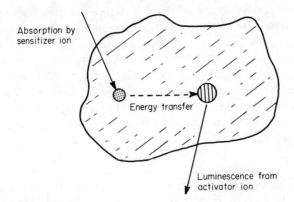

Fig. 1.29. Non-radiative transfer of energy. The sensitizer ion absorbs the radiation and becomes excited. Because of a coupling between sensitizer and activator ions, the sensitizer transmits its excitation to the activator, which becomes excited, and the activator may release the energy as its own characteristic radiation. The activator → sensitizer transfer is *not* a radiative emission and absorption process, rather a non-radiative transfer.

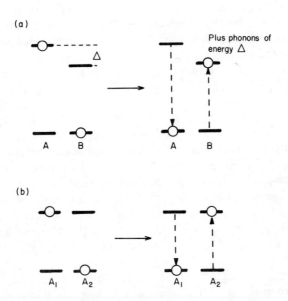

Fig. 1.30. In the transfer of energy between dissimilar ions, the levels will, in general, not be in resonance, and some of the energy is released as a phonon or as phonons. In the case of similar ions the levels should be in resonance, and phonons are not needed to conserve energy.

must take into account not only the interaction between the two ions, but also the interaction of the ions with the lattice. In the transfer between similar ions ($A_1 \rightarrow A_2$ in Fig. 1.30) no excess energy is released, we do not have to invoke the ion–lattice interaction in calculating the probability, and consequently the probability of transfer between similar ions should, in general, be greater than the probability of transfer between dissimilar ions. In singly doped materials, then, energy transfer can occur among the identical dopant ions when the concentration of dopant ions is high so that the average separation between adjacent ions is small, and a sufficiently strong interaction can occur between them. This can have a significant effect upon the quantum efficiency of luminescence at high dopant concentrations.

Sometimes the host material itself may absorb (usually in the ultraviolet) and the energy can be transferred non-radiatively to dopant ions. For example, in $YVO_4:Eu^{3+}$ the vanadate group of the host material absorbs ultraviolet light, then transfers its energy to the Eu^{3+} ions, which emit characteristic Eu^{3+} luminescence.

Clearly many new methods for phosphor fabrication becomes available when the possibilities of double doping and energy transfer are exploited.

1.2.14. Luminescence from Materials with High Dopant Concentrations

One might expect that the higher the concentration of dopant ions the more efficient will be the luminescence. What is found, however, is that increasing the concentration above a certain critical value can lead to a reduction, or quenching, of the luminescence. This is called *concentration quenching*, and it has its origin in the very efficient energy transfer between ions which occurs at high concentrations. The probability of energy transfer to a nearby similar ion can be very much greater than the probability of radiative decay, so that the excitation moves about and "samples", perhaps, a million ions before being emitted as radiation. Now even in the purest and most carefully grown crystals there are always some defects, or traces of other ions, and the excitation can be transferred to these centres. Some of these centres when excited will not transfer the excitation back to the regular dopant ions, but will, instead, decay by multiphonon emission or by infrared emission, and this excitation does not emerge as visible luminescence. These centres are called *sinks* or *quenching traps*, since they draw off the excitation from the regular dopant ions and reduce the quantum efficiency of visible luminescence.

This process is illustrated in Fig. 1.31, where the situation is compared with the optical pumping and luminescence of materials of low concentrations. Let the observed decay time of the excited ions in materials of low concentration be labelled τ_0. In the concentrated material there is now another decay process—energy transfer to sinks and subsequent dissipation

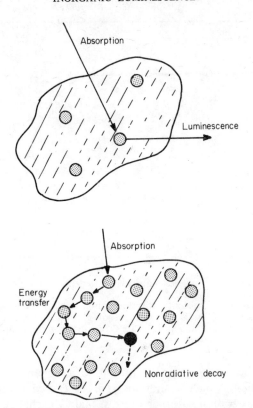

Fig. 1.31. In materials of low dopant concentration absorption of radiation and the subsequent emission of luminescence take place on the same ion. In materials of high dopant concentration the absorbing ion may transfer its excitation to another similar ion rather than emit luminescence. Because of the efficiency of energy transfer between identical atoms, the excitation may be transferred through many ions and may ultimately be trapped at a sink—and dissipated as heat.

of the energy. For a dopant ion adjacent to a sink, the transfer time to the sink will be short, for those farther away the transfer time will be longer, as the exitation has to travel along a chain of dopant ions to the sink. Let the average time of transfer to a sink be τ_x. If we label as τ_{obs} the observed decay time of the excited ions in the concentrated material, then τ_{obs} is given by

$$1/\tau_{\text{obs}} = (1/\tau_0) + (1/\tau_x) \tag{1.51}$$

This reduction in lifetime is much easier to measure than the reduction in quantum efficiency, consequently lifetime measurements play an important part in experimental studies of concentration quenching. Figure 1.32 shows the decay time of ruby luminescence at 77 K as a function of concentration

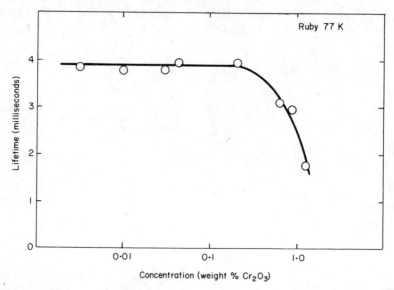

Fig. 1.32. An example of concentration quenching. In ruby the observed decay time of the luminescence at 77 K is independent of chromium concentration up to a concentration of around 0·3 weight % of Cr_2O_3 in the material (unpublished data of J. P. Larkin and G. F. Imbusch).

of Cr^{3+} ions. The lifetime is constant up to a dopant concentration of around 0·3%, above which it shortens. This is caused by transfer of the single Cr^{3+} ion excitation to other centres, mainly exchange coupled pairs of chromium ions, which emit their own characteristic luminescence.

Besides the possibility of energy transfer, a high concentration of dopant ions also means that each such ion is no longer surrounded only by the inert host material—since some of its near neighbours will also be dopant ions. Thus the crystal field at the site of the dopant ion will be changed, and as a result the frequency of luminescence may change with increasing concentration. Because they are less sensitive to the crystal field, we expect that rare-earth ions are less affected by this concentration effect than are transition metal ions. The energy transfer process also should be less efficient for rare-earth ions for the same reason that the coupling between them will not be as strong as that between transition metal ions. Such is indeed found to be the case. For example, the positions of the energy levels of Pr^{3+} in $LaCl_3$ are little changed from the positions of the levels in $PrCl_3$. Further, the quenching of Pr^{3+} luminescence in $PrCl_3$ is apparently unimportant; the luminescence is efficient enough that laser action can be generated in $PrCl_3$ by optical pumping. Laser action can also be generated in HoF_3. On the other hand, $LaCl_3$:Nd exhibits strong Nd^{3+} luminescence, but as the Nd concentration is increased, the interaction between ions manifests itself by the broadening of lines, and by concentration quenching; no luminescence occurs from $NdCl_3$.

The effect of high concentration is much more dramatic for the transition metal ions. Consider, for example, the Mn^{2+} ion which is the active element in many commercial phosphors, and which emits yellow luminescence. Consider its behaviour in MnF_2. This material has an orange colour, indicative of the green and blue absorption bands which are typical of Mn^{2+} ions. Under optical pumping at room temperature, however, no luminescence is emitted by the material. If the material is cooled below 30 K yellow luminescence is obtained, but on examination it is found that this does not originate on the Mn ions of the regular MnF_2 lattice, but rather on Mn^{2+} ions which are adjacent to defects and impurities. These perturbed Mn^{2+} ions have slightly lower energy levels than the regular ions, so they act as shallow traps for the excitation which is being transferred among the manganese ions by energy transfer. They are luminescent traps. When the temperature is high enough—around 30 K in this case—thermal activation can "boil back" the excitation from the perturbed Mn^{2+} ions to the regular Mn^{2+} ions. The excitation is again transferred along chains of manganese ions until it comes into the vicinity of a quenching trap—for example, an Fe^{2+} ion. The manganese excitation may be transferred to the iron ion which is a deep trap and the excitation cannot be returned to the manganese ions by thermal activation. Ultimately the excited Fe^{2+} decays non-radiatively.

In MnF_2 there are, therefore, two kinds of traps: shallow luminescent traps (the perturbed Mn^{2+} ions) and deep quenching traps. At very low temperatures both types of traps can operate; they are fed by the efficient energy transfer along chains of regular Mn^{2+} ions. Some of the absorbed energy is released as light from the shallow luminescent traps. At higher temperatures thermal activation makes the shallow Mn traps inoperative, only the quenching traps are active, and luminescence is quenched. No matter how pure the crystal of MnF_2, the few parts per million of unintentional quenching traps still suffice to quench the luminescence at high temperatures. This indicates how efficient is the energy transfer process in this material. These processes are illustrated schematically in Fig. 1.33.

If MnF_2 is intentionally doped with rare-earth ions, such as Er^{3+} or Eu^{3+}, the excitation can be transferred to these dopant ions, they act as rather shallow luminescent traps, and they emit the characteristic luminescence. At high temperatures, however, thermal activation again depopulates these traps, and all luminescence is again quenched.

At high concentrations we can clearly not regard the optically active ions as being individually isolated from each other—which was the basis for our analysis of luminescence behaviour up to now. For the fully concentrated

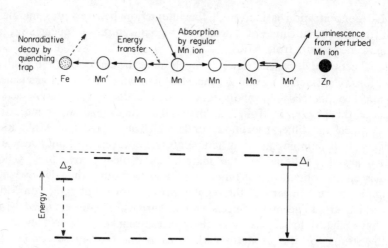

Fig. 1.33. Visualization of the transfer of excitation along chains of exchange coupled Mn^{2+} ions in MnF_2. A Mn^{2+} ion adjacent to a defect, such as a Zn ion in a Mn site, has a lower excited level and may act as a luminescent trap. Fe and Ni ions may draw off the manganese excitation; these can release the energy non-radiatively, consequently they quench the luminescence.

materials the true states of the optically active system are collective states of all the optically active ions—*exciton states*. As far as luminescence is concerned, these exciton states provide the channels whereby excitation is transferred through the material, seeking out traps where either luminescence occurs or the energy is converted into heat. In general, isolated defects and traps determine the luminescence behaviour of concentrated materials.

There is an additional interesting behaviour which is found in concentrated materials of the rare earth and transition metal ions. Because of the unpaired electrons on the optically active ions, these ions behave as elementary magnets. The close proximity of these elementary magnets in the concentrated materials may lead to exchange coupling between the ions, which can result in a magnetic ordering in the material. The occurrence of ferromagnetism and antiferromagnetism can manifest itself in additional luminescence features.

The gap between the ground state and the lowest excited state of MnF_2 is above 2 eV. It is interesting to compare MnF_2 with the large band gap semiconducting material, GaP, which has roughly the same gap between its valence and conduction bands. The nature of these excited states is different in the two materials. In both the absorption of a 2·24 eV photon creates an electron–hole pair at a local centre. In GaP the electron and hole easily become separated from each other and travel about relatively independently of each other, and they contribute to electrical conduction. Absorption at the band gap leads to photoconductivity in GaP. In MnF_2, on the other hand, the electron and hole are created on the same Mn^{2+} ion and much more energy is required to further ionize the ion, that is, to separate the hole and electron from being together in the same ion. The electron and hole pair as a unit can easily move to an adjacent ion. This is just the transfer of excitation which we discussed already in this section. Since the electron and hole move together, there is no net movement of charge, no photoconductivity occurs for band gap transitions in MnF_2, this material remains an insulator.

The luminescence properties of the large gap semiconducting materials are far more important than fully concentrated optically active insulating materials. We shall from now on turn our attention to the luminescence behaviour of large band gap semiconducting materials.

1.3. LARGE BAND GAP SEMICONDUCTOR MATERIALS

1.3.1. Nature of Semiconductors

Semiconducting materials are characterized by a covalent bonding between the atoms of the materials. In this bonding the wavefunctions of the outer electrons on individual atoms overlap each other, and it is no longer possible to associate a particular electron with a particular atom. This is also the situation with metals. We must regard the electron as belonging equally to all atoms in the whole crystal. We would like to solve the Schrödinger equation to obtain the wavefunction of the electron in the electric field of all the atoms. It is not possible to solve such a problem exactly, so we solve a much simpler problem which is a crude approximation to the real problem and which, hopefully, will suffice as a description of the electron's behaviour.

The simplest approximation is to regard the electrons as not interacting with each other, merely moving in a constant potential which is the average potential in the solid. The solution to this problem is a set of plane waves, with spatial dependence given by $\exp(-i\mathbf{k}.\mathbf{r})$. The allowed values of the wavevector, $\mathbf{k}$, can be represented as points in reciprocal space. Although the $\mathbf{k}$ values are discrete, there are so many of them that we can consider that they form a quasi-continuous range of values. $\hbar\mathbf{k}$ can be regarded as the momentum of the electron when it is in the state characterized by this specific value of $\mathbf{k}$. The kinetic energy of the electron is $\hbar^2 k^2/2m$, while the potential energy, being a constant independent of $\mathbf{k}$, can be neglected. So there is also a quasi-continuum of energy values—essentially all values of energy are allowed.

If we attempt to describe the situation a little more realistically, we can replace the constant potential by a potential which reflects simply the periodic structure of the lattice. The effect of this is to split up the quasi-continuum of allowed energy values into bands of allowed energy levels separated by gaps, and the electrons are not allowed to have any of the energy values in the gaps: these are the forbidden gaps. The next stage of sophistication is to take into account that while the electron is on a particular atom, the electron's wavefunction must have many of the characteristics of atomic wave functions. The best wavefunctions combine both an atomic description and a plane wave description; such wavefunctions are not easy to calculate, but are now being computed for many metals and semiconductor materials. From these we obtain the energy band structure of the material.

The results of such band structure calculations for GaAs yield the dependence of energy on **k** shown in Fig. 1.34. This shows how the energy of the electron state varies for the various magnitudes and directions

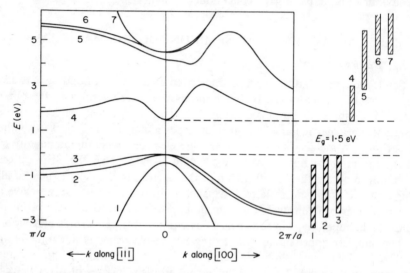

Fig. 1.34. Band structure of GaAs. Each band shows how the energy of the electron state varies for different values of **k** along the [100] and [111] directions. On the right-hand side the energy distribution of the bands is shown. There is an energy gap of 1·5 eV between bands 3 and 4. At low temperatures bands 1, 2, and 3 are filled with electrons, while all the higher bands are empty.

of **k**, when **k** is pointing along the [100] and [111] directions of the unit cell in reciprocal space. Within each E versus **k** curve there is a quasi-continuum of allowed states—the number of different states is comparable to the number of atoms in the crystal. On the right-hand side of the E versus **k** curves we have represented the curves of allowed energy levels solely by their spread in energy. The Pauli principle applies to this electronic system and the lowest energy bands are filled up with one electron per allowed state. In GaAs there are just enough electrons to fill bands 1 to 3. At very low temperatures the electrons will occupy these bands, and there will be no electrons occupying the bands from 4 upwards. The full and empty bands are indicated by heavy and light shading, respectively, in Fig. 1.34. As the figure shows, there is a band gap of 1·5 eV between the top of the highest filled band and the bottom of the next (empty) band. This is the band gap of the material.

Electrons in a filled band cannot contribute to electrical conductivity. The reason is that all the electron states are occupied and there is no net movement of electrons in any direction. The application of a weak external electric field cannot change the distribution of electrons among the states in the filled band, since all the states are already filled, and no net movement of charge can occur. A material in which all the electrons are in filled bands is clearly an electrical insulator, and pure GaAs will be an insulator at lowest temperatures.

As the temperature is raised to room temperature, say, some electrons may acquire sufficient energy to be promoted across the band gap, and they occupy the previously empty band. If this happens, these electrons are then occupying an unfilled band, the distribution of these electrons among the electron states in the band can be changed by the application of a weak external field, and a net flow of electrons can occur. Similarly, since the band from which these electrons were promoted is no longer filled, the application of a weak external electric field can rearrange the distribution of these remaining electrons and lead to conductivity. The conductivity in the previously filled band is most easily discussed in terms of the movement of the holes, or vacancies, left behind by the promoted electrons. These holes have positive charge.

If the gap between the top of the highest band which is normally filled at lowest temperatures and the bottom of the lowest band which is normally empty at lowest temperatures—for example, the 1·5 eV band gap in GaAs—is too large for a detectable amount of electrons to cross it at room temperature, then the material is an insulator at this temperature. If, however, the band gap is not too large, if a detectable amount of electrons can be promoted across the gap at room temperature, the material will exhibit very weak conductivity at room temperature, and this material will be called a *semiconductor*.

The semiconductor material normally used in transistor fabrication, namely silicon, has a band gap of around 1·2 eV. Materials with a larger band gap are of particular use in optical devices.

The electrons in the highest mainly filled band are participating in the covalent bonding between atoms. Consequently, this band is the *valence band*. The mainly empty band just above the band gap, whose few electrons can contribute to electrical conductivity, is called the *conduction band*. The band gap between these two bands is normally labelled E_g.

Besides being promoted from valence to conduction band by thermal energy, electrons may be raised to the conduction band by the absorption of visible light. For GaAs the band gap of 1·5 eV corresponds to a photon of wavelength 0·82 μ, that is, just above the visible. Illuminating GaAs with visible light can promote electrons from the valence to the conduction band, these can contribute to the electrical conductivity, and the material is said to be photoconductive.

For much of our discussion we shall concern ourselves with electron states near the top of the valence band and the bottom of the conduction band, and we shall be interested in their **k** values and energies. The appropriate parts of the GaAs band structure, that is, parts of bands 3 and 4, are shown in Fig. 1.35. In this material the maximum energy of the valence band

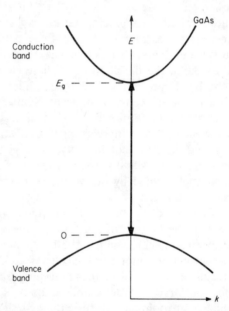

Fig. 1.35. Part of the E versus **k** curves for the top of band 3 (the valence band) and the bottom of band 4 (the conduction band). The shortest transition across the gap between valence and conduction bands (the band gap transition) is indicated by the heavy arrow, and it obeys the selection rule $\Delta\mathbf{k} = 0$. Hence GaAs is a direct gap semiconductor.

occurs at **k** = 0, and the maximum energy of the conduction band also occurs at **k** = 0. Because these extrema occur at the same value of **k**, GaAs is said to be a "direct gap" semiconductor. In GaP, on the other hand, the band gap is around 2·3 eV, the extrema do not occur at the same value of **k**, and this is said to be an "indirect gap" semiconductor. The relevant bands are shown in Fig. 1.36.

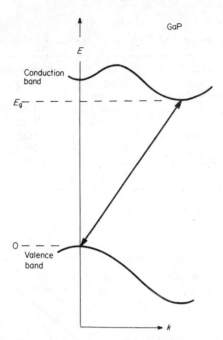

Fig. 1.36. Part of the E versus **k** curves for the valence and conduction bands of GaP. The band gap transition is indicated by the arrow, and there is a change in **k** during this transition. Hence GaP is an indirect gap semiconductor.

The absorption and luminescence processes in these semiconductors will in general involve electronic transitions across the band gap. For luminescence we will see that the photon of maximum energy corresponds roughly to the energy of the band gap. So for visible or near infrared luminescence we require a band gap of around 1·0 eV or greater; 1·7 eV is about the lower limit of visible radiation. A list of such large band gap semiconductors which have interesting luminescence properties is given in Table III, along with the values of the band gaps at low temperatures.

Consider the alloy with the chemical formula $GaAs_{1-x}P_x$. The electrons in such a material, being states of the whole crystal, see an average band gap varying from 1·52 eV when $x = 0$ to 2·3 eV when $x = 1$. Thus the luminescence from $GaAs_{1-x}P_x$ varies with the value of x. The wavelength of the photon corresponding to the band gap as well as the value of the band gap in electron volts is plotted against x in Fig. 1.37. For values of x up to about 0·4, the material is a direct gap semiconductor, but it becomes an indirect semiconductor above $x = 0·4$. Other interesting alloys can be formed from mixtures of InP and AlP, with band gaps varying from 1·42 eV to 2·5 eV. The ability to fabricate materials with specific band gaps is today

80 G. F. IMBUSCH

TABLE III

Band Gaps of Some Semiconductor Materials

Material	Type	Band Gap (eV)
Si	Indirect	1·16
InP	Direct	1·42
GaAs	Direct	1·52
GaP	Indirect	2·3
AlP	Indirect	2·5
SiC	Indirect	3·0

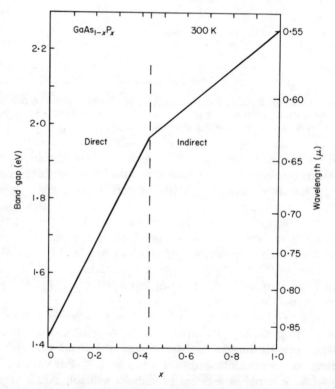

Fig. 1.37. Variation of the band gap and of the wavelength of the band gap transition with the concentration of phosphorus in the alloy $GaAs_{1-x}P_x$ when the material is at room temperature. For values of x up to 0·43 the material is a direct gap semiconductor, while for larger values of x it is an indirect gap semiconductor.

being vigorously exploited in the manufacture of semiconducting technical devices.

1.3.2. Excited States of Pure Semiconductors

Consider the pure semiconductor material at lowest temperatures when the valence band is filled and the conduction band is empty. If an additional electron is added to the material, this electron must occupy one of the states in the conduction band. The wavefunction describing this electron is a wavefunction of the whole crystal. If, instead of adding an extra electron, an electron were raised from the valance band to the conduction band, a hole would be left in the valence band. Let us suppose that the electron and hole drift apart from each other. The wavefunction of this excited electron would be the same as the wavefunction of the electron previously added to the material, since the absence of just one electron from the crystal would hardly affect the crystal wavefunction.

When, however, the electron in the conduction band and the hole in the valence band are in the vicinity of each other, an additional Coulomb interaction can occur between them. The electron–hole pair thus bound together has less energy than when the electron and the hole are far apart, and when the electron is in one of the states of the conduction band. The energy of the bound electron–hole pair is below the bottom of the conduction band. The electron and hole orbit about each other as in a hydrogen-like atom. Since the dielectric constant in covalent crystals is very high, the radius of the orbit is very large, and the electron–hole pair can also move through the material. There is no net movement of charge associated with this movement, only the excitation moves. Such a state of mobile excitation is called an *exciton* state. This loosely bound electron–hole pair in the semiconductor material is called a Wannier exciton. In MnF_2 we saw that the excited state of the Mn^{2+} ion could be considered as a state of excitation in which the excited electron and hole stay on the same atom, although the excitation can travel through the material. Such a mobile but tightly bound electron–hole pair is called a Frenkel exciton. As all these excitons can travel freely through the pure material, they are called free excitons.

Returning to our simple energy level diagram, we see that we must supplement the electronic states previously discussed by including the exciton states. If we use the hydrogen atom model to describe the electron–hole pair, we find that the mutual energy of the electron–hole pair varies as $1/n^2$, where $n = 1, 2, 3, \ldots$. When $n = \infty$ the electron and hole are no longer bound to each other, the electron is now free to move on its own through the crystal, and its energy is appropriate to the bottom of the conduction band. In the simplified energy level diagram for GaAs shown in Fig. 1.6, we showed the free exciton levels as broken lines drawn just below the bottom of the conduction band. The $n = 1, 2, ..., \infty$ labels indicate the spacings of exciton levels on the hydrogen-like model.

1.3.3. Absorption Transitions in a Pure Semiconductor

In Fig. 1.7 we showed the low temperature absorption spectrum of very pure GaAs. This showed two sharp transitions at photon energies below the band gap which have been identified as transitions to the $n = 1$ and $n = 2$ exciton states. The valence→conduction band absorption transitions form an absorption continuum from E_g upwards.

Let us consider the conservation of momentum during these optical transitions. When an electron in some $\mathbf{k}$ state of the valence band is raised to the $\mathbf{k}'$ state of the conduction band, there is a change in electron momentum, $\hbar\Delta\mathbf{k}$. This must be supplied by the photon. The momentum of the photon is $\hbar k_\omega$, but $\hbar k_\omega$ is only around 0·1% of the maximum momentum of the electron, so little momentum can be obtained from the photon. Thus the change in electron state during the absorption transition should obey the selection rule $\Delta\mathbf{k} = 0$, and should be represented by vertical arrows on the E versus $\mathbf{k}$ diagrams.

In GaAs the maximum of the valence band and the minimum of the conduction band occur at the same value of $\mathbf{k}$ (both occur at $\mathbf{k} = 0$), and as a result an optical absorption transition can occur between these two states, that is, at the band gap energy, E_g. In GaP, as Fig. 1.36 shows, $\Delta\mathbf{k}$ is not zero for the band gap absorption, and this transition is not allowed by the selection rule. If we consider an optical absorption transition in which both a photon and a phonon are involved, momentum can be conserved in the band gap transition, the change in $\mathbf{k}$ due to the electron's transfer across the band gap being balanced by the $\mathbf{k}$ value of the phonon emitted or absorbed during the optical transition. In such indirect gap materials, the absorption at the band gap is a phonon-assisted transition and its transition probability is smaller than the band gap absorption in direct gap materials.

Returning to the GaAs absorption transitions shown in Fig. 1.7(a), we notice that the exciton absorption transitions are much broader than the luminescence transitions seen in Fig. 1.7(b). Consequently any fine structure associated with the exciton levels is masked by the relatively large widths of the absorption transitions.

1.3.4. Luminescence from Pure GaAs

Some of the purest semiconductor materials in which low temperature luminescence has been studied are the GaAs crystals examined by Sell; a typical luminescence spectrum obtained by optical pumping of these materials is shown in Fig. 1.7(b), and again in more detail in Fig. 1.38. The broken vertical lines show the position of the band gap and the positions of the $n = 1$ and $n = 2$ exciton levels found in absorption.

At these very low temperatures we expect that optical pumping would result in a large population in the exciton states. From the momentum selection rules only the decay of the $\mathbf{k} = 0$ free exciton is allowed. Because of the sharpness of the luminescence transitions, we might hope to see evidence of fine structure in the exciton levels. This fine structure comes about as follows. For the free exciton state with energy E and with a k value close to zero, there exists a photon state with the same energy and momentum. Any interaction between the radiation and the exciton can couple these resonant states together. The result of the interaction is that two states are formed—slightly separated in energy—each of which contains

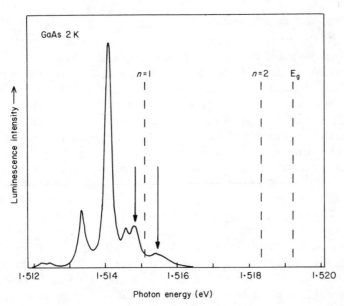

Fig. 1.38. Photoluminescence spectrum of pure GaAs at low temperatures. The positions of the $n = 1$ and $n = 2$ excitons seen in absorption and the position of the band gap are indicated.

some exciton characteristics and some photon characteristics. These are called *polariton* states and each polariton state has two levels slightly split apart in energy. We hope to observe this splitting in the sharp line luminescence decay of free excitons. Sell has identified the luminescence transitions in Fig. 1.38 which are indicated by the heavy arrows (and which occur on either side of the $n = 1$ exciton absorption position) as the luminescence from the split polariton state associated with the $n = 1$ exciton. The positions of these polariton levels are apparently very sensitive to strains in the material. These then are the free exciton luminescence transitions. All the other luminescence features seen in Fig. 1.38 are associated with defects, despite the very high purity of this GaAs material. Once again we see impurity effects dominating the luminescence from very pure materials. The reason for this is found in the high mobility of the exciton. As the exciton moves through the material, it encounters defects and impurities of various kinds, and the exciton tends to decay when in the vicinity of one of these centres. Before proceeding further with an analysis of the luminescence from GaAs and other semiconductors, we must discuss the types of impurities which can occur in semiconductors.

1.3.5. Impurities in Semiconductors

We distinguish three categories of impurities—donors, acceptors, and isoelectronic impurities—and we consider each separately.

Donors. When an impurity atom is incorporated in the semiconductor material, the impurity may release one of its electrons which then goes into one of the conduction band states of the intrinsic semiconductor material. Such an impurity atom is termed a donor as it donates an electron to improve the electrical conductivity of the material. For example, each Ge atom in the material germanium is a Group IV atom with four outer electrons, and these electrons participate in the four covalent bonds required in the germanium crystal. In the growth of the germanium crystal a small amount of As atoms may enter substitutionally for Ge atoms. As is a Group V atom with five valence electrons. Since only four are needed for covalent bonding to the four neighbouring Ge atoms, the fifth electron is not required for bonding and stays loosely bound to the As atom. The binding energy of this electron is small, it is labelled E_D, and its value is around 0·01 eV. If this loosely bound electron of the As atom acquires an energy of 0·01 eV, the electron may leave the As atom and become part of the bulk germanium material. The As atom is now said to be ionized, and E_D is often referred to as the ionization energy. Since the valence band of the germanium is filled, the additional electron donated by the As atom enters the conduction band of the germanium. In this example the As atom is the donor. When the loosely bound electron stays attached to its parent atom we have a neutral donor. When the electron is removed the donor is ionized. If the material is at room temperature almost all donor atoms with ionization energies around 0·01 eV are ionized. The energy level of the electron loosely bound to the neutral donor is shown at an energy E_D below the bottom of the conduction band in Fig. 1.39. Although the donors can place electrons in the conduction

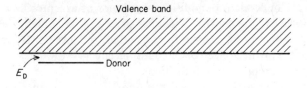

Valence band

Donor

E_D

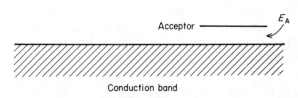

Acceptor —————— E_A

Conduction band

Fig. 1.39. The level of the electron loosely bound to a donor atom is situated an energy E_D below the bottom of the conduction band. The acceptor atom makes available an electron energy level which is E_A above the top of the valence band.

band, these electrons cannot drop to the valence band and emit a photon of light. Since the luminescence process is the coming together of an excited electron *and a hole*, and since the valence band is filled, there are no holes with which these excited electrons can combine, the excited electron stays in the conduction band, and no luminescence results.

Acceptors. In a similar way the germanium material may be doped with Group III atoms such as Ga. With its three valence electrons, Ga can form only three covalent bonds with the neighbouring Ge atoms. The fourth covalent bond is incomplete as it lacks an electron. We can look upon this as a hole attached to the Ga atom. It requires but little energy for one of the electrons in the valence band of the bulk germanium material to occupy this hole. The Ga atom is called an acceptor, as it accepts an electron from the bulk germanium material and leaves a hole in the valence band of the material. The amount of energy required for an electron at the top of the valence band of the bulk semiconductor material to attach itself to the incomplete covalent bond of the acceptor is labelled E_A and is quite small—around 0·03 eV. When such an electron attaches itself to the acceptor, we say that the acceptor is ionized. In the energy level diagram of Fig. 1.39 we show the acceptor as supplying an electronic level which is situated E_A above the top of the valence band.

As long as the density of donors or acceptors is small, these impurities have little effect upon the band structure of the intrinsic semiconductor in which they are imbedded. In addition, the donor or acceptor levels are quite sharp. If, however, the density of donors or acceptors is high, the donor or acceptor wavefunctions can overlap each other, the levels broaden and may even overlap the intrinsic electron bands. Such effects show up as a broadening and shift of the luminescence transitions.

Isoelectronic Impurities. The substitution of N for P in GaP does not contribute to electrical conductivity, but it can have an effect on the luminescence properties of the semiconductor. Particularly in indirect gap semiconductors, they may act as centres where recombination of excited electrons and holes tend to occur.

More than one type of impurity can be incorporated in the same semiconductor. For example, both donors and acceptors may be incorporated in the same semiconductor crystal. Indeed it is impossible to grow semiconductors crystals—or crystals of any type—which are completely free from traces of impurities. In the growth of the purest GaAs there will be traces of Zn and Si, among other impurities, and the Zn and Si atoms act as acceptors and donors, respectively.

In Table IV we show the ionization energies (E_D or E_A) for some impurities which act as donors or acceptors in GaAs. We notice that quite a number of donors have the same E_D value, around 0.006 eV, and quite a number of

TABLE IV

Ionization Energies (E_D of E_A) of Some Impurities in GaAs

Donors		Acceptors	
Element	E_D (eV)	Element	E_A (eV)
S	0·006	Zn	0·029
Se	0·006	Cd	0·030
Te	0·006	Si	0·030
Si	0·006	Be	0·030
Sn	0·006	Mg	0·030
O	0·75, 0·40	Cu	0·15, 0·47
		Fe	0·52
		Ni	0·21

acceptors have the same E_A value, around 0.030 eV. Si can act as a donor— when it substitutes for Ga, and can act as an acceptor—when it substitutes for As. O, Cu, Fe, Ni, and a number of other elements act as very deep traps in GaAs.

The electronic radiative transitions across the band gap in semiconductors are strongly influenced by the presence of the various kinds of impurities which we have enumerated here. We shall list in the next section the different radiative transitions across the band gap which can occur in semiconductors which contain these impurities.

1.3.6. Radiative Transitions across the Band Gap

We represent, in Fig. 1.40, the electron–hole radiative recombination processes across the band gap. These are:

$C \rightarrow V$ *Processes.* Conduction band to valence band transitions which can be seen at high temperatures.

$E \rightarrow V$ *Processes.* This is exciton decay, seen only in very pure materials and at low temperatures—when kT is not larger than the exciton binding energy. Two distinct decay processes are possible. They are (i) the decay of free excitons and (ii) the decay of excitons bound to impurities. In the decay of free excitons it is the polariton states which decay. Transitions due to the decay of the split polariton state associated with the $n = 1$ free exciton in GaAs are shown in Fig. 1.38. They are indicated by arrows, they occur at 1·5148 eV and 1·5155 eV, and are on either side of the position of the $n = 1$ free exciton absorption. Sometimes an exciton stays localized in the vicinity of an impurity. This is a bound exciton. Its energy will be less than the energy of the $n = 1$ free exciton by the amount of the binding energy of the exciton to the impurity. This binding energy is small, being typically around 0·001 eV. The strong luminescence feature seen at 1·5141 eV in the luminescence spectrum of pure GaAs (Fig. 1.38) has been interpreted as

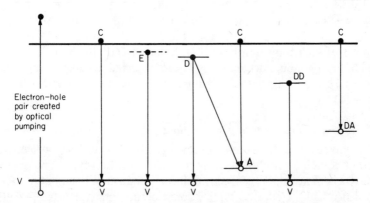

Fig. 1.40. Electron–hole pair recombination processes across the band gap. C → V is the band-to-band transition, E → V is the decay of excitons (free and bound), D → V is the neutral donor–free hole transition, C → A is the free electron–neutral acceptor transition. Deep donors (DD) and deep acceptors (DA) give rise to transitions whose photon energies are well below the band gap (DD → V, and C → DA).

the decay of an exciton bound to a neutral donor. The binding energy is around 0·001 eV and the energy of this luminescence transition is below the energy of the $n = 1$ free exciton by this amount.

$D \rightarrow V$ *Processes.* In this transition the loosely bound electron on the neutral donor recombines with a hole in the valence band. The energy of the transition is $E_g - E_D$. In the case of GaAs, where $E_D = 0·006$ eV for many donors, and where $E_g = 1·5192$ eV at low temperatures, this transition should occur at 1·5132 eV. The luminescence line seen at 1·5133 eV in Fig. 1.38 has been attributed to such a neutral donor-free hole transition. $D \rightarrow V$ transitions on donors with large ionization energies occur well below the energy of the band gap. These are illustrated as $DD \rightarrow V$ processes in Fig. 1.40.

$C \rightarrow A$ *Processes.* Here an electron in the conduction band of the intrinsic semiconductor material drops to an acceptor atom, ionizing the acceptor. This should have an energy of $E_g - E_A$. In GaAs, where many acceptors have $E_A = 0·03$ eV, this transition should occur at around 1·49 eV. Some weak luminescence has been observed by Sell at 1·49 eV in his GaAs materials, and it can be attributed to a free electron–neutral acceptor transition. Transitions from the conduction band to deep acceptor centres will occur well below the band gap. These are shown as $C \rightarrow DA$ processes in Fig. 1.40. Atoms such as Cu, Fe, and Ni form such deep ionization centres in GaAs.

$D \rightarrow A$ *Processes.* If both donors and acceptors are present in sufficient amounts in the same semiconductor material, transitions can occur in which an electron leaves a neutral donor and goes to a neutral acceptor. After the transition both the donor and the acceptor are ionized, and if they are a distance r apart, they have a binding energy $E_b = -e^2/4\pi\varepsilon_0 Kr$. The energy of the transition should then be $E_g - E_D - E_A - E_b$. We expect to see a series of distinct sharp lines for the different allowed values of r. Clear examples of such transition have been found in GaP doped with Si and Te. In addition, phonon sidebands have been observed accompanying some of these sharp transitions. (See Chapter 5 for further examples of $D \rightarrow A$ processes.)

1.3.7. Radiative Processes in p–n Junctions

The luminescence transition in semiconductors takes place when excited electrons combine with holes across the band gap. In n-type semiconductors we have a copious supply of electrons in the conduction band but no holes in the valence band. Hence no luminescence occurs unless holes are created— by optical pumping, for example. Similarly, in p-type semiconductors we have holes in the valence band but no electrons in the conduction band; and no luminescence occurs unless electrons are somehow placed in the conduction band as well. Now by combining n-type and p-type semiconductors in a p–n junction, one can cause recombination of excited electrons (from the n-type) and holes (from the p-type) at the junction without the necessity for optical pumping. We can do this by forward-biasing the p–n junction which has the effect of injecting conduction band electrons into the p-region where they recombine radiatively with the holes. The energy to inject the electrons and holes across the junction comes from the bias battery. This direct conversion of electrical power into luminescence is called *electroluminescence*, and we speak of *injection luminescence*. The energy levels and luminescence transitions in a forward-biased diode were shown in Fig. 1.8. The luminescence takes place in a very shallow region at the junction. The value of the bias voltage, V, is such that eV is roughly the energy of the band gap.

This method of converting low voltage power directly to light is very convenient and finds practical application in the light-emitting diodes (LEDs) which are being widely used in the electronics industry. The process can be quite efficient—laser action has been achieved in a number of direct gap semiconductor diodes. The small size of the LED is another advantage— it is about 0·5 mm across. Further, by using alloys such as $GaAs_{1-x}P_x$, one can vary the band gap, and one can fabricate a diode with luminescence at any colour between red and green. This has given rise to the family of red, yellow, and green LEDs being commercially produced at present. The red LED is made from $GaAs_{1-x}P_x$ with $x \simeq 0·4$, where the material is still a direct gap semiconductor (Fig. 1.37). The green LED is made from almost pure GaP. This, however, is an indirect gap semiconductor, and the radiative transition probability should be small. By doping the semiconductor with nitrogen—which substitutes for some phosphorus atoms—a significant improvement in the luminescence efficiency has been achieved.

Much effort is going into the elucidation of the complex recombination processes occurring at the junctions of light emitting p–n junctions. More advances are expected to be made in the efficiency and range of these devices. It is clear that a new technology combining semiconductor electrical devices and electroluminescence devices is being developed. The recent advances in the production of small low-loss optical fibres for carrying optical information along cables introduces further possibilities in the rapidly growing field of electro-optics.

REFERENCES

Useful textbooks, conference proceedings, and review articles in the general field of inorganic luminescence include:

Ballhausen, C. J. (1962). "Introduction to Ligand Field Theory." McGraw-Hill, London and New York.

Blasse, G. and Bril, A. (1970). *Philips Tech. Rev.* **31**, 304 (I), 314 (II), 324 (III).

Crosswhite, H. M. and Moos, H. W. (eds) (1967). "Optical Properties of Ions in Crystals." Interscience, New York.
Curie, D. (1963). "Luminescence in Crystals." Methuen, London.
Di Bartolo, B. (1968). "Optical Interactions in Solids." Wiley, New York.
Garlick, G. F. J. (1958). "Handbuch der Physik", Vol. 26. Springer-Verlag, Berlin.
Goldberg, P. (ed.) (1966). "Luminescence of Inorganic Solids." Academic Press, New York and London.
McClure, D. S. (1959). "Solid State Physics", Vol. 9 (F. Seitz and D. Turnbull, eds). Academic Press, New York and London.
Williams, F. (ed.) (1970). "Luminescence." North Holland, Amsterdam.
Zeiger, H. J. and Pratt, G. W. (1973). "Magnetic Interactions in Solids." Clarendon Press, Oxford.

The solid-state physics textbook by Elliott and Gibson contains an excellent brief treatment of optical processes in solids. A very recent survey of many aspects of inorganic luminescence is given in the proceedings of the 1974 and 1977 Ettore Majorana summer schools on optical properties of ions in solids.

Elliott, R. J. and Gibson, A. F. (1974). "An Introduction to Solid State Physics." Macmillan, London.
Di Bartolo, B. (ed.) (1975). "Optical Properties of Ions in Solids." Plenum Press, New York and London.
Di Bartolo, B. (ed.) (1978). "Luminescence of Inorganic Solids". Plenum Press, New York and London.

The following are specialist references appropriate to some of the topics covered in the chapter.

The interaction of radiation and matter is discussed in: Bethe, H. A. and Salpeter, E. E. (1957). "Handbuch der Physik", Vol. 35. Springer-Verlag, Berlin.

Dexter, D. L. (1958). "Solid State Physics", Vol. 6 (F. Seitz and D. Turnbull, eds). Academic Press, New York and London.

Group theory appropriate for centres in solids is presented in:

Hamermesh, M. (1961). "Group Theory and its Application to Physical Problems." Addison-Wesley, New York.
Heine, V. (1960). "Group Theory in Quantum Mechanics." Pergamon Press, New York and London.
Tinkham, M. (1964). "Group Theory and Quantum Mechanics." McGraw-Hill, New York.

Transition metal ions are discussed in:

Griffith, J. S. (1961). "Theory of Transition Metal Ions." Cambridge University Press.
Sugano, S., Tanabe, Y. and Kamimura, H. (1970). "Multiplets of Transition Metal Ions in Crystals." Academic Press, New York and London.

Rare earth ions are comprehensively reviewed in the book by Dieke. The spectroscopy of rare earth ions (as well as colour centres) in fluorite crystals is reviewed in the recent book edited by Hayes.

Dieke, G. H. (1968). "Spectra and Energy Levels of Rare Earth Ions in Crystals." Wiley–Interscience, New York.

Hayes, W. (ed.) (1974). "Crystals with the Fluorite Structure." Clarendon Press, Oxford.

F-centres and other similar defect centres are treated in:
Beall Fowler, W. (ed.) (1968). "Physics of Color Centres." Academic Press, New York and London.
Henderson, B. and Wertz, J. E. (1977). "Defects in the Alkaline Earth Oxides." Taylor and Francis, London.
Markham, J. J. (1966). "F-centres in Alkali Halides; Supplement 8 to Solid State Physics" (F. Seitz and D. Turnbull, eds). Academic Press, New York and London.

Lattice dynamics, vibronic transitions and configurational coordinates: these topics are discussed in the books of Markham and Fowler, mentioned previously, as well as in the text by Rebane and in the review article by Maradudin. The proceedings of the 1965 Scottish Universities' Summer School contains some useful papers, in particular an excellent article by M. H. L. Pryce on the effect of lattice vibrations on transitions at point defects. The similarity between the optical and Mössbauer spectra was first pointed out by Schawlow in his reply to a question after his paper at the 1961 Quantum Electronics Conference ("Advances in Quantum Electronics" (J. R. Singer, ed.) (1961). Columbia University Press, New York and London).
Maradudin, A. A. (1966). "Solid State Physics", Vol. 18 (F. Seitz and D. Turnbull, eds). Academic Press, New York and London.
Rebane, K. K. (1970). "Impurity Spectra of Solids." Plenum Press, New York and London.
Stevenson, R. W. H. (ed.) (1966). "Phonons in Perfect Lattices with Point Imperfections." Scottish Universities' Summer School, 1965. Oliver and Boyd, Edinburgh and London.

The treatment of non-radiative transitions in this chapter follows the analysis of Moos and of Weber. A very good review of the various treatments of non-radiative processes is given by F. Auzel in his article in "Luminescence of Inorganic Solids" (B. Di Bartolo, ed.).
Moos, H. W. (1970). *J. Luminescence*, 1, 2, 106.
Weber, M. J. (1973). *Phys. Rev.* 8B, 54.

Energy transfer is thoroughly reviewed by Watts in his article in "Optical Properties of Ions in Solids", previously mentioned. This and the paper by Orbach in the same book give references to previous work. The techniques of fluorescence line narrowing are now being used to study the spectral transfer of excitation among similar ions, that is, the transfer of excitation between near resonant ions in which a slight energy mismatch is taken up by the phonons. The question of how energy transfers between similar ions which are slightly mismatched in energy is fundamental to our understanding of many phenomena involving energy transfer. These matters are discussed by Yen and Orbach in the proceedings of the Lyon conference: "Spectroscopie des Eléments de Transition et des Eléments Lourds dans les Solides", Colloques International du C.N.R.S. No. 255 (1977). The experimental situation has been reviewed by Yen in his paper at the 1978 International Conference on Luminescence (to be published in *J. Luminescence*).

The spectroscopy of magnetic materials is discussed by McClure in his article in "Optical Properties of Ions in Solids", as well as in:

Loudon, R. (1968). *Adv. in Physics*, **17**, 243.

Zeiger, H. J. and Pratt, G. W. (1973). "Magnetic Interactions in Solids." Clarendon Press, Oxford.

Eremenko, V. V. and Petrov, E. G. (1977). *Adv. in Physics*, **26**, 31.

The field of semiconductor luminescence is very clearly treated in a text by Pankove:

Pankove, J. I. (1971). "Optical Processes in Semiconductors." Prentice-Hall, Englewood Cliffs, New Jersey.

This is a field of very active research interest, and new developments are constantly being reported in the research journals and in the periodicals which cater for the electronic and opto-electronic industry. Of particular interest is the recent development of reliable semi-conductor lasers suitable as oscillators for optical communications through glass-fibre cables. In this regard it would seem that optical devices based on glass technology should become of increasing importance. Indeed, we are probably only beginning to appreciate the possibilities inherent in the use of glass, and our technical ability in glass is roughly at a stage comparable to semiconductors in the early fifties. We may see *excitonics* taking its place along with *electronics* as the full capabilities of glass are exploited.

2

Organic Luminescence

MICHAEL D. LUMB

Department of Pure and Applied Physics,
University of Manchester Institute of Science and Technology,
Manchester, England

2.1. THE ORIGINS OF ORGANIC LUMINESCENCE

Luminescence can be described as *cold* light in contrast to light emitted by incandescent bodies which emit light solely because of their high temperature. Luminescence can take on a variety of forms such as photoluminescence, cathodoluminescence, radioluminescence, bioluminescence, thermoluminescence, triboluminescence, sonoluminescence, and electroluminescence. These different forms are concerned with the excitation processes which occur prior to the emission of luminescence. The emission processes are unaffected by these different forms of excitation unless secondary effects such as radiation damage or chemical decomposition occur.

Photoluminescence, the excitation of luminescence by ultraviolet or visible radiation, is the most useful process for studying the fundamental behaviour of luminescent materials because it allows the greatest control over the excitation conditions. The discussions in this chapter will therefore be confined to photoluminescence.

What is so special about organic molecules which causes us to make a distinction between organic and inorganic luminescence? The answer lies in the molecular and atomic structure of these materials. Inorganic materials are held together by ionic or covalent bonds between individual atoms, and thus can be regarded as "atomic" solids. Organic materials are held together by van der Waals forces between molecules and are therefore molecular solids. The consequence of this molecular nature is that the luminescence processes in organic materials are associated with the excited states of molecules, whereas in inorganic materials the luminescence spectra are associated with either defects or impurities in the atomic lattice or with the excited states of the isolated atom or ion. The properties of these two types of luminescence and the theoretical treatment of the excited states differ quite markedly. For example, the spectra of molecular materials in the vapour, liquid, and solid phase, unlike inorganic materials, are similar.

Organic materials are defined as those made up from carbon and hydrogen atoms—the hydrocarbons. Those hydrocarbons which contain double or triple bonds between the carbon atoms, that is the unsaturated hydrocarbons, commonly give rise to strong luminescence emission. This luminescence originates from the excited states of the delocalized π-electrons in these molecules. These π-electrons exist as a consequence of the particular configuration of the carbon atom in the hydrocarbon molecule.

The electronic configuration of the six electrons of the carbon atom in its ground state is denoted by $1s^2 2s^2 2p^2$. Prior to forming a bond with a neighbouring carbon atom one of the 2s electrons is promoted to the 2p level to give an electronic configuration of carbon "prepared for binding" as $1s^2 2s^1 2p^3$. In order to explain the structure of saturated hydrocarbon molecules such as methane, ethane, etc., a technique known as the "hybridization" of wave functions must be employed to produce four equivalent orbitals from the $2s^1 2p^3$ electronic configuration; this technique gives rise to sp^3 hybridization and results in the tetrahedral methane molecule. For larger molecules such as ethane, H_3C-CH_3, sp^3 hybridization is again employed and gives rise to a σ-bond between the carbon atoms. Recent work on the saturated hydrocarbons and in particular the paraffin series has shown that weak luminescence also occurs in these materials due to excitation of the σ-electrons. Other hybridized configurations are possible such as sp^2 and sp. In sp^2 hybridization, one of the original p-orbitals is left unchanged and the sp^2 hybridization gives rise to three hybrid orbitals lying in the same plane and at $120°$ to each other. These hybrid orbitals form the σ-bonds between carbon–carbon and carbon–hydrogen atoms and account for the ring-structure of molecules such as benzene. The remaining p-orbital extends above and below the molecular plane in a direction perpendicular to it, with a node (zero electron density) in this plane. This p-orbital overlaps with the p-orbital of a neighbouring carbon atom to produce a π-bond as shown for ethylene, $H_2C=CH_2$, in Fig. 2.1. The degree of overlap of the π-bond is much less than that of σ-bonds and this is reflected in the binding energy of the bonds. For example, $6\cdot33$ eV is required to break a single σ-bond and only $3\cdot98$ eV to break a double $\sigma\pi$-bond into a single σ-bond.

sp hybridization accounts for the triple-bonds in molecules like acetylene, $HC\equiv CH$, and gives rise to one σ- and two π-bonds.

It is the excited states of the π-electron systems which are of interest to the organic luminescence spectroscopist, and in particular the double-bonded molecules such as the polyenes, aromatic hydrocarbons, nucleic, and amino acids, etc.

Studies of the simple polyenes and aromatic hydrocarbons enable the fundamental processes of organic luminescence to be understood and research in this area has been intense over the past two decades. Figure 2.2 shows the π-electron orbitals for benzene. The p-orbital for each atom is in close contact with both its neighbours; thus the π-orbital extends completely around the benzene ring, producing a delocalized orbital with a node along

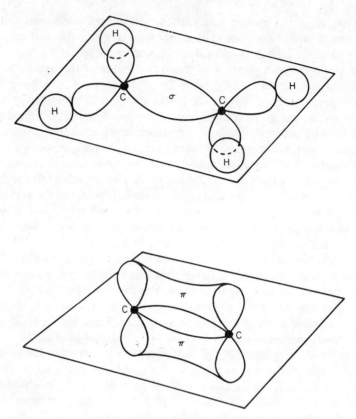

Fig. 2.1. Ethylene molecule.

the σ-bonded molecular skeleton. The overlap between the σ- and π-orbitals is therefore very small and this makes it possible to treat the π-electron system independently of the rest of the molecule. As might be expected, these π-electrons are less tightly bound to their parent carbon nuclei than the localized σ-electrons and thus require less energy to excite them. Absorption in the ultraviolet and visible regions of the spectrum, therefore, readily occurs for these molecules. This excitation energy cannot easily be lost as vibrational energy to the main molecular skeleton, and the system loses most of its energy by the luminescence emission of light.

As mentioned in the Preface, the theoretical treatment of the excited states of aromatic molecules will not be discussed in detail as whole texts have been devoted to this area. The free-electron model is discussed at length by Platt (1964), and the applications of molecular-orbital theory, Hückel theory, linear combination of atomic orbitals, etc., to organic

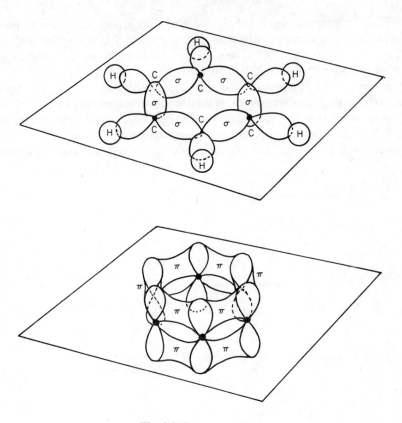

Fig. 2.2. Benzene molecule.

molecules are discussed in several publications to which the reader is referred at the end of this chapter. Chapter 4 also includes the theory of magnetic interactions and the excited states of aromatic molecules.

2.2. LUMINESCENCE PROCESSES

All π-electron systems contain an even number of π-electrons since there are two electrons per π-bond. The ground state of any π-electron system is therefore a singlet state since the electron spins will be paired off.

If the π-electrons are excited without change of spin the resultant excited states S_1, S_2, S_3, etc., are also singlet states. However, if the excited π-electron suffers a spin reversal between the ground and excited state, the resultant excited state is a triplet state. Under the application of an external magnetic field this triplet state splits into three Zeeman levels. The singlet states are unaffected by an external field.

Electronic transitions between a singlet and triplet state are formally forbidden by quantum mechanical selection rules. The absorption process, therefore, occurs principally between the ground state S_0 and the singlet states S_1, S_2, S_3, etc. Excitation to levels higher than S_3 usually results in excitation of the σ-electrons, and thus the excited states of the σ-electrons compete with the π-electron states for the excitation energy. In general, the absorption by the σ-states obscures the π-electron absorption at this level.

Figure 2.3 shows the luminescence processes which can occur in the π-electronic system of organic molecules. It should be noted that each

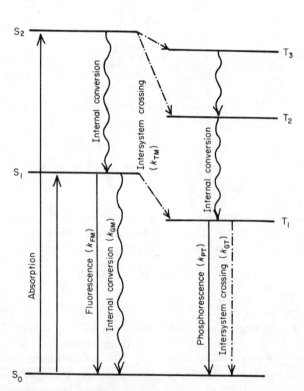

Fig. 2.3. π-Electron luminescence processes (rate parameters in parentheses): —— absorption; ——, emission (radiative process); ∿, internal conversion (non-radiative); — · — ·, intersystem crossing (non-radiative process).

triplet state lies below the corresponding singlet state—a situation dictated by Hund's rule. Figure 2.3 also shows the rate parameters which are listed in Table I.

Superimposed on each of the electronic levels are vibrational sub-levels due to the vibrational energy of the main molecular skeleton (see Fig. 2.6).

Luminescence may be produced by absorption into any of the excited singlet states $S_1, S_2, S_3, ...,$ etc. However, the primary fluorescence emission, in general, occurs from the lowest excited singlet state S_1 irrespective of the initial state excited. The radiative transitions from higher excited states $S_2, S_3, ...,$ etc. are very weak due to the rapid and efficient non-radiative process of internal conversion between $S_2, S_3, ...,$ etc. and the lowest excited state S_1 (Kasha, 1950).

TABLE I

Rate Parameters

$S_1 \to S_0$	fluorescence	k_{FM}
$S_1 \to S_0$	internal conversion	k_{GM} ⎫ k_{IM}
$S_1 \to T_1$	intersystem crossing	k_{TM} ⎭
$T_1 \to S_0$	phosphorescence	k_{PT}
$T_1 \to S_0$	intersystem crossing	k_{GT}

The radiative lifetime of S_1 is of the order of 10^{-9} s which is long compared with the period of molecular vibrations ($\approx 10^{-12}$ s), so that the molecule normally reaches thermal equilibrium before the fluorescence emission occurs from the lowest vibrational state of S_1. Thus for a given molecule the emission spectrum, the fluorescence lifetime, and the fluorescence quantum efficiency (the ratio of the number of fluorescence photons emitted to the number of photons absorbed) are independent of the state initially excited. The exceptions to this rule, which is known as Vavilov's law, will be discussed later.

Non-radiative processes play an important role in the luminescence processes which can take place in an excited organic molecule and in addition to internal conversion a non-radiative process known as intersystem crossing takes place. Intersystem crossing is the non-radiative transfer of energy from the singlet manifold to the triplet manifold and vice versa. Although singlet–triplet transitions are formally forbidden by quantum mechanics, there is still a finite rate constant ($\approx 10^6$ s^{-1}) for intersystem crossing from the excited singlet to lowest triplet state T_1.

The rate constant for intersystem crossing and internal conversion decreases with the increase in energy gap between the respective electronic states. Thus the rate constant for internal conversion between S_3 and S_2 or S_2 and S_1 is very much greater than that between S_1 and S_0 and hence fluorescence is readily observed between S_1 and S_0 but not between S_3 and S_2 or S_2 and S_1, etc. Similarly T_3 to T_2 and T_2 to T_1 internal conversion rates are also high. Intersystem crossing rate constants apart from being formally forbidden are higher if the gap between the singlet and triplet states is small. For example, in Fig. 2.3 the S_2 to T_3 intersystem crossing rate constant should be much larger than that from S_2 to T_2. However, as mentioned earlier, internal conversion occurs in the triplet manifold and a molecule excited into a higher triplet state, say T_3, will rapidly internally convert to the lowest triplet state T_1. The probability of radiative emission from T_1 to S_0 will then compete with the non-radiative intersystem crossing from T_1 to S_0. Intrinsically, the radiative process is the more probable and in the absence of external perturbations phosphorescence emission of light occurs. Since transitions, radiative or non-radiative, between T_1 and S_0 are formally forbidden, the lifetime of T_1 is usually of the order of seconds and decreases as the energy gap decreases. The long-lived phosphorescence emission in contrast with the short-lived fluorescence emission has a very special meaning in organic luminescence as it also implies a different emission spectrum at longer wavelengths (lower energy) than the corresponding fluorescence emission. Figure 2.4 shows the fluorescence and phosphorescence of benzene at low temperatures.

Another type of long-lived luminescence known as delayed fluorescence also occurs in organic molecular systems, and although originating from the triplet state it has an emission spectrum identical to that of the normal

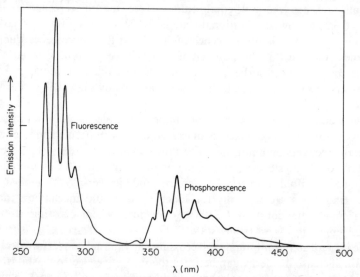

Fig. 2.4. Fluorescence and phosphorescence emission spectrum of benzene dilute solid solution at 77 K.

prompt fluorescence spectrum. Delayed fluorescence occurs when the long-lived triplet state T_1 can be thermally depopulated via S_1 or when pairs of excited triplet molecules interact to produce an excitation energy greater than or equal to the S_1 energy.

Figure 2.5 illustrates the origins of these two types of delayed fluorescence which will be explained in more detail later.

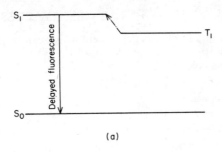

(a)

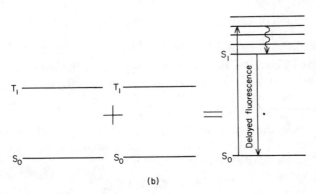

(b)

Fig. 2.5.(a) P-type delayed fluorescence; (b) E-type delayed fluorescence: ——, radiative process; ~~~, non-radiative processes.

Although the luminescence processes in most organic molecules involve transitions from the ground π-electron state to an excited π^*-state an important type of transition known as the n–π^* transition should be briefly mentioned. This transition occurs in molecules which contain a single non-bonding electron known as the n-electron. The n-electron is associated with substituents such as oxygen or nitrogen in the molecule. In general, the lowest excited singlet n–π^* states lies below the lowest excited π–π^* state and hence n–π^* transitions have an important role in the luminescence processes in these molecules. These n–π^* states differ quite markedly from the $\pi-\pi^*$ states in that they have a low extinction coefficient, low fluorescence yield, but have a high intersystem crossing rate and hence high phosphorescence yield. π^*-n phosphorescence may even occur in the liquid phase at room temperature (see Section 2.4.2). Polar solvents affect the position of the $n-\pi^*$ state more significantly than the $\pi-\pi^*$ state and may even reverse the position of the lowest excited state from an $n-\pi^*$ state of a $\pi-\pi^*$ state.

2.3. EXPERIMENTAL OBSERVABLES

The properties of any luminescence system which can be measured directly by experimental observation are:
(1) Absorption spectra.
(2) Emission spectra.
(3) Excitation spectra.
(4) Lifetimes and rate parameters.
(5) Quantum efficiency and yields.

From these measurements the rate parameters of other processes such as non-radiative processes and energy transfer can be deduced.

2.3.1. Absorption Spectra

The absorption of a photon by an atom is governed by the Einstein relation

$$hv = hc/\lambda = E_2 - E_1 \qquad (2.1)$$

where E_2 and E_1 are the electronic energy levels of the atom, h is Planck's constant, and v, λ, and c are the frequency, wavelength, and velocity of the incident photon. This relation represents the quantization condition for the absorption or emission of light by an atom.

In a molecule, the vibrational and rotational energies of the atoms which are also quantized are superimposed on the electronic energy levels of the atoms and so the Einstein relation becomes

$$hv = \Delta E_e + \Delta E_{vib} + \Delta E_{rot} \qquad (2.2)$$

where ΔE_e, ΔE_{vib}, and ΔE_{rot} are the respective changes in the electronic, vibrational, and rotational energies. In polyatomic molecules at room temperature, the rotational energy states which are very closely spaced ($\approx 10^{-4}$ eV separation) tend to smear out and simply broaden the vibrational bands ($\approx 10^{-1}$ eV separation).

Figure 2.6 shows a typical absorption spectrum for a $S_0 \rightarrow S_1$ transition. The peaks in the spectrum correspond to the vibrational energy levels. The strength of the absorption is determined by the degree of overlap of the wavefunction of the lowest vibrational level of the ground state S_{00} and the wavefunctions of the vibrational levels of the first singlet state $S_{10} - S_{1n}$ (see Fig. 2.9).

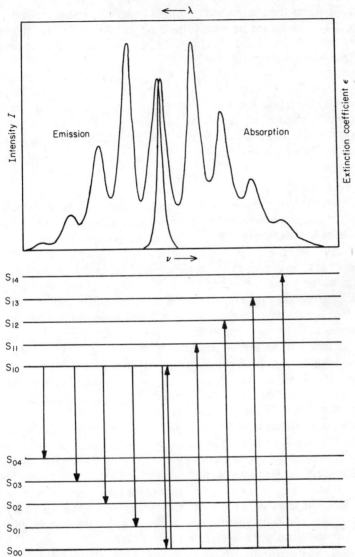

Fig. 2.6. Vibrational levels of S_1 and S_0 showing origins of vibrational structure in emission and absorption spectra.

In practice, the absorption spectrum is plotted in terms of the molecular extinction coefficient ε against frequency ν, wavelength λ, or wavenumber $\bar{\nu}(= 1/\lambda)$.

The molecular extinction coefficient ε is obtained from the measurement of the optical density of a sample of known concentration c and thickness d.

Figure 2.7 shows the attenuation of a light beam on passage through such a sample. The transmitted intensity I_t falls off exponentially with thickness d and for convenience base 10 is chosen rather than base e for the definition of ε. Thus ε is given by

$$\varepsilon = (1/cd)\log_{10}I_0/I_t \tag{2.3}$$

By measurement of the transmission, I_t/I_0, by conventional absorption spectrometry, the optical density and hence ε can be found. It should be noted that ε is a molecular property of the material in question. Figure 2.8 shows ε versus $\bar{\nu}$ for benzene in solution.

2.3.2. Emission Spectra

An insight into the processes of absorption and the consequent emission of light can be gained by considering the potential energy curves of a diatomic molecule as shown in Fig. 2.9. The S_0 curve represents the molecule in its ground state, and the S_1 curve the excited state.

At ordinary temperatures the molecule will be in the lowest vibrational level gg' of the ground state and on the absorption of visible or ultraviolet light will be raised to the excited state S_1. This transition occurs too rapidly ($\approx 10^{-15}$ s) for the atomic nuclei to alter their distances apart and is therefore represented by the vertical line ge (Franck–Condon principle). Since the bond length of an excited molecule is usually greater than the ground-state molecule, the molecule at e finds itself in a highly compressed state. This excess vibrational energy is rapidly lost to neighbouring molecules (except in the case of the isolated gas molecule) and the molecule finds itself in the lowest vibrational level of the excited state, e'. The molecule now has the possibility of dropping to the ground state with the emission of light. The time it spends in the excited state will depend on the mean life τ of the molecule. Again, because of the Franck–Condon principle this transition is represented by a vertical line $e'g'$ in Fig. 2.9. Because of the loss of vibrational energy, the emitted photon (with the exception in some cases of the $S_{00} \rightleftharpoons S_{10}$ transitions) has less energy than the absorbed photon and therefore the emission spectrum is shifted to longer wavelengths with respect to the absorption spectrum (Stokes' law). The intensity of these transitions, whether in absorption or emission, is determined ultimately by the overlap of the vibrational wavefunctions of the ground and excited states which are also shown in Fig. 2.9.

From Fig. 2.9 and Fig. 2.6 which show the vibrational levels of the ground, S_0, and excited state S_1, associated with the absorption and emission spectra, it is obvious that absorption spectra give data about the vibrational levels of the excited state and the emission spectra yield data about the vibrational levels of the ground state.

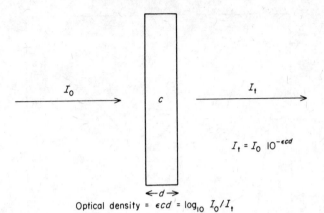

$$I_t = I_0 \, 10^{-\epsilon cd}$$

Optical density $= \epsilon cd = \log_{10} I_0/I_t$

Fig. 2.7. Absorption of light by a sample of thickness d cm and concentration c mole/litre^{-1}.

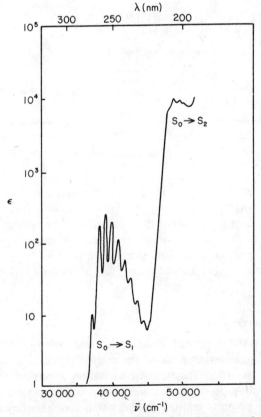

Fig. 2.8. Absorption spectrum of benzene in n-hexane solution.

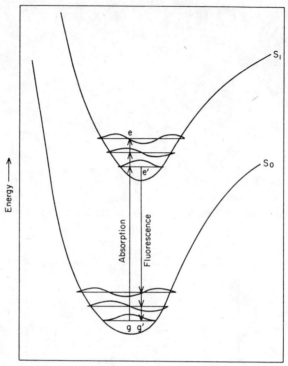

Interatomic distance, r

Fig. 2.9. Potential energy curve for a diatomic molecule, For polyatomic molecule the interatomic distance r is replaced by a configurational coordinate.

It should be noted that in Fig. 2.6 the emission spectrum is the mirror image of the absorption spectrum. This situation frequently arises in organic luminescence and is indicative of the similarity of the respective vibrational wavefunctions in the excited and ground states. If mirror symmetry is not observed in a molecular system, then some hindrance to the relaxation processes in the excited or ground states must be occurring (Birks and Dyson, 1963).

2.3.3. Excitation Spectra

In addition to the study of the absorption and emission spectra, a further means of gaining knowledge about the excited states of organic molecules is from their excitation spectra. An excitation spectrum is obtained by observing the variation in the luminescence quantum intensity (the ratio of the number of photons emitted to the number of photons incident) with change in excitation wavelength.

For most organic molecules the quantum yield is independent of excitation wavelength because of the very efficient process of internal conversion from higher excited states to the lowest excited singlet state. However, molecules such as benzene, toluene, etc., do show changes in quantum intensity for excitation into higher excited state (see Section 2.6.2) and thus excitation spectra of concentrated solutions, in which all the incident radiation is absorbed, are useful in studying the radiationless transitions from higher excited states.

Even for those molecules for which the quantum intensity is invariant with excitation wavelength, excitation spectra provide a powerful tool for measuring the absorption spectra of molecules which are present at too low a concentration for detection by conventional absorption spectrometry. This technique is possible because for a dilute solution the quantum intensity I is proportional to the extinction coefficient ε since

$$I = I_0(1 - 10^{-\varepsilon cd})q \tag{2.4}$$

$$\simeq I_0 2 \cdot 3 \varepsilon cdq \quad \text{for } \varepsilon cd \ll 1 \tag{2.5}$$

where I_0 is the incident excitation intensity, q is the quantum efficiency; and $(1 - 10^{-\varepsilon cd})$ is the fraction of photons absorbed for a concentration c and sample thickness d. Thus the excitation spectrum will be identical to the absorption spectrum provided $\varepsilon cd \ll 1$. The measurable quantum intensity Φ is limited by the sensitivity of the spectrofluorimeter and depends ultimately on the intensity of the excitation source, whereas in very dilute solution absorption spectrometry the ultimate sensitivity is governed by the accuracy within which one can measure the intensity of the sample beam with respect to the almost equal reference beam. Parker (1968a) estimated that concentrations as low as 10^{-12} M can be detected by excitation spectrometry compared with a minimum concentration of 10^{-8} M by absorption spectrometry.

Excitation spectrometry is also used to determine the quantum efficiency of energy transfer between donor and acceptor molecules. By observation of the acceptor luminescence quantum yield, Φ_D when the donor is alone excited, and the luminescence quantum yield, Φ_A, when the acceptor is directly excited, then the energy transfer efficiency is given by Φ_D/Φ_A.

2.3.4. Lifetimes and Rate Parameters

Following the absorption of radiation, the molecules may re-emit this excitation energy as fluorescence. The spontaneous emission transition probability as shown by Einstein for any excited state is directly proportional to the corresponding absorption probability. Thus the radiative transition probability, or the probability per unit time, k_{FM}, that an electron in a particular excited state will fall to the ground state, can be deduced from the absorption spectrum if the mean wavelength of the emission spectrum is known and the mirror-image relationship applies between emission and absorption spectra.

This relation can be expressed in the form

$$k_{\text{FM}} = 2 \cdot 88 \times 10^9 n^2 \langle \bar{\nu}^{-3} \rangle_{\text{av}}^{-1} \int \frac{\varepsilon(\bar{\nu}) \, d\bar{\nu}}{\bar{\nu}} \qquad (2.6)$$

where n is the refractive index of the medium in which the molecule is situated, $\varepsilon(\bar{\nu})$ is the extinction coefficient as a function of wavenumber, $(\bar{\nu}^{-3})_{\text{av}}^{-1}$ is associated with the mean wavenumber of the fluorescence spectrum $F_{\text{M}}(\bar{\nu})$ and is given by

$$(\bar{\nu}^{-3})_{\text{av}}^{-1} = \int F_{\text{M}}(\bar{\nu}) \, d\bar{\nu} \bigg/ \int F_{\text{M}}(\bar{\nu}) \, d\bar{\nu}/\bar{\nu}^3 \qquad (2.7)$$

The expression for k_{FM} in its earlier form assumed that the absorption and fluorescence bands were sharp and that the emission occurred at the same wavelength as the absorption. Strickler and Berg (1962) extended the use of the formula to broad-band molecular spectra. A further refinement (Birks and Dyson, 1963) suggests that the square of the refractive index should be replaced by n_f^3/n_a where n_a and n_f are the refractive indices at the absorption and emission wavelengths respectively.

If k_{FM} is the probability per unit time that the electron will fall to a lower energy state, then the average length of time that the electron spends in the excited state will equal the reciprocal of k_{FM}. This average time is known as the radiative lifetime, τ_0, of the excited state.

The observed fluorescence lifetime, τ, however, will depend on the competition between the radiative and non-radiative processes. If k_{IM} is the sum of the non-radiative rate parameter for internal conversion (k_{GM}) and intersystem crossing (k_{TM}) from the lowest excited state S_1 then the total probability per unit time for the transition between S_1 and S_0 will be $k_{\text{FM}} + k_{\text{IM}}$. Thus the observed lifetime will be given by

$$\tau = \frac{1}{k_{\text{FM}} + k_{\text{IM}}} \qquad (2.8)$$

If the concentration of excited singlet-state molecules is M* moles/litre, then the rate of change of M* after excitation by a δ-pulse for a spontaneous transition is given by

$$\frac{dM^*}{dt} = -(k_{\text{FM}} + k_{\text{IM}}) \, [\text{M*}] \qquad (2.9)$$

and since the fluorescence quantum intensity, I, at any instant is given by

$$I = k_{FM}[M^*] \qquad (2.10)$$

then integration of Eq. (2.9) gives an exponential decay of the form

$$I = I_0 \exp - (k_{FM} + k_{IM}) t = I_0 \exp(-t/\tau) \qquad (2.11)$$

Thus observation of the decay of fluorescence intensity with time after excitation by a δ-pulse or square pulse gives the fluorescence lifetime and hence the sum of the rate parameters k_{FM} and k_{IM}.

2.3.5. Quantum Efficiency and Yields

The quantum efficiency of fluorescence, q_{FM}, is defined as the ratio of the number of fluorescence photons emitted to the number of photons absorbed and is given by

$$q_{FM} = \frac{k_{FM}}{k_{FM} + k_{IM}} = \frac{\tau}{\tau_0} \qquad (2.12)$$

If q_{FM} and the fluorescence lifetime τ are measured, the radiative lifetime τ_0 and rate parameters k_{FM} and k_{IM} can be determined. This measurement of q_{FM} and τ constitutes the major means of obtaining information about the fluorescence processes in organic molecules.

It is important to distinguish at some stage the difference between the terms quantum efficiency and quantum yield. The term quantum yield is used to define the fraction of excited molecules which may follow a given pathway whether radiative or non-radiative. For example, the triplet quantum yield Φ_{TM} is given by

$$\Phi_{TM} = \frac{k_{TM}}{k_{TM} + k_{GM} + k_{FM}} = k_{TM}\tau \qquad (2.13)$$

and gives the fraction of molecules in the excited state S_1 which will cross to T_1 by intersystem crossing. The phosphorescence quantum efficiency, q_{PT}, is given by

$$q_{PT} = \frac{k_{PT}}{k_{PT} + k_{GT}} \qquad (2.14)$$

which gives the fraction of photons emitted from T_1 if the triplet state is excited directly. The total phosphorescence quantum yield Φ_{PT} is thus given by

$$\Phi_{PT} = \Phi_{TM} q_{PT} \qquad (2.15)$$

Therefore

$$\Phi_{PT} = \frac{k_{TM}}{k_{TM} + k_{GM} + k_{FM}} \frac{k_{PT}}{k_{PT} + k_{GT}} \qquad (2.16)$$

TABLE II

Quantum yields of fluorescence and triplet formation at room temperature for dilute solutions and phosphorescence quantum yields and lifetimes at 77 K in EPA glass

Compound	Solvent	Φ_{FM} (= q_{FM})	Φ_{TM}	Φ_{PT}	τ_p (s)
Benzene	Cyclohexane	0·06[a]	0·24[a]	0·18[b]	6·3[b]
Naphthalene	Ethanol	0·21[c]	0·80[c]	0·51[d]	2·3[d]
Anthracene	Ethanol	0·30[c]	0·72[c]	0·0002[e]	0·05[e]
Phenanthrane	Ethanol	0·13[c]	0·85[c]	0·11[f]	3·9[f]
Pyrene	Ethanol	0·65[g]	0·38[g]	0·019[g]	0·5[g]
Biphenyl	Cyclohexane	0·15[h]	0·81[i]	0·24[i]	4·6[e]
p-Terphenyl	n-Hexane	0·77[h]	0·11[j]	0·07[j]	2·3[k]

[a] Cundall, R. B., Pereira, L. C., and Robinson, D. A. (1972). *Chem. Phys. Lett.* **13**, 253.
[b] Li, R. and Lim, E. C. (1972). *J. Chem. Phys.* **57**, 605.
[c] Horrocks, A. R. and Wilkinson, F. (1968). *Proc. Roy. Soc.* A, **306**, 257.
[d] Emolaev, V. L. and Svitashev, O. (1965). *Opt. Spectrosc.* **7**, 399.
[e] Langelaar, J., Rettschnick, R. P. H., and Hoytink, G. J. (1971). *J. Chem. Phys.* **54**, 1.
[f] Masetti, F., Mazzucato, U., and Galiazzo, G. (1971). *J. Luminescence*, **4**, 8.
[g] Medinger, T. and Wilkinson, F. (1966). *Trans. Faraday Soc.* **62**, 1785.
[h] Birks, J. B. (1970). "Photophysics of Aromatic Molecules", p. 130. Wiley–Interscience, London.
[i] Sandros, K. (1969). *Acta Chem. Scand.* **22**, 2815.
[j] Heinzelmann, W. and Labhart, H. (1969). *Chem. Phys. Lett.* **4**, 20.
[k] Kellog, R. E. and Bennett, R. G. (1964). *J. Chem. Phys.* **41**, 3042.

and represents the fraction of photons emitted from T_1 if S_1 was excited directly. Thus the term quantum efficiency is used for the fundamental emission processes and quantum yields for any other processes. Table II lists some typical values of quantum yields. It can be seen that for many molecules $\Phi_{FM} + \Phi_{TM}$ equals unity, hence implying that k_{GM} in general is small.

2.4. ORGANIC COMPOUNDS

A complete chapter could be devoted to a description of fluorescent organic compounds; however, it is the purpose of this section to give a brief summary of typical organic molecules and the effects of substituents and conjugation on the fluorescence spectra.

Any molecule containing delocalized π-electrons has a reasonable chance of being fluorescent. Thus the linear polyenes and aromatic hydrocarbons fall into this category. With the recent advances in low-intensity and far-ultraviolet spectroscopy, even non-aromatic molecules such as the paraffins are found to be weakly fluorescent (Hirayama and Lipsky, 1969). However, the aromatic hydrocarbons are by far the most prolific emitters of luminescence and this discussion will in the main be confined to these molecules.

2.4.1. Aromatic Molecules

The simplest aromatic molecule is benzene and Fig. 2.10 shows an emission spectrum of benzene in dilute solution. This molecule emits at the shortest wavelengths of any of the aromatic compounds. Its fluorescence quantum efficiency is low, $q_{FM} = 0.06$, and the lifetime $\tau = 29$ ns. Substitution of a methyl group on benzene giving toluene shifts the spectrum to slightly longer wavelengths and the quantum efficiency is increased. This spectrum is also shown in Fig. 2.10. Large alkyl groups, C_2H_5, also shift the spectrum

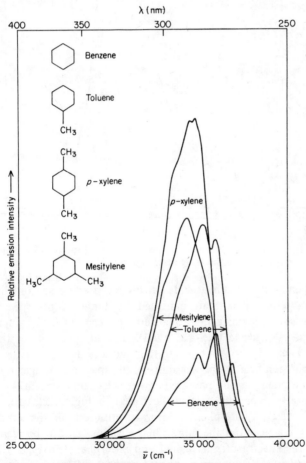

Fig. 2.10. Emission spectra of dilute solutions of benzene, toluene, p-xylene, and mesitylene. The areas under the spectra are proportional to the quantum efficiencies.

to longer wavelengths, as shown in Fig. 2.11. However, as the alkyl groups become larger the shift to longer wavelengths becomes progressively less compared to the effect of substituting additional alkyl groups at other positions on the ring as shown in the case of *p*-xylene and mesitylene in Fig. 2.10. Increasing the size of the alkyl group as before only produces a small shift to longer wavelengths.

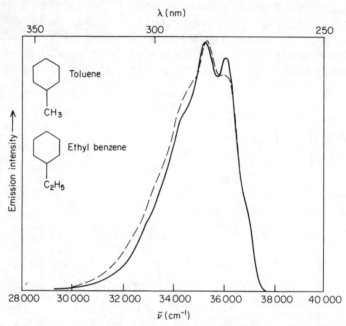

Fig. 2.11. Emission spectra of dilute solutions of ethyl benzene, - - -, compared with toluene, ——, normalized to the same maximum intensity.

The influence of the substituents on quantum efficiency and lifetime is, however, more complex than indicated by these simple spectral shifts, *p*-xylene for example has a high quantum efficiency of 0·33 compared with the other alkyl benzene (see Table III). The lifetimes of all these alkyl substituted benzenes, however, are in the region of 30 ns.

Similar but significantly greater shifts of the spectra to longer wavelengths occur when a phenyl group having its own π-electronic system is substituted on the benzene ring. This fact is illustrated in Fig. 2.12(a) for biphenyl, *p*-terphenyl, and *p*-quaterphenyl. Meso-substitution on the linear phenyl chain tends to smear out the vibrational structure as can be seen by comparison of the spectra of *p*-terphenyl and *m*-terphenyl in Figs. 2.12(a) and 2.12(b). Further substitutions of phenyl groups shift these spectra to even longer wavelengths (Fig. 2.12(b)(ii)).

TABLE III

Quantum Efficiencies and Fluorescence Lifetimes of Some Typical Organic
Molecules in Dilute Cyclohexane Solution at Room Temperature

Compound	q_{FM}	τ (ns)
Benzene	0·058	29
Toluene	0·14	34
p-Xylene	0·33	30
Mesitylene	0·14	36·5
Naphthalene	0·19	96
Anthracene	0·30	4·9
Tetracene	0·17	6·4
Phenanthrene	0·11	57·5
Pyrene	0·65[a]	450[a]
Perylene	0·78	6·4
Biphenyl	0·15	16
p-Terphenyl	0·77	0·95
p-Quaterphenyl	0·74	0·8
Diphenylhexatriene	0·65	12·4
Diphenyloctatetraene	0·07	6·2
Tetraphenylbutadiene	0·48	1·76
2,5-Diphenyloxazole	0·83	1·4
Azulene ($S_2 \to S_0$)	0·20	1·4
Rhodamine-B[†]	0·97[b]	3·2

† In ethanolic solution.
All values of τ are taken from Berlman (1971) with the exception of reference (a).
The values of q_{FM} are taken from Birks (1970d) with the exception of references (a)
and (b).
  a Birks, J. B., Dyson, D. J., and Munro, I. H. (1963), *Proc. Roy. Soc.* A **275**, 575.
  b Weber, G. and Teale, F. W. J. (1957). *Trans. Faraday Soc.* **53**, 646.

For the condensed aromatic hydrocarbons the effect of adding benzene
rings produces very pronounced spectral shifts without loss of vibrational
structure (Fig. 2.13). The position of the substituent is also significant, since
"straight" chains of benzene molecules emit at longer wavelengths than
"bent" molecules as can be seen by comparing anthracene in Fig. 2.13 with
phenanthrene in Fig. 2.14.

Further condensation of the benzene rings occurs in a molecule such as
pyrene (Fig. 2.14) with a large shift to longer wavelengths and this sequence
is followed by perylene and other larger molecules.

A special class of aromatic molecules are the scintillator solutes which
have very high quantum efficiencies and consequently short decay times
(see Table III). In general these molecules have rather complex structures;
however, p-terphenyl and tetraphenylbutadiene are both good scintillator
solutes. The oxazole group is also prevalent in many scintillator solutes.

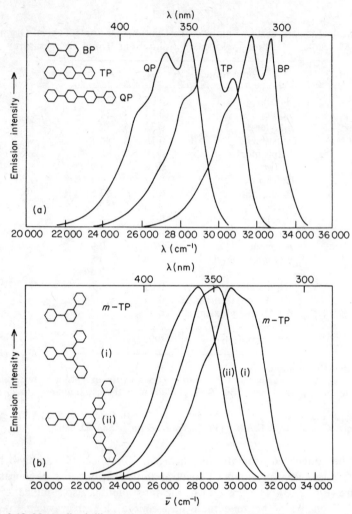

Fig. 2.12. Normalized dilute solution emission spectra: (a) biphenyl (BP), p-terphenyl (TP), p-quaterphenyl (QP), (b) m-terphenyl, 1,3,5-triphenyl benzene (i) and 1,3,5-tri-(biphenyl) benzene (ii) (after Berlman, 1971).

The molecules which combine the molecular structure of the linear polyenes with phenyl groups are also interesting. In these molecules the π-electrons extend along the whole length of the polyene chain as well as circulating in the adjoining benzene rings. Molecules of this type play an important part in the visual pigments of the retina.

The spectra of some linear polyenes are shown in Fig. 2.15 as well as the spectra of some typical scintillator solutes.

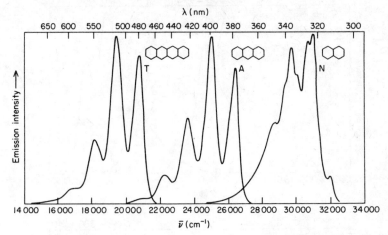

Fig. 2.13. Normalized dilute solution emission spectra of naphthalene (N), anthracene (A), and tetracene (T).

Invariably in any discussion of aromatic molecular fluorescence the exceptional qualities of the molecule azulene must be mentioned. This particular molecule is exceptional, apart from containing a five-membered ring, in that the principal fluorescence emission (Fig. 2.16) occurs from the second excited state to the ground state, $S_2 \rightarrow S_0$, and not from the lowest excited state as is the case for all other aromatic molecules. However, Rentzepis *et al.* (1970), Huppert *et al.* (1972a, b), Hirayama *et al.* (1973) and Gregory *et al.* (1973) have shown that fluorescence from higher excited

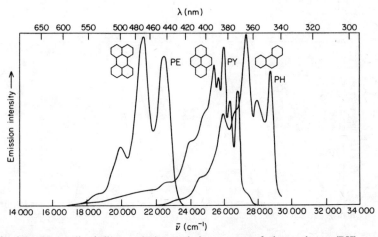

Fig. 2.14. Normalized dilute solution emission spectra of phenanthrene (PH), pyrene (PY), and perylene (PE) (after Berlman 1971).

116

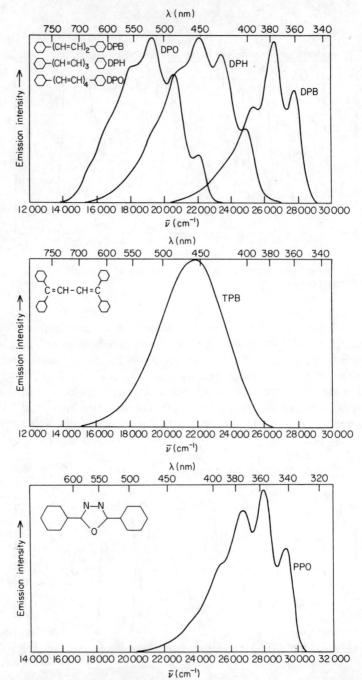

Fig. 2.15. Phenyl polyene and scintillator solute dilute solution emission spectra: diphenylbutadiene (DPB), diphenylhexatriene (DPH), diphenyloctatetraene (DPO), tetraphenylbutadiene (TPB), and 2,5-diphenyloxazole (PPO).

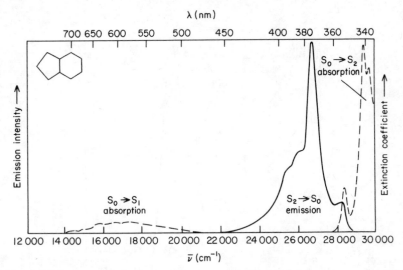

Fig. 2.16. Azulene emission spectrum, ——, and absorption spectrum, - - - - - (after Berlman, 1971).

states of aromatic molecules is not uncommon and does occur but with a very low quantum efficiency ($\leqslant 0.1\%$) compared with the $S_1 \rightarrow S_0$ fluorescence.

Fluorescent dyes form another special group of aromatic compounds which emit light in the visible region of the spectrum. Thus these dyes, which are even more complex in their molecular structure than the scintillator solutes, absorb light in the visible region. Figure 2.17 shows the fluorescence spectra and molecular structure of a typical dye. Unfortunately, most dyes tend to dimerize or polymerize under irradiation by light so that the fluorescence quantum efficiency decreases with time.

Although this brief survey of aromatic molecules leaves out a vast number of basic fluorescent molecules as well as a whole variety of substituents such as nitrogen, hydroxyl groups, halogen groups, etc., it is hoped that it will give some idea about typical fluorescent spectra, quantum efficiencies, and lifetimes.

2.4.2. Aliphatic Molecules

Most aliphatic compounds either are non-luminescent or are emitters of low-intensity luminescence. The most notable exception to this rule is the molecule biacetyl which exhibits an intense green phosphorescence even in liquid solution. The molecule is of the form:

$$H_3C-C-C-CH_3$$
$$\overset{\|}{O} \ \overset{\|}{O}$$

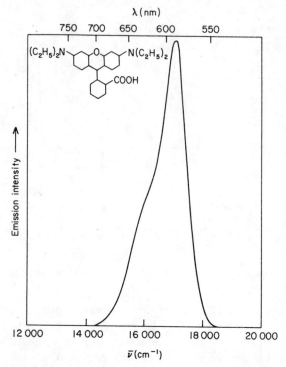

Fig. 2.17. Dilute solution emission spectrum of a typical organic dye (rhodamine B) (after Berlman, 1971).

This phosphorescence is associated with excitation of the carbonyl group (C=O) in the biacetyl which has a non-bonding electron—n-electron—and this gives rise to an n–π* transition (Parker, 1968b). Most molecules, including aromatics, in which n–π* transitions are present are intensely phosphorescent and weakly fluorescent.

The other group of aliphatic molecules of interest are the linear paraffins which until recently were thought to be non-fluorescent. Studies by Hirayama and Lipsky (1969), Hirayama et al. (1970) and Rothman et al. (1973) show that these materials can be excited in the vacuum ultraviolet region at around 170 nm and emit in the near ultraviolet region at about 210 nm. The quantum yields are, however, of the order of 10^{-3} and lifetimes are in the region of a few nanoseconds. As these paraffins contain no π-electrons, excitation occurs via the σ-electron states. In aromatic molecules the higher π-electron states overlap the lower σ-electron states and these studies on the paraffins are possibly relevant to the vacuum ultraviolet excitation studies of aromatic molecules.

2.5. GAS, LIQUID, RIGID MEDIA, AND CRYSTALLINE PHASES

2.5.1. Gas Phase

The luminescence of molecules in the gas phase has been reviewed by Stockburger (1973) and therefore only a brief summary of the fundamental effects of the gas-phase environment will be given here. The organic molecule in the vapour phase at low pressures can be regarded as an isolated molecule. Thus if a molecule is excited into a high vibronic state of the lowest excited state S_1, and if the pressure is sufficiently low and the containing vessel sufficiently large that the mean free time between collisions is longer than the fluorescence lifetime, then fluorescence emission will occur from this high vibronic state to the ground state. Excitation into S_2 can produce even more interesting possibilities such as:

(a) Direct emission from a vibronic state of S_2 to the ground state.
(b) Internal conversion from S_2 to a high vibronic state of S_1 and subsequent emission to S_0.

It can be seen, therefore, that whether (a) or (b) applies depends on the efficiency of the internal conversion process compared with the emission efficiency of the $S_2 \rightarrow S_0$ process. However, (a) tends to increase relative to (b) with increase of excitation energy (Baba et al., 1971).

An increase in pressure or the introduction of a foreign gas leads to a vibrational deactivation of the fluorescent molecule; the excess vibrational energy above the zero point energy being transferred to the neighbouring molecules and emission occurring from the zero point energy of the lowest excited state S_{10}.

Since a molecule in a high vibronic state corresponds to an effective temperature much higher than the ambient, then internal thermally activated quenching processes such as intersystem crossing to the triplet manifold may lead to a lowering of the fluorescence quantum efficiency at low pressures. Increase of pressure or addition of a foreign molecule (other than known quenchers of luminescence) may, therefore, lead to an increase in the quantum efficiency.

As would be expected, the spectra of molecules in the gas or vapour phase are much sharper than in the liquid or solid phase.

2.5.2. Liquid Phase

Most studies of the fluorescence of organic molecules are made in liquid solution.

Apart from the convenience of handling solutions and the ease of observation of the spectra compared with the gas and solid phases, it is also a more convenient medium for assaying materials which are inherently fluorescent or can be made fluorescent by chemically reacting the material with a fluorescent compound (Guilbault, 1973).

The principal fluorescence emission in solution, with only a few exceptions, takes place from the zero point level of the lowest excited singlet state S_{10} to the vibronic levels of the ground state. Excitation into higher vibronic states of excited singlet states leads to dissipation of the excess vibrational energy to neighbouring molecules until thermal equilibrium is established. As in the high pressure or high density gas situation, the quantum efficiency in solution should be higher than for the isolated molecule, since the temperatures of the molecules are equilibrated with the surroundings prior to emission and the thermally activated internal quenching processes are reduced.

Since the refractive index and dielectric constant are higher for liquids than for gases, all spectra undergo a solvent shift to longer wavelengths due to these changes in the dielectric medium. A corresponding but smaller shift occurs in the absorption spectrum although large shifts in the polyenes have been observed (Hudson and Kohler, 1973). Solvent solute interactions also broaden the vibrational bands of the spectra. The degree of broadening is indicative of the strength of the solvent–solute interaction. For example, perfluor-hexane (C_6F_{14}), a unique solvent, shows very little solvent interaction with the alkyl benzenes and the spectra almost resemble the gas-phase spectra (Lawson et al., 1969).

The environment of neighbouring molecules also means that if any of these molecules is a non-fluorescent quencher of luminescence, dissolved oxygen for example, then the quantum efficiency and fluorescent lifetime are reduced. However, these quenching processes are controlled by diffusion and hence a quenching molecule in a viscous solution will produce less reduction in fluorescence emission than in a low viscosity solution.

At high solute concentrations, self-quenching of fluorescence can occur. This is a diffusion-controlled process and is often accompanied by the appearance of a broad structureless new fluorescence emission at longer wavelengths. This behaviour is attributed to the formation of an excited-state dimer or "excimer" formed by the association of an excited solute molecule and a ground-state solute molecule.

If a solution contains two fluorescent solute molecules X and Y, then provided the first excited singlet state of Y lies just below that of X then energy transfer from X to Y can occur. Thus as the concentration of Y is increased the fluorescence emission of Y also increases at the expense of the fluorescence of X which reduces in a manner similar to that for a quenching process. Again this energy transfer process is diffusion-controlled. It should be noted, however, that although both self-quenching and energy transfer occur readily in the liquid phase, these processes also occur in the gas phase.

2.5.3. Rigid Media and Crystalline Solids

(a) *Rigid Inert Media.* The most important application of rigid inert media is in the study of phosphorescence. In a rigid medium the long-lived triplet state has the opportunity of emitting to the singlet ground state without being quenched by traces of impurities. These impurities cannot diffuse to the excited molecule and deactivate it, although in some cases weak diffusion can occur even in rigid media. Reduction of temperature to 77 K or lower also reduces diffusion and any thermal deactivation of the triplet state. Thus low temperature rigid glasses such as EPA† or plastic media are particularly useful for phosphorescence studies.

Another interesting application of rigid glasses is used in the study of the quasilinear spectra of the aromatic hydrocarbons. This work was initiated by Shpol'skii (1952) when it was observed that the fluorescence spectra of many aromatic molecules in a frozen solution of a normal paraffin at 77 K exhibited highly resolved fine structure. The optimum condition for observing this fine structure is obtained when the length of the paraffin molecule matches the long axis length of the aromatic molecule.

Recently aromatic molecules in rare-gas matrices (Hallam, 1970) have been used for luminescence studies in the vacuum ultraviolet region. The aromatic vapour is mixed with a rare gas and then solidified at 77 K. This technique is particularly useful, since the rare-gas matrix is transparent to the vacuum ultraviolet radiation and the aromatic molecule can be treated as an isolated molecule at a low temperature. The effect of krypton, neon, argon, and xenon matrices on the molecular emission processes can also be studied; these studies being particularly useful as the phosphorescence emission is enhanced by the heavy atom effect (see Section 2.7.1).

(b) *Organic Crystals.* The molecules in organic aromatic solids are held together by van der Waals' forces; however, since these forces are weak compared with the Coulombic and exchange forces in ionic or covalent crystals, the luminescence processes in aromatic crystals can still be linked directly with the molecular excited states. Thus organic solids are regarded as molecular crystals.

A molecule in the environment of similar neighbouring molecules is effected by two factors:

(a) A "solvent" red shift due to the change in the dielectric and optical properties of the crystal from the vapour or liquid environment, and the interaction of the excited molecules with equivalent molecules (exciton shift). This shift also reflects the anisotropy of the dielectric and optical properties. The red shift ($\simeq$ few hundred cm^{-1}) for the singlet states is two orders of magnitude larger than for triplet states.

† A mixture of ethyl ether, iso-pentane, and ethyl alcohol, volume ratio 5 : 5 : 2 respectively.

(b) The Davydov splitting of the unperturbed energy state. This occurs in crystals where there is more than one inequivalent molecule per unit cell and therefore includes most organic crystals. The energy state splits into as many components as there are inequivalent sites. This Davydov splitting which is a few hundred cm^{-1} for singlets and $\simeq$ tens of cm^{-1} for the triplet states.

Thus the energy states, E_c, for the crystals can be written as

$$E_c = E_0 - S \pm D$$

where E_0 is transition energy of the isolated molecule, S is the solvent shift energy, and D is half the Davydov splitting energy.

Apart from these spectral shifts and anisotropic changes which make polarization studies of these crystals interesting, a most important property of organic crystals is that of migration of the excitation energy. A molecule in an excited state can very easily transfer its excitation energy to similarly oriented equivalent molecules. Thus the excitation energy becomes delocalized and leads to the concept of excitons which are neutral non-conducting entities which can travel through the crystal carrying the excitation energy with them. The lifetime and hence mean free paths of these excitons depend on the lifetime of the particular molecular excited state. Thus singlet excitons have short lifetimes ($\simeq$ ns) and mean free paths of a few hundred ångströms whereas triplet excitons with longer lifetimes have mean free paths of a few microns. However, the longer lifetimes of the triplet excitions do make them more susceptible to quenching by impurities or exciton traps.

A recent development in the spectroscopy of organic crystals has been in the use of magnetic fields. As mentioned earlier, the application of a magnetic field removes the degeneracy of the triplet state and hence has a pronounced effect on the intensity of prompt and delayed fluorescence. This behaviour is attributed to triplet–triplet exciton fusion and singlet exciton fission processes as shown in Fig. 2.18. Chapter 4 deals with this subject in more detail.

In certain instances the spectra in the crystalline phase may show more structure than in the liquid or rigid-glass environment. For example, p-terphenyl in solution fluoresces with a broad emission band, whereas crystalline p-terphenyl gives a well-structured emission spectrum (Birks and Cameron, 1959). These differences are attributed to the phenyl groups in the crystal having set orientations to one another whereas in the solution phase the phenyl groups are randomly oriented.

As in the case of gases and liquids, energy transfer, quenching, etc., can readily occur in the crystalline phase. However, instead of the physical diffusion of an excited molecule towards an acceptor molecule prior to energy transfer, the delocalized excitation energy (in the form of excitons

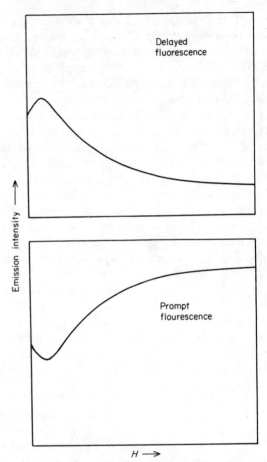

Fig. 2.18. Variation of delayed fluorescence (triplet exciton fusion) and variation of prompt fluorescence (singlet exciton fission) with increase in magnetic field H (after Geacintov and Swenberg, 1978).

in the crystal environment) can travel through the crystal until an acceptor molecule is reached; thus energy migration plays an important role in energy transfer in crystals.

2.6. RADIATIONLESS PROCESSES

2.6.1. Introduction

A radiationless process is a transition which takes place between energy states without the emission of light. Radiationless processes are, therefore, in competition with the radiative processes responsible for luminescence emission and can be crudely regarded as the dissipation of energy in the form of heat. It would be expected that this heat energy should be detectable as infrared radiation; however, to avoid the multiple collisions which occur between molecules and which distribute this dissipated energy over a large number of molecules, experiments would have to be carried out at very low pressures and in very large containers in order really to isolate the molecule. These experimental conditions are extremely difficult to achieve. These experimental difficulties also confuse the issue as to whether radiationless processes are confined to the individual molecule or are dictated by external conditions. The evidence so far (Kistiakowsky and Parmenter, 1965; Anderson and Kistiakowsky, 1969) points to the intramolecular property, that is, the radiationless processes are an inherent property of the molecule but can, however, be strongly influenced by the molecular environment.

Very strong evidence exists for the existence of radiationless processes in organic molecules. The fact that the principal fluorescence emission occurs from the lowest excited singlet state irrespective of excitation into higher excited states represents strong evidence for internal conversion between higher excited states and the lowest excited singlet state. The existence of phosphorescence emission at longer wavelengths than the fluorescence emission indicates that some radiationless loss of energy (intersystem crossing) must occur after excitation and prior to emission. Most quantum efficiencies are less than unity and hence radiationless processes compete with the radiative processes.

The measurement of the quantum yields and lifetimes of the fluorescence and phosphorescence emission coupled with the measurement of triplet quantum yields enables the rate parameters for intersystem crossing, k_{TM} and k_{GT} and internal conversion, k_{GM}, to be evaluated. Thus measurement of q_{FM} gives $k_{FM}/(k_{FM}+k_{GM}+k_{TM})$ and τ_M gives $1/(k_{FM}+k_{GM}+k_{TM})$ and hence the radiationless rate parameter $(k_{GM}+k_{TM})$ can be found. Measurement of the triplet quantum yield $\Phi_{TM}(=k_{TM}\tau_M)$ gives the absolute value of k_{TM} and hence k_{GM} (Wilkinson, 1975). Similarly, measurement of the phosphorescence quantum yield Φ_{PT} and phosphorescence lifetime τ_p coupled with the measurement of Φ_{TM} enable the intersystem crossing rate parameter k_{GT} to be evaluated (see Section 2.3.5 and Eqs (2.15) and (2.16)). The direct evaluation of k_{GT} from the phosphorescence quantum efficiency q_{PT} and lifetime τ_p is experimentally not easy due to the difficulty of directly exciting T_1 from S_0.

For excitation into higher excited states than S_1 or T_1, apart from the experimental difficulties of observing the emission from higher excited states, the evaluation of the radiationless rate parameters becomes more complex. However, recent measurements have produced values for the radiationless rate parameter for the $S_2 \rightarrow S_0$ transition from the measurement of quantum efficiencies and lifetimes of the S_2 to S_0 fluorescence (Rentzepis *et al.*, 1970; Deinum *et al.*, 1971).

2.6.2. Theory of Radiationless Processes

An excellent review of the theory of radiationless processes has been given by Henry and Siebrand (1974). It will be sufficient, therefore, briefly to summarize the current ideas about radiationless transitions. Figure 2.19 shows the potential energy curves for the lowest singlet and triplet excited states and the ground state. Considering first of all internal conversion between S_1 and S_0, it was thought in the earlier theories that crossing of potential curves was essential for radiationless transitions between S_1 and

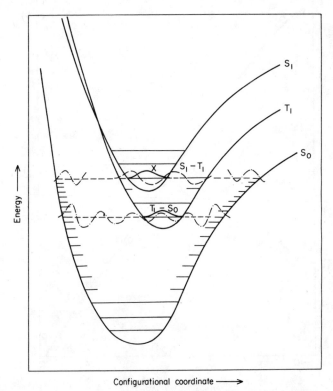

Configurational coordinate ⟶

Fig. 2.19. Potential energy curves for a polyatomic molecule showing overlap of wavefunctions which give rise to intersystem crossing radiationless transitions: ——, wavefunctions for zeroth vibrational level of S_1 and T_1; — · — ·, wavefunctions for high vibrational levels of S_0; – – –, wavefunctions for high vibrational level of T_1. The $S_1 \rightarrow S_0$ internal conversion process would correspond to the overlap of —— and — · — · at X.

S_0, later Ross and co-workers (Hunt *et al.*, 1962) introduced the idea of tunnelling between states which could occur at points other than the crossing points of the curves. Recently, it was realized that curve crossing was not necessary and indeed unlikely in large aromatic molecules and that radiationless transitions between states could be explained in terms of the overlap of vibrational wavefunctions (Franck–Condon factors) in the same manner as radiative transitions. Thus both internal conversion and intersystem crossing for most molecules can be explained in terms of the Franck–Condon factors. The agreement between theory and experiment for intersystem crossing based on the Franck–Condon factor for the energy gap between the appropriate electronic states is excellent (see Fig. 2.20). It can be seen that

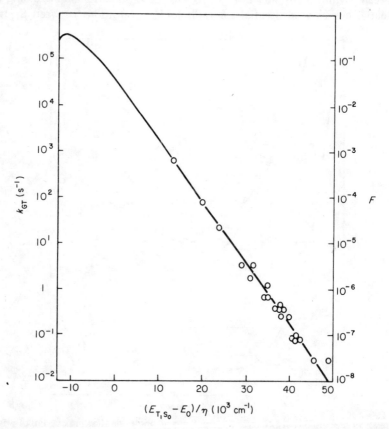

Fig. 2.20. Plot of the radiationless intersystem crossing rate parameter k_{GT} between T_1 and S_0 and Franck–Condon factor F for planar aromatic hydrocarbons against $(E_{T_1 S_0} - E_0)/\eta$, where η is the relative number of hydrogen atoms in the molecule. $E_{T_1 S_0}$ is the triplet energy above the ground state S_0 and E_0 is a constant ($\simeq 5000\ cm^{-1}$) associated with CH and CC stretching modes (after Henry and Siebrand, 1974).

because the vibrational overlap of wavefunctions decreases as the vibrational energy increases, the Franck–Condon factor decreases as the energy gap between the electronic states increases.

For internal conversion, theory and experiment are more difficult to compare because of the lack of accurate experimental values of k_{GM}. However, extrapolation of the data from Fig. 2.20 enables Franck–Condon factors for the S_1–S_0 gap to be evaluated and hence theoretical values of k_{GM} can be predicted. The agreement between theory and experiment is good for large organic molecules, but significant differences occur for small molecules such as benzene and its alkyl derivatives.

Benzene, in particular, deserves some mention in that the internal conversion rate k_{GM} is unusually high, $\Phi_{FM} + \Phi_{TM} \neq 1$ (see Table II). The quantum efficiency in the gaseous, liquid and solid phases also depends on the excitation wavelength (Braun et al., 1963) and becomes progressively less for excitation into S_2 and S_3. Most organic molecules have a quantum efficiency which is independent of excitation wavelength and thus in benzene some unusual radiationless process is occurring for excitation into excited states greater than S_1. This has been attributed to the formation of a transient isomeric state of benzene which can revert back to normal benzene by collision with neighbouring molecules (reviewed by Cundall and Ogilvie, 1975).

2.7. DIFFUSION-CONTROLLED PROCESSES

From the point of view of organic luminescence the most important diffusion-controlled processes are those of impurity quenching, concentration quenching, and energy transfer. In solution, a reaction or effective collision between one molecule and another has a high probability of taking place when the molecules are in close proximity to one another. If the activation energy of the reaction is low then the probability that a reaction will occur for each collision is very high, so that the diffusion of the reacting molecules towards one another becomes the rate-determining step. This type of reaction is known as a diffusion-controlled process.

In solution, collisional interactions between molecules are controlled by diffusion, that is, they depend fundamentally on the diffusion coefficients of the molecules involved. Thus viscosity and temperature play significant roles in diffusion-controlled processes.

The diffusion-controlled rate parameter k_{diff} is related (Sveshnikoff, 1935, 1937; Umberger and La Mer, 1945) to the diffusion coefficients and radii of the molecules involved in the collisional interaction by the relation:

$$k_{diff} = 4\pi N' r_{AB} D_{AB} p \left(1 + \frac{r_{AB}}{\sqrt{(D_{AB} \tau_M)}}\right) \qquad (2.17)$$

where N' is the number of molecules per millimole, r_{AB} represents the closest distance of approach of two molecules A and B and D_{AB} $(= D_A + D_B)$ represents the simultaneous diffusion of A and B. τ_M is the lifetime of the luminescent molecule A in the absence of B. p is the probability per collision that a reaction will take place and thus varies between 0 and 1. In many cases the transient term $\{1 + [r_{AB}/\sqrt{(D_{AB}\tau_M)}]\}$ can be ignored when $D_{AB}\tau_M \gg r_{AB}^2$.

For large spherical solute molecules moving amongst smaller solvent molecules, the diffusion D_{AB} can be equated to the Stokes–Einstein diffusion coefficient D_{SE} for spheres of radius r moving in a viscous liquid (Sutherland, 1905) and hence

$$D_{AB} = D_{SE} = kT/6\pi\eta r \qquad (2.18)$$

where k is Boltzmann's constant, T is the absolute temperature, and η is the macroscopic viscosity. Combining Eqs. (2.17) and (2.18) and neglecting the transient term gives the modified Debye (1942) equation:

$$k_{diff} = (8RT/3000\eta)p \qquad (2.19)$$

For solute molecules which have radii comparable to the solvent molecules then

$$k_{diff} = (8RT/2000\eta)p \qquad (2.20)$$

Recently, in a review of diffusion-controlled processes by Alwattar et al. (1973) it has been shown that where the solute molecules are smaller than the solvent molecules then $k_{diff} \gg (8RT/2000\eta)p$. In many instances, however, the Debye equation has predicted values for k_{diff} which have agreed with experimental observations, but these equations should nevertheless be used with some caution because of their dependence on the relative radii of the solute and solvent molecules (Debye, 1942).

2.7.1. Quenching Processes

In solution most quenching processes are diffusion-controlled. The fluorescence quantum yield of an organic molecule is reduced in the presence of a quenching molecule such as oxygen or a halogen. If the magnitude of the quenching depends linearly on the concentration of the quencher, then the quenching is said to obey Stern–Volmer kinetics and follows the Stern–Volmer (1919) equation which is of the form:

$$\Phi_0/\Phi_q = 1 + K_q[Q] \qquad (2.21)$$

where K_q is the Stern–Volmer quenching constant, Φ_0 is the fluorescence quantum yield when no quencher is present and Φ_q the quantum yield in the presence of a quencher concentration of Q moles/litre.

The kinetic scheme for quenching can be written as follows:

	Process	*Rate parameter*
Fluorescence	$M^* \to M + h\nu$	k_{FM}
Internal conversion	$M^* \to M$	k_{GM}
Intersystem crossing	$M^* \to M$	k_{TM}
Quenching	$M^* + Q \to M + Q$	$k_{QM}[Q]$

The quantum yield in the presence of the quencher is, therefore, given by

$$\Phi_q = \frac{k_{FM}}{k_{FM} + k_{GM} + k_{TM} + k_{QM}[Q]} \tag{2.22}$$

and in the absence of the quencher,

$$[Q] = 0,$$

$$\Phi_0 = \frac{k_{FM}}{k_{FM} + k_{GM} + k_{TM}} \tag{2.23}$$

Hence

$$\frac{\Phi_0}{\Phi_q} = \frac{k_{FM}(k_{FM} + k_{GM} + k_{TM} + k_{QM}[Q])}{k_{FM}(k_{FM} + k_{GM} + k_{TM})}$$

$$= 1 + \frac{k_{QM}[Q]}{k_{FM} + k_{GM} + k_{TM}}$$

$$= 1 + \tau_M k_{QM}[Q] \tag{2.24}$$

where τ_M is the fluorescence lifetime in the absence of Q.

The Stern–Volmer quenching constant K_q is thus equal to $\tau_M k_{QM}$ and if the process is diffusion controlled then k_{QM} equals k_{diff}. The Stern–Volmer equation, therefore, can be written in the form

$$\Phi_0 / \Phi_q = 1 + \tau_M k_{diff}[Q] \tag{2.25}$$

provided the quenching process is a collisional process and is diffusion-controlled.

Quenching can take place in a variety of ways; the most common quenching process is that produced by dissolved oxygen in solution. The detailed mechanics of the process are rather complicated involving the possible formation of an oxygen organic excited state complex (exciplex). However, the net result is that oxygen catalyses the non-radiative processes of the organic fluorescent molecule. These processes can be represented by the equations:

(i) $S_1 + {}^3O_2 \to T_1 + {}^3O_2$ (catalysed intersystem crossing)   (2.26)

and

(ii) $S_1 + {}^3O_2 \rightarrow S_0 + {}^3O_2$ (catalysed internal conversion) (2.27)

Process (i) is common in low viscosity solutions and process (ii) is common in high viscosity and rigid solutions (Birks, 1970b).

The quenching of fluorescence by halogens, the so-called "heavy-atom effect", is a much simpler process than oxygen quenching. Most aromatic molecules with a halogen substituent have low fluorescence quantum yields, since the perturbation produced by the halogen atom increases the intersystem crossing rate. This effect is known as the internal heavy-atom effect. Similarly, an external heavy-atom effect occurs when a halogen derivative in the form of bromine gas or carbon tetrachloride strongly quenches the fluorescence by increasing the intersystem crossing rate. Both the internal and external heavy-atom effects increase the intersystem crossing rate by increasing the spin–orbit coupling between triplet and singlet states.

Quenching can also occur by the process of energy transfer, a donor excited molecule can have its energy transferred to a fluorescent or non-fluorescent acceptor molecule; thus reducing the fluorescence intensity of the donor emission with increase of concentration of the acceptor. Once again Stern–Volmer kinetics apply to such quenching processes.

Quenching studies can be used to estimate the lifetime of fluorescent molecules in solution (Berlman, 1971). If the fluorescence intensities of aerated and deoxygenated solutions are compared for different solutes and if $k_{QM}[Q]$ is assumed to be constant then

$$\Phi_0/\Phi_q = 1 + \tau_M k_{QM}[Q] = 1 + k'\tau_M \qquad (2.28)$$

and hence from the known values of τ_M and a graph of the form shown in Fig. 2.21 the unknown lifetime of a solute can be estimated from the magnitude of the oxygen quenching. The above method, of course, assumes that the oxygen concentration and diffusion coefficient are constant in the aerated solution and that the diffusion coefficient of the much larger organic molecule is small compared with the diffusion coefficient of oxygen.

2.7.2. Concentration Quenching (Excimer Formation)

Concentration quenching or self-quenching is also a diffusion-controlled process and in this case the solute fluorescence is decreased with increase of solute concentration. Once again the Stern–Volmer law can be applied and written as

$$\Phi_0/\Phi_q = 1 + k[c] \qquad (2.29)$$

where k is the Stern–Volmer coefficient of concentration quenching and c is the solute concentration.

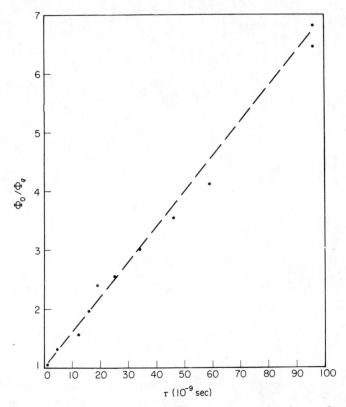

Fig. 2.21 Plot of the relative intensity Φ_0/Φ_q of deoxygenated and aerated solutions against fluorescence lifetime for various solutes (after Berlman, 1971)

The physical basis of concentration quenching was discovered when a detailed study of the fluorescence of pyrene solutions was carried out by Förster and Kasper (1955). They observed that the quenching of the characteristic structured violet emission of pyrene with increase of pyrene concentration was accompanied by the appearance of a broad structureless emission at longer wavelengths. This long wavelength emission increased in intensity as the concentration increased (see Fig. 2.22). The absorption spectrum did not show any corresponding changes with concentration, thus showing that the molecular interaction responsible for the long wavelength emission occurred only in the excited state and not in the ground state. Förster and Kasper attributed the short wavelength structured emission to the normal fluorescence of the unassociated monomer molecules and the long wavelength structureless emission to the fluorescence of transient dimers formed by the association of excited and unexcited molecules. The term

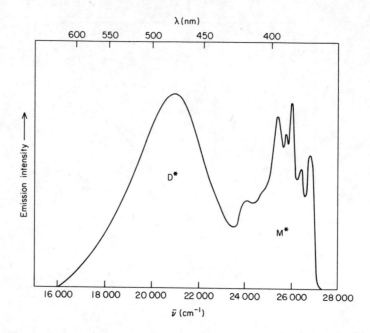

Fig. 2.22. Emission spectrum of 10^{-3} M solution of pyrene in cyclohexene.

"excimer" was proposed by Stevens and Hutton (1960) in order to distinguish this dimer which only exists in the excited state from the normal type of dimer which exists in the ground state. The excimer, on the emission of the long wavelength emission, dissociates into a pair of unexcited ground-state molecules. The fluorescence, formation, and dissociation of the excimer process can best be illustrated by the scheme shown in Fig. 2.23, where M represents a monomer molecule in the ground state S_0; M* and ^{3}M* represent monomer molecules in the excited singlet S_1 and triplet T_1 states. Similarly D* represents an excited excimer molecule and ^{3}D* represents the triplet state of the excimer. The concentration rate parameter $k_{DM}c$ is the excimer formation rate parameter and k_{MD} the dissociation rate parameter. The other rate parameters for fluorescence etc. follow the notation described earlier. Table IV lists the values of these parameters for some typical excimer-forming molecules.

The quantum intensity of monomer fluorescence follows the Stern–Volmer law and is of the form:

$$\Phi_M = \frac{q_{FM}}{1 + c/c_h} \qquad (2.30)$$

where c_h is the half-value concentration at which $\Phi_M = \frac{1}{2}q_M$ and can be shown to be equal to

$$[k_M(k_D+k_{MD})/k_{DM}k_D]; \quad k_M = k_{FM}+k_{IM} \quad \text{and} \quad k_D = k_{FD}+k_{ID}.$$

The c_h value is the reciprocal of the Stern–Volmer concentration quenching constant k. Similarly the quantum intensity of excimer fluorescence is related to c_h by

$$\Phi_D = \frac{q_{FD}}{1+(c_h/c)} \tag{2.31}$$

From Eqs. (2.30) and (2.31) the ratio of the excimer to monomer intensities is found to be proportional to c and is given by the expression:

$$\frac{\Phi_D}{\Phi_M} = \frac{k_{FD}k_{DM}c}{k_{FM}(k_D+k_{MD})} \tag{2.32}$$

The time dependence of the monomer and excimer emission can be found (Birks $et\ al.$, 1963) by solving the rate equations for the monomer excimer system, after excitation by a δ-pulse.

$$\frac{dM^*}{dt} = -(k_{FM}+k_{GM}+k_{TM}+k_{DM}c)[M^*]+k_{MD}[D^*] \tag{2.33}$$

$$\frac{dD^*}{dt} = -(k_{FD}+k_{GD}+k_{TD}+k_{MD})[D^*]+k_{DM}c[M^*] \tag{2.34}$$

which give a time dependence of intensity of the form

$$i_M(t) = c_M[\exp(-\lambda_1 t)+A\exp'(-\lambda_2 t)] \tag{2.35}$$

and

$$i_D(t) = c_D[\exp(-\lambda_1 t)-\exp(-\lambda_2 t)] \tag{2.36}$$

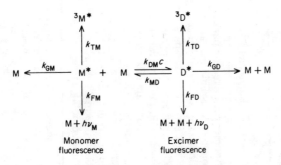

Fig. 2.23. Excimer kinetic scheme.

TABLE IV

Excimer Parameters at Room Temperature

Parameter		Benzene[a,b] in cyclohexane	Naphthalene[a] in ethanol	Pyrene[d,e] in cyclohexane
Monomer radiative rate parameter	k_{FM}	1.8×10^6 s^{-1}	2.3×10^6 s^{-1}	1.5×10^6 s^{-1}
Monomer non-radiative rate parameter	k_{IM}	28×10^6 s^{-1}	17×10^6 s^{-1}	0.7×10^6 s^{-1}
Monomer quantum efficiency	q_{FM}	0·06	0·12	0·65
Excimer formation rate parameter	k_{DM} c	—	—	6.7×10^9 M^{-1} s^{-1}
Excimer radiative rate parameter	k_{FD}	0.8×10^6 s^{-1}	0.9×10^6 s^{-1}	11.6×10^6 s^{-1}
Excimer non-radiative rate parameter	k_{ID}	35×10^6 s^{-1}	1.5×10^6 s^{-1}	4×10^6 s^{-1}
Excimer quantum efficiency	q_{FD}	0·023	0·32	0·75
Half-value concentration	c_h	4·8 M	1 M	5×10^{-4} M
Excimer dissociation rate parameter	k_{MD}	—	—	6.5×10^6 s^{-1}
Monomer lifetime ($c \to 0$)†	τ_M	34 ns	52 ns	450 ns
Excimer lifetime ($c \to \infty$)†	τ_D	28 ns	380 ns	65 ns
Excimer binding energy	B	0·35 eV	0·27 eV	0·40 eV

† τ_M and τ_D defined as $1/k_{FM}+k_{IM}$ and $1/k_{FD}+k_{ID}$ respectively.

a Cundall, R. B. and Pereira, L. C. (1973). Chem. Phys. Lett. 18, 371.
b Cundall, R. B. and Robinson, D. A. (1972). J. Chem. Soc. Faraday Transactions II, 68, 1133.
c Selinger, B. K. (1966). Austral. J. Chem. 19, 825.
d Birks, J. B., Dyson, D. J., and Munro, I. H. (1963). Proc. Roy. Soc. A, 275, 575.
e Birks, J. B., Lumb, M. D., and Munro, I. H. (1964). Proc. Roy. Soc. A, 280, 289.

where c_M, c_D, A, λ_1, and λ_2 are functions of the rate parameters and con- concentration c. From analysis of the decay curves of the type shown in Fig. 2.24 over a range of concentrations the rate parameters for the excimer–monomer system can be found. However, other measurements (see Section 2.6.1) are required to separate the non-radiative processes k_{GM} and k_{TM} for the monomer and k_{GD} and k_{TD} for the excimer.

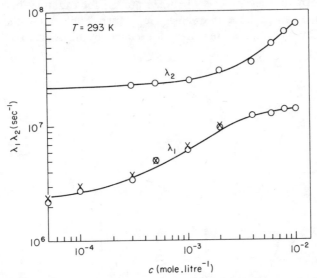

Fig. 2.24. Plot of rate parameters λ_1 and λ_2 for excimer formation against concentration (after Birks *et al.*, 1963).

Since the first observation of excimer formation over twenty years ago, it has been found that excimer formation is the rule rather than the exception with aromatic hydrocarbons. The planar structure of the benzene rings lends itself to easy association with a neighbouring molecule and a resonance interaction takes place between the two molecules with their planes separated by a perpendicular distance of about 3–4 Å (Murrell and Tanaka, 1964). Thus not only pyrene, but benzene and its alkyl derivatives, naphthalene and its derivatives, etc., form excimers.

Excimer formation is reduced by the effects of substituents on the aromatic ring system. For example, the xylenes show almost no excimer emission or self-quenching at room temperature, but on reduction of temperature excimer bands can be seen (Lumb and Weyl, 1967). In anthracene, because of the molecular geometry, ground-state dimer formation is favoured in preference to excimer formation. However, excimers and dimers are formed between 9-methyl anthracene molecules (Birks and Aladekomo, 1963; Barnes and Birks, 1966).

The excimer can be regarded as a molecular entity having singlet and triplet states just like a normal organic molecule; the only difference being that it has a repulsive ground state. Figure 2.25 shows the potential energy diagram for a pair of excimer-forming molecules. The repulsive ground state can be clearly seen.

Excimer formation also occurs in the crystal and in the gas phase. In fact, excimer formation even occurs between atoms; for example, helium in its excited state can produce an excimer He_2. More details about these atomic excimers can be found in the review of excimers by Birks (1975).

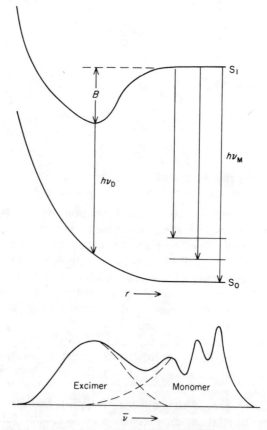

Fig. 2.25. Potential energy curve for a pair of excimer-forming molecules and corresponding emission spectra. B represents the binding energy of the excimer.

2.8. THE TRIPLET STATE

2.8.1. Population of the Triplet State, Phosphorescence, and Delayed Fluorescence

All organic molecules have an even number of π-electrons, and hence in the ground state all the electron spins are paired off and thus the ground state is a singlet state. However, when one of the electrons is excited it can undergo a spin reversal and its spin will be orientated in the same direction as the electron remaining in the ground state. This situation creates a triplet state, since the total spin vector can take up three different orientations in space. In the absence of an external magnetic field or strong spin–orbit interaction this triplet state is degenerate. However, application of a strong magnetic field removes the degeneracy and gives rise to three states of slightly different energy. As mentioned previously, the triplet states always lie somewhat lower in energy than the corresponding singlet levels.

Population of the excited triplet states directly from the singlet ground state is not easy due to the spin-forbiddeness of transitions between states of different multiplicity. S_0 to T_1 absorption is therefore very weak with extinction coefficients of the order of 10^{-3}. Similarly, the reverse process from T_1 to S_0 is spin-forbidden and hence the lowest triplet state T_1 can be regarded as a metastable state having a lifetime in the range of milliseconds to seconds.

The triplet state is responsible for the phosphorescence and delayed fluorescence processes in organic molecules. However, it is indeed surprising that emission from the triplet state should be observed at all if population of the triplet state can only occur via direct but weak absorption from S_0 to T_1. In fact, the lowest triplet state, T_1, is populated almost entirely by inter-system crossing from the excited singlet state S_1. Although this process is a spin-forbidden transition, the transition probability (Franck–Condon factor) increases as the energy gap between the respective states decreases; thus the intersystem crossing rate between S_1 and T_1 is 10^6 times greater than that between S_0 and T_1. This intersystem crossing rate k_{TM} between S_1 and T_1 is in direct competition with the fluorescence emission process between S_1 to S_0; however, the fluorescence and internal conversion rate parameters k_{FM} and k_{GM} are comparable with k_{TM} and thus the triplet quantum yield $\Phi_{TM}\,(=k_{TM}/k_{FM}+k_{TM}+k_{GM})$ is of the same order of magnitude as the fluorescence quantum yield.

The lowest triplet state can thus have a similar population to the lowest singlet state. It would be expected, therefore, that all fluorescent organic molecules are also phosphorescent. However, because of its long lifetime, the triplet state is very susceptible to quenching by trace impurities, and in particular oxygen. In any medium where significant diffusion can occur, that is in the gas and liquid phases, quenching of the phosphorescence occurs. Thus phosphorescence is most commonly seen in rigid media at low temperatures.

A further emission phenomenon associated with the triplet state is that of delayed fluorescence. Delayed fluorescence can occur in two separate ways, by thermal depopulation of T_1 to S_1 or by triplet–triplet annihilation. The former process is known as E-type delayed fluorescence and the latter as P-type delayed fluorescence (Parker and Hatchard, 1961, 1962).

E-type delayed fluorescence, which was first observed in eosin, is simply the reverse of the $T_1 \rightarrow S_1$ intersystem crossing process, and it occurs when the T_1–S_1 energy gap is of the order of kT (see Fig. 2.5(a)). The lifetime of this delayed fluorescence is the same as the phosphorescence lifetime. E-type delayed fluorescence is observed mainly in dye molecules, and does not normally occur in aromatic molecules because the singlet–triplet energy gap is relatively large.

P-type delayed fluorescence, first observed in pyrene, is observed for many aromatic hydrocarbons in fluid solution, concentrated rigid solution, and in the crystal phase. It is due to the interaction between pairs of triplet-excited molecules which have a total energy greater than S_1 and represented by the equation

$$T_1 + T_1 \longrightarrow S_1 + S_0 \qquad (2.37)$$

Obviously, these triplet–triplet interactions are most likely to occur when a high concentration of excited triplet molecules is present. In fluid solution it is believed that the triplet-excited molecules interact by a diffusion-controlled process, whereas in rigid media and in crystals triplet exciton migration occurs.

The lifetime of the P-type fluorescence is normally half the triplet lifetime, and the intensity of the delayed fluorescence depends on the square of incident light intensity (I_0). However, for high excitation intensities and in excimer-forming concentrated solutions, the observation of the delayed fluorescence shortly after excitation shows a non-exponential decay and the intensity varies linearly with the excitation intensity (Birks, 1970a). If sufficient time elapses, however, between excitation and the observation of delayed fluorescence, that is a low enough concentration of excited triplets is reached, then the decay becomes exponential with a lifetime of half the triplet lifetime and the intensity once again depends on I_0^2. This behaviour is consistent with the excimer kinetic scheme proposed by Birks (1970a).

2.8.2. Quenching of the Triplet State

As mentioned previously, the long lifetime of the triplet state makes it very susceptible to quenching. Thus all measurements must be carried out on deoxygenated ultra-pure solutions, even if the solutions are to be frozen later. The quenching mechanism, apart from the process of energy transfer to an acceptor molecule, is usually due to an increase in the radiationless intersystem crossing rate parameter k_{GT} produced by a paramagnetic molecule such as oxygen, or by a heavy atom such as a halogen which increases the spin–orbit interaction and thus catalyses the intersystem crossing rate. As with fluorescence quenching, the halogen atom can be external or internal to the molecular system.

2.8.3. Higher Excited Triplet States

Excitation into higher excited states of the singlet manifold results in rapid internal conversion of the excitation energy to the lowest excited singlet state; in an identical manner excitation of higher triplet states results in equally rapid internal conversion to the lowest excited triplet state. Thus phosphorescence always occurs from the lowest excited triplet state T_1. Intersystem crossing can also occur between higher excited singlet and triplet states. How can these higher excited triplet states be observed? Certainly, emission from higher excited singlet states is difficult enough to observe for the singlet manifold and direct excitation of triplet states from the S_0 ground state has a very low extinction coefficient. Ordinary absorption spectrometry would be inadequate since $S_0 \rightarrow S_n$ absorption would be in the same spectral region as the $S_0 \rightarrow T_n$ absorption. This problem was resolved by Porter and Windsor (1958) and Porter (1974) using the technique of flash photolysis.

The principle of this technique is to populate the lowest triplet state, T_1, using an intense flash of light which also depletes the S_0 ground state. After a time greater than the singlet excited state lifetime but less than the triplet state lifetime, the transient T_1–T_n absorption spectra is measured using a second pulsed light source. This absorption spectrum is thus used to identify and to obtain measurements of the extinction coefficients of these higher triplet states provided that the T_1 concentration can be estimated. Figure 2.26

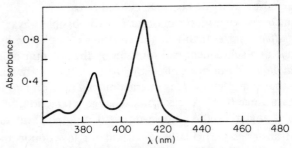

Fig. 2.26. Triplet–triplet absorption spectrum of naphthalene in an n-hexane measured by the flash-photolysis technique (after Wilkinson, 1974).

shows a triplet absorption spectra obtained using this technique. The decay kinetics of these triplet states can also be studied by flash photolysis. With the advent of pulsed lasers it is now also possible to use this technique for measuring $S_1 \to S_n$ absorption spectra (Porter and Topp, 1970).

2.9. ENERGY TRANSFER AND MIGRATION

2.9.1. Introduction

If a molecule Y is in the vicinity of an excited molecule X* and the energy of an excited state of Y lies below that of X*, then *energy transfer* from the donor molecule X* to the acceptor molecule Y can occur. This process is illustrated in Fig. 2.27 for singlet–singlet energy transfer. Energy transfer can also occur from a donor excited triplet state T* to an acceptor molecule Y which has either a triplet state or an excited singlet state of lower energy than T*. Singlet–triplet transfer processes can also occur, but are less efficient processes.

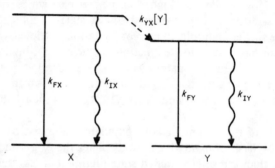

Fig. 2.27. Energy transfer from a donor molecule X to an acceptor molecule Y: ——, radiative transitions; ⌇⌇⌇, non-radiative transitions.

The use of the term *energy migration* is confined to describing energy transfer between molecules of the same kind. Energy migration therefore causes the excitation energy to become delocalized.

Energy transfer or migration can occur in two possible ways:

(a) by *radiative transfer* involving the emission of a photon by the donor molecule and subsequent reabsorption by the acceptor molecule;

(b) by *radiationless transfer* caused by an interaction between the donor and accepter molecules. This interaction can take on two forms:

    (i) short-range (6–15 Å) electron exchange interaction,

    (ii) long-range (20–60 Å) dipole–dipole (Coulombic) interaction.

Radiative transfer is particularly important for low donor and acceptor concentrations when the distances between acceptor and donor molecules are too large even with the help of diffusion for radiationless transfer to occur. The efficiency of transfer, naturally, depends on the overlap of the emission spectrum of the donor with the absorption spectrum of the acceptor. This radiative process has been described as trivial—a title which implies simplicity rather than a process to be ignored, since it plays an important role in energy transfer in solid solutions.

At high enough molecular concentrations radiationless energy transfer takes over from the radiative process. This transition from radiative to radiationless transfer has been illustrated very clearly by the work of Birks and Kuchela (1961) for the scintillator solutes p-terphenyl and TPB, and is shown in Fig. 2.28.

The two radiationless processes, electron exchange and dipole–dipole interaction as competitive energy-transfer processes have led to, and still create, a great deal of controversy in the interpretation of energy-transfer processes. In solution the excited donor and unexcited acceptor molecules are mobile, and their movement is therefore influenced by diffusion. Thus, energy transfer in solution is produced by collisional interaction, and this may take place by diffusion-assisted dipole–dipole interaction which can occur at effective radii greater than the dipole interaction radius of 20–60 Å or by exchange interaction at a radius of 6–15 Å.

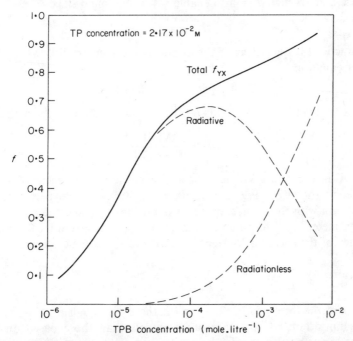

Fig. 2.28. Radiative and radiationless energy transfer for the system p-terphenyl (TP)–tetraphenylbutadiene (TPB) in toluene solution (after Birks and Kuchela, 1961).

In rigid media, the diffusion element is almost eliminated, and the interpretation is less confused since the molecules concerned are stationary and it is therefore easier to distinguish between the appropriate interactions.

An important parameter in the energy-transfer process is the energy-transfer quantum efficiency, f_{YX}, which gives the fraction of the molecules of X which transfer their energy to Y. f_{YX} can be found experimentallly from the excitation spectrum of the X–Y system. By observing the emission intensity of Y for different excitation wavelengths and provided the system absorbs all the incident radiation, then

$$f_{YX} = \Phi_{YX}/\Phi_{YO} \qquad (2.38)$$

where Φ_{YX} is the quantum yield of Y when excited via X, and Φ_{YO} is the quantum efficiency of Y when excited directly. Alternatively, the quantum yields of X in the presence and absence of Y give

$$\Phi_{XY}/\Phi_{XO} = 1 - f_{YX} \qquad (2.39)$$

where Φ_{XY} is the quantum yield of X when quenched by Y, and Φ_{XO} is the quantum efficiency of X alone. Lifetime measurements τ_{XY} and τ_{XO} in the presence and absence of Y also enable f_{YX} to be found since

$$\tau_{XY}/\tau_{XO} = 1 - f_{YX} \qquad (2.40)$$

In practice the quenching of the donor by the acceptor follows the Stern–Volmer equation

$$\Phi_{XO}/\Phi_{XY} = 1 + K[Y] \qquad (2.41)$$

and this means that f_{YX} depends on the concentration Y in the manner

$$f_{YX} = \frac{K[Y]}{1 + K[Y]} \qquad (2.42)$$

where K is the Stern–Volmer quenching constant. In terms of the rate parameters shown in Fig. 2.27, this energy-transfer rate parameter is in competition with the fluorescence rate parameter k_{FX} and the non-radiative rate parameter k_{IX}. The parameters k_{FY} and k_{IY} only determine the quantum efficiency of the acceptor Y. Hence in terms of this kinetic scheme

$$f_{YX} = \frac{k_{YX}[Y]}{k_{FX} + k_{IX} + k_{YX}[Y]} \qquad (2.43)$$

which is of the form of Eq. (2.42) and K therefore equals $\tau_{XO} k_{YX}$.

It is important to note that other models than those based on Stern–Volmer kinetics for the dependence of f_{YX} on acceptor concentration such as Förster kinetics (Förster, 1959) and the active sphere model (Perrin, 1924) do not produce large variations in the value of f_{YX} for a given concentration and thus the Stern–Volmer plot using Eq. (2.42) cannot within the experimental errors distinguish between these alternative models.

2.9.2. Electron Exchange Interaction

The electron exchange interaction occurs via a short-lived intermediate complex (exciplex) and is analogous to the quenching processes of oxygen or external heavy-atom effects which involve the approach of the acceptor molecule or quencher to distances only slightly greater than the molecular radius of the donor molecule. This interaction can be expressed in the form

$$^1X^* + {}^1Y \longrightarrow {}^1(XY)^* \longrightarrow {}^1X + {}^1Y^* \quad \text{(singlet–singlet)} \quad (2.44)$$

and

$$^3X^* + {}^1Y \longrightarrow {}^3(XY)^* \longrightarrow {}^1X + {}^3Y^* \quad \text{(triplet–triplet)} \quad (2.45)$$

The electron exchange interaction requires overlap of the electronic wavefunctions of X and Y, and it is therefore short-range. Spin must also be conserved during this transition (Dexter, 1953). Thus electron exchange can occur for singlet–singlet migration or transfer and triplet–triplet migration or transfer.

2.9.3. Dipole–dipole Energy Transfer

The theory of dipole–dipole or Coulombic interaction has been developed by Förster (1959) and produces an expression for the energy transfer probability k_{YX}' of the form

$$k_{YX}' = \frac{9000\,(\ln 10)\,K^2}{128\pi^5 n^4 N r^6 \tau_X} \int_0^\infty F_X(\bar{\nu})\,\varepsilon_Y(\bar{\nu})\frac{d\bar{\nu}}{\bar{\nu}^4} = \frac{1}{\tau_X}\left(\frac{R_0}{r}\right)^6 \quad (2.46)$$

where N is Avogadro's number, n is the refractive index of the medium, K^2 is an orientation factor (approximately 2/3) and τ_X is the lifetime of the donor in the absence of the acceptor. $F_X(\bar{\nu})$ is the emission spectrum of the donor plotted against wavenumber $\bar{\nu}$ and normalized so that the quantum efficiency of the donor Φ_{FX} equals $\int_0^\infty F_X(\bar{\nu})\,d\bar{\nu}$ and $\varepsilon_Y(\bar{\nu})$ is the absorption spectrum of the acceptor.

As can be seen from this expression, k_{YX}' depends significantly on the overlap of the emission and absorption spectra of the donor and acceptor molecule and is strongly influenced by the strength of the optical transitions of X and Y. However, spin does not have to be conserved in this interaction and hence triplet–singlet as well as singlet–singlet and triplet–triplet transfer is feasible for this type of interaction.

Equation (2.46) also shows that k_{YX}' is related to a critical radius R_0 which is defined as the distance between donor and acceptor molecules which gives a 50% probability of energy transfer. R_0 thus represents the distance at which there is an equal probability of an excited X molecule being de-excited by fluorescence and internal quenching or by transfer to Y. Hence, R_0 can be calculated from the expression

$$R_0{}^6 = \frac{9000\,(\ln 10)\,K^2}{128\pi^5 n^4 N} \int_0^\infty F_X(\bar{\nu})\,\varepsilon_Y(\bar{\nu})\,\frac{d\bar{\nu}}{\bar{\nu}^4} \qquad (2.47)$$

The values for R_0 calculated from Eq. (2.47) are in the region of 16–60 Å, and thus confirm that dipole–dipole transfer is a long-range process.

A distinction between electron exchange and dipole–dipole transfer can be made experimentally by comparing the energy-transfer rate parameter, k_{YX}, with the rate parameter, k_{QX}, for collisional quenching, using a quencher such as carbon tetrabromide for which $p = 1$ (see Eq. (2.21)). If k_{QX} is equal to k_{YX} over a reasonable temperature range, then it can be assumed that the energy-transfer process is a short-range process, and hence possibly operates via the electron exchange mechanism.

2.9.4. Energy Transfer in Solution

In solution, diffusion strongly influences collisional energy transfer, and the study of energy transfer in solution is particularly important because of its application to an understanding of the energy-transfer process in liquid scintillators. A liquid scintillator consists of an aromatic solvent such as toluene which transfers its excitation energy to a fluorescent solute. The solvent represents the donor molecule X and the solute the acceptor molecule Y.

Both electron exchange and dipole–dipole transfer can occur in solution; however, diffusion enables the acceptor and donor molecules to come close enough for efficient transfer to take place. If the energy-transfer process is diffusion-controlled, then the diffusion-controlled rate parameter k_{diff} can be calculated if the diffusion coefficients of X and Y are known. If the collisional separation distance is equated to the sum of the molecular radii $(R = R_X + R_Y)$, then $k_{diff} = 4\pi N' R (D_X + D_Y)$. In practice, it is found that k_{YX} is greater than k_{diff}, suggesting that either

(i) D_X is effectively increased by a quantity Λ due to energy migration in the solvent molecules, hence

$$k_{YX} = 4\pi N'(R_X + R_Y)(D_X + D_Y + \Lambda) \qquad (2.48)$$

and/or

(ii) the range R of solvent–solute energy transfer is beyond the collisional separation distance $(R_X + R_Y)$.

It would appear that (ii) suggests that the Förster dipole–dipole mechanism should be applicable to this form of energy transfer, and R_0

calculations (see Eq. (2.47)) give radii which are in fact greater than R. However, quenching studies where the dipole–dipole interaction cannot occur indicate that $k_{YX} = k_{QX}$ implying that the process is short-range. The results of Birks *et al.* (1971) indicate either that the interaction radius is three times the molecular radii due to the influence of excimers, or that energy migration increases the effective diffusion coefficients by threefold. However, further experimentation is required before these mechanisms can be finally resolved into an effective increase in D or an increase in R or both.

Triplet–triplet energy transfer occurs for most aromatic molecules in rigid solution at low temperatures. However, it can also occur at room temperature in fluid solution for a molecule like biacetyl which is unique in that it phosphoresces in room temperature fluid solutions. Triplet–triplet energy transfer is consistent with the electron exchange model. However, triplet–triplet migration follows the Förster dipole–dipole mechanism (see Birks, 1970c).

Observations of triplet–singlet transfer in solution have been confined to low-temperature solid solutions. These measurements are consistent with Förster transfer by dipole–dipole interaction, which also accounts for singlet–triplet transfer in thin films.

2.9.5. *Energy Transfer in Crystals*

Singlet–singlet transfer in crystals readily occurs between neighbouring identical molecules and leads to a rapid migration of excitation energy throughout the crystal lattice due to dipole–dipole interactions. This migration is known as exciton migration. If the exciton mean free path is greater than the crystal lattice spacing, then the exciton band model (Rice and Jortner, 1967), can be applied to this excitation wave; if the mean free path is less than the lattice spacing then the "hopping" model is applicable (Trilifaj, 1959). If a crystal such as anthracene contains a small quantity of fluorescent acceptor molecules such as tetracene which has its first excited singlet state, S_1, below that of anthracene, then very efficient energy transfer will occur and the crystal will emit the characteristic tetracene emission. The anthracene excitation energy rapidly migrates to a tetracene site where energy is transferred.

Triplet–triplet exciton migration and transfer also occurs readily in the crystal phase. The mean free path of triplet excitons is 10^3 times the mean free path of singlet excitons and although the mean free path is reduced due to a shortening of the triplet lifetime by triplet–triplet annihilation, the transfer process is still very efficient.

Triplet–triplet energy transfer in mixed crystals is thought to occur by the electron exchange mechanism with the excitation energy reaching the acceptor molecule by triplet–exciton migration between donor molecules.

Triplet–singlet transfer also occurs in the crystalline phase but as in the case of solid solutions this only occurs to any significant extent at low temperatures.

2.10. CONCLUSION

Luminescence spectroscopy of organic molecules provides an important means of investigating the excited states of molecules. It is hoped, in this rather brief chapter, that the reader gains some basic understanding of these molecular processes and an insight into the techniques required to observe experimentally those parameters which reflect these molecular properties. Some areas of the subject have not been discussed, but it is hoped that the reader will find adequate coverage of these in the general list of books and review articles on this vast subject at the end of this chapter.

REFERENCES

Books and Review Articles on Organic Luminescence and Related Subjects

Berlman, I. B. (1971). "Handbook of Fluorescence Spectra of Aromatic Molecules." Academic Press, New York and London.

Birks, J. B. (1970). "Photophysics of Aromatic Molecules." Wiley–Interscience, London.

Birks, J. B. (ed.) (1973). "Organic Molecular Photophysics", Vol. I. Wiley–Interscience, London.

Birks, J. B. (ed.) (1975). "Organic Molecular Photophysics", Vol. II. Wiley–Interscience, London.

Cundall, R. B. and Palmer, F. (1974) (Review). "Luminescence spectroscopy." *Ann. Rep. Prog. Chem. A, Phys. Inorg. Chem.* **70**, 31.

Guilbault, G. G. (1973). "Practical Fluorescence". Dekker, New York.

McGlynn, S. P., Azumi, T., and Kinoshita, M. (1969). "Molecular Spectroscopy of the Triplet State." Prentice-Hall, New York.

Murrell, J. N. (1963). "The Theory of the Electronic Spectra of Organic Molecules." Methuen, London.

Parker, C. A. (1968). "Photoluminescence of Solutions". Elsevier, Amsterdam.

Platt, J. R. (1964). "Systematics of the Electronic Spectra of Conjugated Molecules." Wiley, New York.

Specific References

Alwattar, A. H., Lumb, M. D., and Birks, J. B. (1973). "Organic Molecular Photophysics" (J. B. Birks, ed.), Vol. I, pp. 453–454. Wiley–Interscience, London.

Anderson, E. M. and Kistiakowsky, G. B. (1969). *J. Chem. Phys.* **51**, 182.

Baba, H., Nakajima, A., Aoi, M., and Chihora, K. (1971). *J. Chem. Phys.* **55**, 2433.

Barnes, R. L. and Birks, J. B. (1966). *Proc. Roy. Soc.* A **291**, 570.

Berlman, I. B. (1971). *In* "Handbook of Fluorescence Spectra of Aromatic Molecules", p. 58. Academic Press, New York and London.

Birks, J. B. (1970a). *In* "Photophysics of Aromatic Molecules", p. 378. Wiley–Interscience, London.

Birks, J. B. (1970b). "Photophysics of Aromatic Molecules", p. 499. Wiley–Interscience, London.

Birks, J. B. (1970c). "Photophysics of Aromatic Molecules", pp. 591–593. Wiley–Interscience, London.

Birks, J. B. (1970d). "Photophysics of Aromatic Molecules", pp. 122–123. Wiley–Interscience, London.

Birks, J. B. (1975). *Rep. Prog. Phys.* **38**, 903.

Birks, J. B. and Aladekomo, J. B. (1963). *Photochem. Photobiol.* **2**, 415.

Birks, J. B. and Cameron, A. J. W. (1959). *Proc. Roy. Soc.* A **249**, 297.

Birks, J. B. and Dyson, J. E. (1963). *Proc. Roy. Soc.* A **275**, 135.

Birks, J. B. and Kuchela, K. N. (1961). *Proc. Phys. Soc.* **77**, 1083.

Birks, J. B., Dyson, D. J., and Munro, I. H. (1963). *Proc. Roy. Soc.* A **275**, 575.

Birks, J. B., Najjar, H. Y., and Lumb, M. D. (1971). *J. Phys.* B **4**, 1516.

Braun, C. L., Kato, S., and Lipsky, S. (1963). *J. Chem. Phys.* **39**, 1645.

Cundall, R. B. and Ogilvie, S. McD. (1975). "The Photophysics of Benzene in Fluid Media, Organic Molecular Photophysics" (J. B. Birks, ed.), Vol. II, pp. 33–93, Wiley–Interscience, London.

Debye, P. (1942). *Trans. Electrochem. Soc.* **82**, 265.

Deinum, T., Warkhoren, C. J., Langelaar, J., Rettschnick, R. P. H., and Van Voorst, J. D. W. (1971). *Chem. Phys. Lett.* **12**, 189.

Dexter, D. L. (1953). *J. Chem. Phys.* **21**, 836.

Förster, Th. (1959). *Disc. Faraday Soc.* **27**, 7. (Note that π^6 in this paper should be replaced by π^5.)

Förster, Th. and Kasper, K. (1955). *Z. Elektrochem.* **59**, 976.

Geacintov, N. E. and Swenberg, C. E. (1978). "Luminescence Spectroscopy" (M. D. Lumb, ed.), Chapter 4. Academic Press, London and New York.

Gregory, T. A., Hirayama, F., and Lipsky, S. (1973). *J. Chem. Phys.* **58**, 4697.

Guilbault, G. G. (1973). *In* "Practical Fluorescence", pp. 277–348. Dekker, New York.

Hallam, H. E. (ed.) (1970). "Vibrational Spectroscopy of Trapped Species, Infra-red and Raman Studies of Matrix-isolated Molecules, Radicals and Ions." Wiley, London.

Henry, B. R. and Siebrand, W. (1974). *In* "Radiationless Transitions, Organic Molecular Photophysics" (J. B. Birks, ed.), Vol. I, pp. 153–237. Wiley–Interscience, London.

Hirayama, F. and Lipsky, S. (1969). *J. Chem. Phys.* **51**, 3616.

Hirayama, F., Rothman, W., and Lipsky, S. (1970). *Chem. Phys. Lett.* **5**, 296.

Hirayama, F., Gregory, T. A., and Lipsky, S. (1973). *J. Chem. Phys.* **58**, 4696.

Hudson, B. S. and Kohler, B. E. (1973). *J. Chem. Phys.* **59**, 4984.

Hunt, G. R., McCoy, E. F., and Ross, I. G. (1962), *Australian J. Chem.* **15**, 591.

Huppert, D., Rentzepis, P. M., and Jortner, J. (1972a). *Chem. Phys. Lett.* **13**, 225.

Huppert, D., Rentzepis, P. M., and Jortner, J. (1972b). *J. Chem. Phys.* **56**, 5826.

Kasha, M. (1950). *Disc. Faraday Soc.* **9**, 14.

Kistiakowsky, G. B. and Parmenter, C. S. (1965). *J. Chem. Phys.* **42**, 2942.

Lawson, C. W., Hirayama, F., and Lipsky, S. (1969). *J. Chem. Phys.* **51**, 1590.

Lumb, M. D. and Weyl, D. A. (1967). *J. Molec. Spectrosc.* **23**, 465.

Murrell, J. N. and Tanaka, J. (1964). *Molec. Phys.* **7**, 363.

Parker, C. A. (1968a). *In* "Photoluminescence of Solutions", p. 21. Elsevier, Amsterdam.

Parker, C. A. (1968b). *In* "Photoluminescence of Solutions", p. 30. Elsevier, Amsterdam.

Parker, C. A. and Hatchard, C. G. (1961). *Trans. Faraday Soc.* **57**, 1894.

Parker, C. A. and Hatchard, C. G. (1962). *Proc. Roy. Soc.* A **269**, 574.

Perrin, F. (1924). *C. R. Acad. Sci. Paris*, **178**, 1978.

Porter, G. (1974). *Proc. Roy. Inst.* **47**, 143.

Porter, G. and Topp, M. R. (1970). *Proc. Roy. Soc.* A **315**, 163.

Porter, G. and Windsor, M. W. (1958). *Proc. Roy. Soc.* A **245**, 238.

Rentzepis, P. M., Jortner, J. and Jones, R. P. (1970). *Chem. Phys. Lett.* **4**, 599.

Rice, S. A. and Jortner, J. (1967). "Physics and Chemistry of the Organic Solid State", Vol. 3 (D. Fox, M. M. Labes, and A. Weissburger, eds) p. 199. Interscience, New York.

Rothman, W., Hirayama, F., and Lipsky, S. (1973). *J. Chem. Phys.* **58**, 1300.

Shpol'skii, E. V., Illina, A. A., and Klimova, L. A. (1952). *Dokl. Akad. Nauk SSSR*, **87**, 935.

Stern, O. and Volmer, M. (1919). *Phys. Z.* **20**, 183.

Stevens, B. and Hutton, E. (1960). *Nature*, **186**, 1045.

Stockburger, M. (1973). "Fluorescence of Aromatic Molecular Vapours, Organic Molecular Photophysics" (J. B. Birks, ed.), Vol. I, pp. 58–102. Wiley–Interscience, London.

Strickler, S. J., and Berg, R. A. (1962). *J. Chem. Phys.* **37**, 814.

Sutherland, W. (1905). *Phil. Mag.* **9**, 781.

Sveshnikoff, B. (1935). *Acta Physiochim. URSS*, **3**, 257.

Sveshnikoff, B. (1937). *Acta Physiochim. URSS*, **7**, 755.

Trilifaj, M. (1959). *Czech. J. Phys.* **6**, 533.

Umberger, J. Q. and La Mer, V. K. (1945). *J. Am. Chem. Soc.* **67**, 1099.

Wilkinson, F. (1975). "Organic Molecular Photophysics" (J. B. Birks ed.), Vol. II, p. 99. Wiley–Interscience, London.

3

Luminescence Instrumentation

T. D. S. HAMILTON

The Schuster Laboratories, University of Manchester, Manchester, England

I. H. MUNRO

Daresbury Laboratory, Daresbury, Cheshire, England

and

G. WALKER

Department of Pure and Applied Physics,
University of Manchester Institute of Science and Technology,
Manchester, England

3.1. INTRODUCTION

This chapter is concerned with the techniques and instrumentation used in the production, observation, and measurement of luminescence. In order to restrict its length the techniques described are mainly limited to those of optical spectroscopy; electrical and magnetic measurements are specifically excluded. Experimental techniques used in the optical detection of magnetic resonance are described in Chapter 5. Since a discussion of the preparation and purification of the various types of organic and inorganic phosphors would fill a large book in itself, only a few general comments are made at the end of the chapter. Optical measurements fall conveniently into two main categories: the measurement of spectra and the decay of luminescence intensity with time. However, these categories are not mutually exclusive and measurements of spectra at different times during the luminescence decay can be made using the methods of time-resolved spectroscopy.

Since the terms "fluorescence" and "phosphorescence" are often defined in different ways, a brief explanation may make for clarity.

These terms can be unambiguously defined for organic materials since the corresponding emission spectra are different, i.e. different radiative transitions are involved. However, for inorganic materials usually (but not always) the same radiative transition is involved in prompt emission and afterglow and the term phosphorescence is often applied to a process similar to that which would be called "delayed fluorescence" in organics (see Chapter 2). However, it is common practice to ascribe the term phosphorescence to all long-lived luminescence (i.e. with decay times longer than about a millisecond) irrespective of the mechanisms involved.

The chapter begins with a basic description of the main types of spectral measurements which are carried out on luminescent materials. A full discussion follows of the main elements of experimental optical spectroscopy, namely, excitation sources, spectrometers, detectors, and signal recovery techniques, with particular emphasis on the advantages and limitations of the instrumentation described. Techniques used in the measurement of luminescence decay and in the measurement of time-resolved spectra are then discussed in Section 3.5 and the chapter ends with a few comments on a variety of other important topics.

3.1.1. Absorption Spectra

Absorption of light by a homogeneous absorbing medium or a medium in which an absorbing species is uniformly distributed follows the Beer–Lambert law:

$$I(\nu) = I_0(\nu)\,e^{-k(\nu)d} \tag{3.1}$$

where $I_0(\nu)$ and $I(\nu)$ are the incident and transmitted intensities respectively, d is the length of the light path in the medium and $k(\nu)$ is the coefficient of absorption of the medium. However, for convenience the law is often rewritten

$$I(\nu) = I_0(\nu)\,10^{-\alpha(\nu)d} \tag{3.2}$$

where $\alpha(\nu)$ is the decadic absorption coefficient, often confusingly referred to simply as the absorption coefficient, although of course the difference between α and k is fairly trivial since $k(\nu) = 2\cdot3\alpha(\nu)$.

If the absorbing species are molecules in a solution or ions or other localized centres in a solid then $\alpha(\nu) = \varepsilon(\nu)\,c$, where c is the concentration of the absorbing species, usually expressed in moles litre^{-1}, and $\varepsilon(\nu)$ is called the molar extinction coefficient and is a fundamental property of the absorbing centres.

Although the transmission $(I(\nu)/I_0(\nu))$ of the sample is the parameter measured experimentally, spectra are often plotted in terms of absorbance or optical density which is defined as $\log_{10}(I_0(\nu)/I(\nu))$. More properly, however, absorption spectra should be plotted in terms of ε (or α in the case of crystal lattice absorption) against wavelength, wavenumber or frequency, since this is the fundamental parameter.

Clearly,

$$\varepsilon(\nu) = \frac{1}{cd}\log_{10}\left(\frac{I_0(\nu)}{I(\nu)}\right) \tag{3.3}$$

The measurement of an absorption spectrum in a particular spectral region requires a source with a continuous spectral emission in that region, a spectrometer or monochromator to select the wavelength or frequency and a detector to measure the transmitted monochromatic light (see Fig. 3.1). A high intensity source is not usually required; a hydrogen or deuterium lamp is often used for the ultraviolet, and a simple tungsten filament lamp for the visible and near infrared.

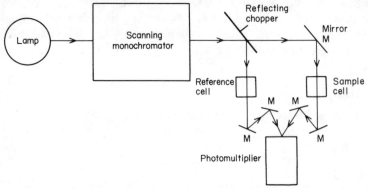

Fig. 3.1. Block diagram of apparatus for determination of absorption spectra.

Similarly, unless very low transmissions are to be measured, the spectrometer does not have to be of very high luminosity (see Section 3.3.4) although the stray light level should be reasonably low (see Section 3.3.5). A photomultiplier is used as a detector in the ultraviolet and visible regions and usually a lead sulphide cell in the near infrared.

Measurements using a simple single-beam instrument must be corrected for variations in lamp output, spectrometer efficiency, and detector sensitivity with wavelength (see Section 3.3.6). However, in most commercial instruments, the necessity for such corrections is eliminated by using a dual or split beam. The monochromatic beam emerging from the spectrometer is split into two, one passing through the absorbing sample, the other by-passing it or passing through a reference cell. The detector alternately samples the two beams and the ratio of the two signals yields the transmission $(I(\nu)/I_0(\nu))$. The same detector is used to monitor both beams since no two photodetectors have exactly the same spectral sensitivity even though they may be of the same type (see Fig. 3.1).

If the sample is in the form of a solution then the absorption is easily varied by using different concentrations. However, if the absorption spectra of solids are to be measured then the magnitude of the absorption coefficient will determine the specimen thickness. Very thin crystal plates may be necessary if the absorption coefficient is high and since high absorption often results in a high reflectivity, corrections for reflective losses may be necessary. Light scattering by the sample can also be a problem which can be overcome by polishing the sample surface or, where this is not possible, by mounting the sample between glass slides in a transparent oil of the same average refractive index. In cases where the absorption coefficient is very high ($\sim 10^5$ cm^{-1} or higher, e.g. fundamental band-to-band absorption in insulators and semiconductors and second or third absorption bands in aromatic crystals), the determination of absorption spectra by transmission measurements becomes very difficult. In these circumstances reflection spectra are measured although the evaluation of the absorption coefficient from such spectra is more difficult than in simple transmission measurements (see, for example, Garbuny 1965).

$$\text{Reflectivity } R = \frac{(n-1)^2 + K^2}{(n+1)^2 + K^2} \quad \text{(normal incidence)} \tag{3.4}$$

where n and K are the real and imaginary parts of the refractive index. K, which is often referred to as the extinction coefficient, is related to the absorption coefficient k by

$$k = 4\pi K/\lambda \tag{3.5}$$

Both n and K can be evaluated from reflection measurements using the Kramers–Kronig relations (Phillips and Taft, 1959; Garbuny, 1965).

If the absorbing solid is not isotropic then the absorption spectrum will be different for different planes of polarization of the incident light relative to the crystal axes. For electric dipole transitions, two different absorption spectra may be distinguished in uniaxial (tetragonal, trigonal, or hexagonal) crystals and three in biaxial (orthorhombic, monoclinic, or triclinic) crystals, when light is polarized along each of the indicatrix axes. Polarized absorption spectra can be valuable, for example, in the determination of the site symmetry of absorbing ions in crystals.

The light incident on the crystal is passed through a polarizer and the crystal is oriented with respect to the plane of polarization. For visible and infrared measurements on very small crystals, polarizing microscopes with a universal stage have been used, one in the sample beam (with crystal) and one in the reference beam.

The measurement of absorption spectra of microcrystalline powder samples presents special problems. There are basically two approaches: firstly, the powder grains are suspended in a transparent oil or liquid of about the same refractive index or embedded in pressed potassium bromide pellets; secondly, the diffuse reflectance spectrum of a layer of powder can be measured with suitable optical adaptations to the specimen chamber. In the latter measurements the diffuse reflectance of an MgO screen is usually taken as a reference. For further details regarding reflection spectroscopy the reader is referred to the monographs by Wendlandt and Hecht (1966) and Kortüm (1969).

3.1.2. Excitation Spectra

An excitation spectrum is determined by measuring the luminescence emission intensity, either totally integrated or in a specific wavelength range, as the wavelength or frequency of excitation is scanned (Fig. 3.2). The emitted luminescence intensity I_L (photons/s) can be written as

$$I_L(v) = I_0(v) Q(1 - 10^{-\alpha(v)d}) \tag{3.6}$$

where $I_0(v)$ is the intensity of the incident monochromatic light falling on a sample of absorption coefficient $\alpha(v)$, thickness d, and luminescence quantum efficiency Q.† For samples of very low absorbance such as very dilute solutions or crystals with a very low concentration c of absorbing centres, this expression becomes

$$I_L(v) = 2·3 I_0(v) Q \varepsilon(v) cd \tag{3.7}$$

since $\alpha(v) = \varepsilon(v) c$ where $\varepsilon(v)$ is the molar extinction coefficient.

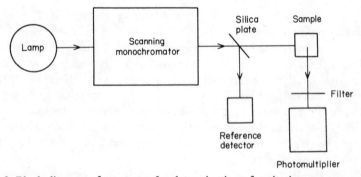

Fig. 3.2. Block diagram of apparatus for determination of excitation spectra.

If an incident light monitor is used so that the variations in $I_0(v)$ with frequency are compensated for (see Section 3.3.7) and Q is assumed to be frequency independent then $I_L(v)$ is proportional to $\varepsilon(v)$ with an error of less than 4% for $\alpha(v) d < 0·02$. When an excitation spectrum is measured under these circumstances it will have a profile which is identical to that of the absorption spectrum. When very thin or only weakly absorbing samples are obtainable, a study of the excitation spectrum may be the only means of measuring the absorption spectrum in any detail and with high resolution.

† Q is defined as the ratio of the number of photons emitted to the number of photons absorbed by the luminescent species (see Section 3.6.2).

It may also be possible to measure the absorption spectrum of a weakly absorbing species which is luminescent in the presence of strongly absorbing non-luminescent species by means of excitation spectroscopy. For concentrations within the dilute solution range, $I_L(\nu)$ is also proportional to c, a relationship which is of considerable importance in filter fluorometry. Alternatively, when the sample is sufficiently thick or concentrated for practically all the incident light to be absorbed at each frequency then

$$I_L(\nu) = I_0(\nu)\,Q \tag{3.8}$$

In this case of strong absorption, if Q is known to be frequency independent for the material then $I_L(\nu)$ is directly proportional to $I_0(\nu)$ and such materials can be used as quantum counters (see Section 3.3.7). Alternatively, if variations in $I_0(\nu)$ with frequency are automatically compensated for, then it can be established whether or not Q is frequency dependent. A wide range of materials display a quantum efficiency which is independent of excitation wavelength. In such measurements, practically all the light should be absorbed in a thin layer of the sample from which the emission is recorded. If the geometry of the emitting layer alters appreciably then the angle of collection of light by the photocathode changes and a spurious variation in emission intensity may consequently be recorded; this effect will be more noticeable if emission is viewed perpendicular to the direction of excitation.

In the very weak absorption case, emission will necessarily also be weak and therefore a high intensity of excitation will be required for a good signal-to-noise ratio. A tunable dye laser may be ideal as a source of excitation (Section 3.2.3) if the range of excitation wavelength required can be accommodated. Alternatively, a high intensity continuum source such as a xenon arc (Section 3.2.1) and a high luminosity monochromator (Section 3.3.4) will provide excitation over a large spectral range.

When the emission is very weak, excitation light reflected or scattered by the sample may be detected by the photomultiplier and may often be of very much larger intensity than the emission. Stray light from the monochromator may also be troublesome for very weakly emitting samples (see Section 3.3.5). A cut-off filter in front of the photomultiplier which absorbs the scattered excitation light but transmits the emission may suffice. However, sometimes a more sophisticated technique is required employing some form of time resolution such as a double-chopper technique (Section 3.5.6) to separate the emission from extraneous scattered light.

In the weak absorption limit, polarized excitation spectra of single anisotropic crystals will obviously yield the same information as polarized absorption spectra (Section 3.1.1).

3.1.3. Emission Spectra

A luminescence emission spectrum is a plot of luminescence intensity $I_L(\nu)$ against wavelength or frequency for a given intensity of excitation. The excitation may be broad-band or narrow-line ultraviolet, visible or sometimes near infrared (photoluminescence), electrical (electroluminescence) or by bombarding particles such as electrons (cathodoluminescence).

Often the emission is the result of a single radiative electronic transition between a low-lying excited state and the ground state of the molecule or centre (e.g. $S_1 \rightarrow S_0$ in organic molecules, $^4T_1(G) \rightarrow {}^6A_1(S)$ in Mn^{2+}-activated phosphors, etc.). Excitation into higher states usually results in the same emission, since the system normally relaxes rapidly to its emitting state. However, systems such as rare-earth phosphors may give emission due to several transitions at once depending on the excitation energy, and, of course, an inorganic solid may contain more than one type of luminescence centre. With the exception of rare-earth, Cr^{3+}, and exciton emission, the luminescence spectra of liquids and solids are generally broad-band spectra although at very low temperatures these broad bands may break up into quasi-line spectra or have some discernible structure.

The measurement of emission spectra of efficient phosphors such as ZnS phosphors is relatively easy requiring fairly modest equipment. However, if very much weaker emissions are to be analyzed with good wavelength resolution then an intense excitation source, a high-luminosity spectrometer, a low-noise detector, and a good signal-recovery system will be required (Fig. 3.3). If the emission is particularly weak then either laser beam or

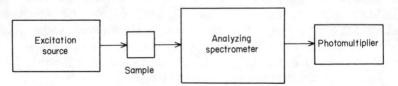

Fig. 3.3. Block diagram of apparatus for determination of emission spectra.

electron beam excitation may have to be used. Moreover, some materials require the energy or frequency of the excitation necessary in order to produce emission to be so high that an electron beam may be the only convenient form of excitation (see Section 3.2.5). Except in special circumstances, the frequency of emission is lower than the excitation frequency or energy (Stokes' law). In some cases the minimum excitation frequency required to produce emission may be very much higher than the frequency of emission.

The intense mercury lines in the ultraviolet and blue regions of the spectrum are often convenient for excitation purposes. However, light scattered into the spectrometer from longer wavelength lines may interfere with the emission spectrum and some filtering may be necessary (see Section 3.3.1). The best type of filter is, of course, a second spectrometer or mono-chromator to isolate the excitation wavelength. If such a monochromator is available then an intense continuum source such as a xenon arc may be used as in excitation spectroscopy (see Sections 3.1.2 and 3.2.1). The problem of background scattered light does not arise, however, if electron-beam or laser-beam excitation is used (see Sections 3.2.3 and 3.2.5).

Although it is possible to determine a spectrum by point-by-point measurements at different wavelengths, most systems employ automatic wavelength scanning (Section 3.3.2). However, the output from the photo-detector traced out on a pen-recorder or digitally recorded on tape will not be the "true" emission spectrum; the spectrometer transmission and detector sensitivity usually vary considerably with wavelength and calibration of these variables is therefore necessary so that spectral corrections can be applied (Section 3.3.7).

In some systems such corrections can be applied automatically once the calibration has been carried out.

It is usual to allow the detector to view the luminescence from the same surface of the sample upon which the excitation is incident. Self-absorption of the emission, which occurs when the emission and absorption bands overlap, is thereby minimized. Such self-absorption effects result in the truncation of the short wavelength edge of the emission spectrum. If other non-luminescent absorbing centres are present in the sample then of course further absorption of the emission may also occur.

In general, emission from liquids, isotropic solids, and randomly orientated microcrystalline powder grains will be unpolarized when excited by electrons or unpolarized light. However, emission from anisotropic crystals will generally be polarized to some degree. Spectrometer spectral transmission curves are often very dependent on the state of polarization of the light entering the spectrometer (see Section 3.3.8) and therefore care must be taken to ensure that the state of polarization of this light is known. If the emission spectra of anisotropic crystals are to be accurately determined then the light entering the spectrometer must have the same polarization as that used in calibration procedures, e.g. the emitted light should be effectively "de-polarized" before entering the spectrometer (see Section 3.3.8) if the spectrometer has been calibrated for unpolarized light.

Time-resolved emission spectra can be measured using high intensity pulsed or modulated excitation, e.g. a laser, electron beam or flash tube (see Section 3.5.6). The spectrum of the emission is scanned at a particular time interval after excitation using a sampling technique which samples the luminescence intensity of the appropriate part of the decay curve after each excitation pulse. Alternatively, the excitation is modulated at a given frequency using, say, a square wave and the spectrum of the emission modulated at the same frequency is measured using lock-in amplification techniques (Section 3.4.3). In the latter case, only emission with a decay time short compared with the reciprocal of the modulation frequency will be measured, since emission with a much longer decay time will not be significantly modulated. Such techniques are useful for separating emission bands with appreciably different decay times.

3.2. EXCITATION SOURCES

3.2.1. Arc Light Sources

The type of light source required in luminescence spectroscopy will primarily depend on whether excitation or emission spectra are being measured. For the former a high-intensity smooth continuous source is required covering the range of excitation wavelengths—sharp spectral lines are undesirable. For emission spectra, however, it is often possible to use a single excitation wavelength, in which case a source with a predominant spectral line may be more effective both in excitation intensity and reduced stray light.

Incandescent sources produce pure continua but are poor ultraviolet generators and so are of little use below about 450 nm, though they can be of use for calibration (Marette, 1976). The most generally used sources are high pressure xenon and mercury arc lamps with some use of low pressure hydrogen or deuterium for lower intensities (Fig. 3.4). The high-pressure lamps operate at high temperature and pressures up to 100 atmospheres, so that the normal atomic line spectra are broadened out to give a continuum. The mercury lamp, while somewhat more efficient in the conversion of input power to light output, has rather more line structure than xenon, so the latter is more commonly used. When used as an ultraviolet source one disadvantage of both mercury and xenon is the high output in the visible and infrared regions which must be efficiently discriminated against so as not to interfere with detection of the often weak luminescence emission. (The electrodes run white hot and so are a contributor to the longer wavelength emission). In some cases it is preferable to use deuterium lamps which, though less bright, produce good ultraviolet continua with little visible emission. However, sometimes, particularly when transition-metal or rare-earth excitation spectra are required, lamp emission in the visible region is an invaluable asset.

Arc lamps are available in a wide range of types and power ratings up to 20 kW (Osram (Wotan), Mazda, Hanovia, PEK). However, it is most important to realize that power rating is not a deciding factor in spectro-scopic applications. More important is the amount of light from the source that can be accepted by the optical system and passed into the sample. In this circumstance it is brightness or luminance that is the determining factor. Some care must be exercised in interpreting the manufacturer's specifications as the quoted values refer to the average over some specified area of the source and are usually given in photometric (e.g. candelas/m^2) rather than

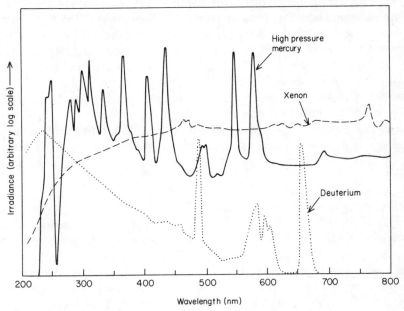

Fig. 3.4. Comparison of spectral outputs of arc light sources.

radiometric units (watts/m^2 . steradian). Photometric units are based on the response of a standard eye and hence have little relevance to luminescence measurements. However, for the same type of lamp (e.g. xenon) which may be expected to produce similar spectral distributions, the relative photo-metric brightness should be a fair guide to relative radiometric brightness.

3.2.2. Light Source Stabilization

In fluorimetry it is necessary to keep the source intensity constant or to correct for the variations. Intensity fluctuations arise as a result of power supply variations, changes in arc or discharge conditions, and geometry or sputtering of the envelope with electrode material. Gas discharge lamps operated in either the discharge or arc modes are approximately constant voltage elements and so it is necessary to regulate the current rather than the voltage for constant light output. A number of circuits have been described for this purpose (Redfield, 1961; Scouler and Mills, 1964; Schiff, 1968) or for stabilizing the light output directly (Schurer and Stoelhorst, 1967; Langelaar et al., 1969). Since arc lamps are rather subject to arc wander, which can lead to large effective intensity variations in focused optical systems (Breeze and Ke, 1972), it is preferable to correct for intensity variations than stabilize the current. This should not be applied to the light direct from the lamp but rather to the output of the excitation monochromator since different spectral regions may vary in differing ways. The fluctuations may also be monitored and the luminescence signal divided by this to compensate for the variations (Aspnes, 1967; Langelaar et al., 1969; Hodby, 1970; Topp and Schmid, 1971). Accurate analogue dividing over a wide dynamic range is difficult however, so that source fluctuations should in any case be minimized. Arc lamps operate at large currents which can present a problem when smoothing a dc supply. The use of a three-phase rather than single-phase supply will give an advantage of a factor 10 using the same smoothing capacitor. It may also be noted that the magnetic fields due to these heavy currents in the leads to the lamp can affect the position of the arc.

Operating conditions for arc lamps are particularly important and should be carefully observed. Conditions are usually specified for orientation, mechanical mounting, operating temperatures and cooling, cleaning of the envelope, polarity and ripple of the power supply, and range of operating currents. As the currents are large good low resistance connections are necessary. The high intensity ultraviolet radiation is injurious, particularly to the eyes, and produces ozone which is poisonous above rather small concentrations and so should be removed efficiently (Neely et al., 1975). The lamps also operate at high pressures, e.g. 8 atmospheres cold and 25 atmospheres or more hot, so should be handled with great care and mounted in an explosion-proof housing.

In many applications it is necessary or desirable to pulse the light source, as for example in flash photolysis or the measurement of delayed emissions. This can also give a great increase in peak power for a given average lamp power rating. Modulators for such applications are described in De Sa and Gibson (1966), Hodgson and Keene (1972), Hviid and Nielsen (1972), Taylor et al. (1972), Fünfschilling and Zschokke-Gränacher (1974), but if high frequencies are required there may be difficulties with acoustical resonances (Gallo and Courtney, 1967).

3.2.3. Lasers

Lasers offer the prospect of an excitation source of exceedingly high intensity compared with classical light sources. The high degree of collimation of laser light enables utilization of almost 100% of the available emission. The angular divergence of a laser beam is defined ultimately by diffraction effects; in practice a divergence of less than 0·5 mrad can be achieved. Classical light sources, including xenon arcs which most nearly approach a point source, emit over a solid angle of 4π and therefore the utilization of the emitted light can rarely exceed more than about 15% without complex light collection systems. In addition, at any one time often only a small proportion of the output of a classical source is used owing to the polychromatic nature of the emission. Obviously line sources are better than continuous sources in this respect but the entire output of a laser can be contained in a very narrow line. A further advantage of lasers is the low source noise compared with classical arc sources. Lasers can also provide pulsed light sources with extremely short pulse widths.

Laser operation requires an excess population of the upper excited state over the lower state involved in the stimulated emission transition,[†] and the different ways of achieving such a population inversion between levels of differing separations are embodied in the various laser devices now available.

The amplifying medium (i.e. the medium in which population inversion occurs) forms part of an optical resonator which is defined by two terminal mirrors; at least one of these mirrors is not totally reflecting, thereby allowing the laser emission to escape from the cavity. Plane parallel mirrors can be used although a confocal concave mirror arrangement has much smaller diffraction losses. Thus stimulated emission is amplified by passing it through the gain medium many times. Mirrors external to the amplifying medium can be used provided reflections at the medium boundaries are suppressed. One method used to suppress these unwanted reflections is to employ rod-ends or windows inclined at the Brewster angle; light polarized in the plane of incidence is then not reflected at such boundaries and hence is preferentially amplified. Light from such a laser cavity will therefore be polarized even though the emission from the amplifying medium may not be expected to be polarized.

Since there are many excellent texts on lasers (e.g. Lengyel, 1966, 1971), we shall restrict the discussion here to a résumé of lasers and laser techniques which may be useful for the excitation of luminescence.

Some lasers are pulsed giving very large intensities for very short times; others can be continuously operated (continuous wave, CW). Some are of fixed frequency or wavelength; others are to some extent tunable.

Usually, solid-state lasers based on transition-metal ions such as Cr^{3+} or

---

[†] This is, of course, a necessary but not sufficient condition for laser action since the excess population of the upper state must yield sufficient amplification to offset losses; i.e. there is a threshold condition for the build-up of oscillations (see, for example, Lengyel, 1971).

rare-earth ions are pulsed owing to the fact that they are optically pumped using a xenon flash tube. However, solid-state CW-operated lasers are available which use a very intense continuously operating pump source having its radiation efficiently focused on to the laser medium; e.g. CW ruby lasers and Nd:YAG lasers have been constructed which operate at room temperature (Evtahov and Neeland, 1965; Geusic et al., 1968). However, there are often problems caused by the large heat dissipation in CW operation.

Population inversion is much more easily achieved in four-level laser systems. If the gap between the lower excited state involved in the stimulated emission transition and the ground state is large enough, then the former will not be appreciably populated even at room temperature—hence the importance of trivalent neodynium lasers. Laser emission can be obtained from this ion in a variety of host structures such as $CaWO_4$, YAG (Yttrium Aluminium Garnet), and certain glasses. Harmonics of Nd laser emission ($\lambda = 1\cdot06$ μm) are often useful for luminescence-excitation purposes or as a pump for dye lasers and optical parametric amplifiers (see below).

Semiconductor lasers constructed using p–n junction diodes of materials such as GaAs (LEDs) require only low voltage excitation. The injection of current-carriers across the junction results in recombination radiation, and at very high current densities an amplifying state can be achieved. Such lasers are extremely small in size, the resonator cavity being formed by the parallel, polished ends of the junction in a direction perpendicular to the current flow. Pulsed operation is usual although CW operation is possible at low temperatures (Rieck, 1970). However, GaAs–GaAlAs heterostructure diodes can be CW operated at room temperature (Kressel, 1971). The emission wavelength of most diode lasers, which is in the red or infrared, can be tuned over a limited range using a variety of techniques (Melngailis and Mooradian, 1975); these techniques usually exploit the variation in band gap with temperature, hydrostatic pressure, or magnetic field. Since the spectrum of recombination emission is usually broad, it is also possible to tune within this emission band by having one end of the diode anti-reflection coated and using a grating as a cavity mirror (Rossi et al., 1973). However, to be useful for luminescence excitation purposes, emission from diode lasers requires frequency doubling (see below). Another disadvantage of such lasers is the rather large beam divergence.

Electron-beam excitation of semi-conductor materials has also been shown to produce laser emission in some cases, e.g. laser emission has been obtained from CdS_xSe_{1-x} at low temperatures in the wavelength range 490 nm (CdS) to 690 nm (CdSe) (Hurwitz, 1966).

Gas lasers, many of which can be continuously excited by an electrical discharge, provide fixed-frequency CW emission at a variety of wavelengths. In addition to the ubiquitous helium–neon laser ($\lambda = 633$ nm), argon ion and helium–cadmium lasers are now in common use. The latter can provide CW emission at 442 nm and 325 nm, and argon ion lasers can be tuned to several lines in the green, blue, and near ultraviolet. Nitrogen lasers, which are also commercially available, can provide high energy pulsed ultraviolet emission at 337 nm and are often used as a pump for dye lasers (see below). Recently, the development of rare-gas halide lasers has provided laser emission in the far ultraviolet (Searles and Hart, 1975; Searles, 1976). Such systems are undergoing rapid development at the present time.

Ultraviolet emission, particularly pulsed emission, can also be obtained by harmonic generation using lasers which have their fundamental frequency in the visible or near infrared. Second-harmonic generation (SHG), or frequency doubling, is now a well-established technique which uses certain birefringent, non-centrosymmetric, crystals that behave optically in a non-linear fashion when the intensity of a transmitted beam of light is very high. The effect can be understood by considering the addition of non-linear terms to the usual linear relationship between the polarization P and the electric field E in one dimension

$$P = a_1 E + a_2 E^2 + a_3 E^3 + \ldots \qquad (3.9)$$

For a given E the non-linear effects obviously depend on the magnitude of the coefficients a_2, a_3, etc. If the electric field E is oscillating at an angular frequency ω then it is easy to show that the second term will give rise to a polarization oscillating at a frequency 2ω. This conclusion still holds when we extend the situation to that of a non-linear, three-dimensional, anisotropic crystal although the scalar coefficient a_2 then becomes a third-rank tensor relating the second-order polarization vector to a function of electric field components.

However, for efficient SHG the phase-matching condition $2k = k'$ must be satisfied where $k = \omega n/c$ and $k' = 2\omega n'/c$, n and n' being the refractive indices at the frequencies ω and 2ω respectively. This condition can be met by orienting the crystal such that the fundamental and the second-harmonic travel with the same velocity in the crystal (i.e. $n = n'$), e.g. one as the o-wave and the other as the e-wave. Because the normal to the wavefront and the ray direction are not in general parallel for the e-wave, the extra-ordinary beam diverges from the ordinary beam causing a reduction in SHG efficiency. This walk-off can, however, be avoided by ensuring that both rays travel perpendicular to the optic axis. Because the birefringence is usually temperature dependent, i.e. one of the principal refractive indices is more temperature dependent than the other, it is sometimes possible to phase-match in this orientation by varying the crystal temperature. 90° phase-matching is also less sensitive to incident-beam divergence than other phase-matching directions (Zernicke and Midwinter, 1973a).

Commonly used non-linear crystals include ADP (ammonium dihydrogen phosphate) and KDP (potassium dihydrogen phosphate); such crystals have good transmission characteristics in the ultraviolet, visible and near infrared, non-linear coefficients of suitable magnitude, and a large enough birefringence to allow phase-matching. Using a pulsed laser with the crystal outside the laser resonator, SHG efficiency is usually in the range 10–40%.[†] However, very efficient SHG has been obtained for CW laser emission by placing the crystal inside the resonator cavity. In fact, it has proved possible to convert all the fundamental CW output of an Nd:YAG laser, linearly polarized by an intra-cavity plate mounted at the Brewster angle, to second-harmonic output using an intracavity barium sodium niobate crystal (Geusic *et al.*, 1968). SHG is in fact a special case of the more general property of frequency mixing in non-linear crystals. Substitution for E in Eq. (3.9) of two oscillating fields at frequencies ω_1 and ω_2 will yield from the second term a polarization oscillating at the combination frequencies $\omega_1 + \omega_2$ and $\omega_1 - \omega_2$ in addition to $2\omega_1$ and $2\omega_2$. Thus sum and difference frequency generation may occur, but the frequencies which are generated efficiently will be those for which phase matching occurs.

For sum frequency generation

$$\omega_1 + \omega_2 = \omega_3 \qquad (3.10)$$

and the phase-matching condition is [‡]

$$k_1 + k_2 = k_3. \qquad (3.11)$$

This condition may be written

$$n_1 \omega_1 + n_2 \omega_2 = n_3 \omega_3 \qquad (3.12)$$

where n_1, n_2, and n_3 are the appropriate refractive indices at ω_1, ω_2, and ω_3 respectively. For example, for a negative uniaxial crystal, n_3 is the refractive index for the e-ray at the generated frequency ω_3, and n_1 and n_2 could be the ordinary indices at ω_1 and ω_2 respectively (type I phase-matching); alternatively, n_1 could be the index for the e-ray at ω_1 and n_2 the index for the o-ray at ω_2 (type II phase-matching). It is easy to show from Eq. (3.12) that a necessary constraint is that the value of n_3 must lie between n_1 and n_2. The special case of SHG, which we have already discussed briefly, is obtained when $\omega_1 = \omega_2$ and $n_1 = n_2 = n_3$ for phase-matching.

[†] The maximum theoretical conversion efficiency is 100% for a truly monochromatic, plane-wave fundamental. However, if all the fundamental had been converted to second harmonic the process would in theory then reverse to reproduce the fundamental if the crystal was long enough.

[‡] This is really a vector equation but here we assume that the wavevectors are collinear.

Precisely what frequencies can or cannot be phase-matched for any particular crystal can be ascertained from dispersion curves such as those in Fig. 3.5.

Equation (3.11) also expresses the phase-matching condition for difference frequency generation if, say, ω_2 is the frequency generated from initial frequencies ω_3 and ω_1.

In the foregoing discussion we have only considered the effects associated with the second term in Eq. (3.9). Third-term effects are usually much smaller and phase-matching conditions are more difficult to achieve in practice.

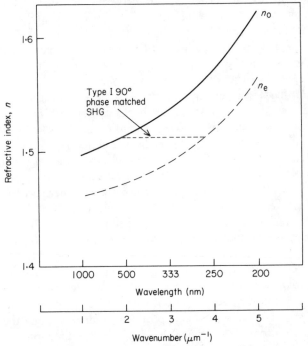

Fig. 3.5. Dispersion curves for ADP at room temperature.

However, recently, efficient phase-matched, third-harmonic generation of the fundamental of an Nd:YAG laser has been achieved in an alkali-metal vapour suitably mixed with an inert gas (Bloom et al., 1975). Phase matching is possible on account of the anomalous dispersion of the alkali vapour (Miles and Harris, 1973).

A non-linear crystal may also be used as an optical parametric generator or oscillator to produce a frequency-tunable light source (Zernicke and Midwinter, 1973b). The crystal is "pumped" by a laser beam of fixed frequency ω_p and two beams of lower frequencies ω_s and ω_i (called the "signal" and the "idler") are produced such that $\omega_p = \omega_s + \omega_i$. The frequencies ω_s and ω_i are determined by the phase-matching condition $k_p = k_s + k_i$ and can be varied by changing the crystal temperature. The process can be considered as a difference-frequency generation between the pump ω_p and a frequency ω_s which is present in the noise. If the crystal is placed in a cavity resonant at either ω_s or ω_i (or both) then both ω_s and ω_i will be amplified in accordance with the Manley–Rowe relation (Zernicke and Midwinter, 1973c). However, efficient generation of ω_s and ω_i has been obtained in a single pass of the fourth harmonic of an Nd laser through an ADP crystal giving an output tunable between about 420 nm and 720 nm (Yarborough and Massey, 1971).

In all these frequency-mixing techniques, it is usually desirable to remove the initial frequencies from the output beam leaving the generated frequency required. Sometimes a simple filter will suffice, but often the frequencies are separated spatially by a polarizing Glan prism.

If efficient harmonic generation is to be achieved with the crystal outside the resonant cavity, then obviously a short pulse of high peak intensity will be better than a longer pulse of lower intensity. For this and other reasons, solid-state lasers are often Q-switched to produce regularly spaced, well-defined, giant pulses. For Q-switching an electro-optic or fast mechanical shutter is placed in the cavity between the laser medium and one of the reflectors. Excitation with the shutter closed can then be built up well beyond the threshold level which is operative when the shutter is open. When such a high level of excitation is reached the shutter is opened, resulting in a giant pulse of short duration as the excess excitation is dissipated. A Kerr cell, Pockels cell, rotating mirror, and a bleachable filter have all been used as a shutter for Q-switching; Pockels cells, in particular, are used for very rapid switching. (A Pockels cell is an electro-optic device which exploits the linear variation of birefringence of crystals such as ADP with applied electric field. The field is applied longitudinally along the optic axis and the light is also passed through the crystal in this direction.† The uniaxial crystal becomes biaxial and a retardation proportional to the applied voltage is produced. A $\lambda/2$ retardation will effectively rotate the plane of polarization through 90° and therefore allow a Pockels cell to be used as a shutter when a suitably orientated polarizer is incorporated). Q-switched pulses are, however, limited to a minimum width of about 10 ns because of the necessary build-up time.

Nevertheless, it is possible to produce trains of extremely short, regularly spaced pulses from either solid-state or gas lasers by a technique known as mode-locking (De Maria et al., 1969; Smith, 1970; Von der Linde, 1973).

† Transverse-field devices are also used (see Ley et al., 1970).

Usually the relative phases of the longitudinal modes of an optical cavity of length L, which are spaced in frequency by $c/2L$, fluctuate in a random fashion causing the output to vary in an uncontrolled way. In a mode-locked laser all such modes are phase-locked to give a pulse every $2L/c$ s. The ratio of pulse separation to pulse width is about equal to the number of oscillating modes N. The number of such modes, and hence the ultimate pulse width, will be limited by the gain bandwidth of the laser medium; pulses of several picoseconds width have been obtained from Nd lasers.

Some lasers will self mode-lock under certain conditions, but mode-locking can be induced by internal modulation using some form of loss modulator driven at the mode-spacing frequency. The most usual technique for inducing very short-pulse mode-locking uses a saturable absorber (i.e. a bleachable filter) inside the cavity. In such an absorber, which is usually in the form of an organic dye, the absorption decreases with an increase in intensity beyond a certain level, and therefore a short, high power pulse will suffer less absorption than a longer pulse of the same energy.

Perhaps the most important development in lasers for luminescence excitation has been the organic dye laser. Such lasers, which are frequency-tunable, can be operated in CW or pulsed mode and most closely approach the ideal light source for excitation. Organic dye lasers are essentially four-level systems in which the emission is a result of a transition from the lowest vibrational level of the first excited singlet state to one of the vibrational levels of the ground state. Pumping takes place from the lowest vibrational level of the ground state to any vibronic level in excited singlet states.

The output wavelength of a dye laser can be continuously varied within the broad emission band of the dye which has a typical width of a few tens of nanometres. The dye, which is in solution, is rapidly circulated through the excitation chamber so that the problems of heat dissipation and photo-decomposition are minimized. Pulsed dye laser emission can be achieved with flash tube excitation but at present CW emission requires a second powerful CW laser for excitation, such as an argon ion laser, in order to achieve the necessary radiation density. Pulsed dye lasers may also be pumped by an auxiliary pulsed laser such as a nitrogen laser or a harmonic of a Q-switched Nd laser.

In the absence of wavelength selective elements, a dye laser will self-tune according to conditions dependent on such factors as cavity-mirror reflectivity and the concentration of the dye solution. More usually, however, the laser is extrinsically tuned by the use of wavelength selective reflectors or dispersive elements in the cavity. For example, a cavity mirror may be replaced by a diffraction grating or a prism may be introduced into the cavity; birefringent filters have also been used in the cavity and linewidths of less than 1 Å are typically achieved.

The most efficient laser dyes are the rhodamine dyes (6G or B) which emit in the yellow-red region. For emission in the blue-green region coumarin dyes are normally used and well-known scintillators such as p-terphenyl have been used to produce near ultraviolet emission. However, frequency doubling of pulsed dye lasers using rhodamine or other dyes can also produce a tunable emission in the ultraviolet (Dunning et al., 1972). SHG using an intra-cavity, lithium formate crystal has also been obtained using a CW dye laser (Gabel and Hercher, 1972).

An additional feature of dye lasers is the ability to produce wavelength-tunable picosecond-width pulses by mode-locking (Shank and Ippen, 1973; Von der Linde, 1973). The frequency bandwidth of a typical dye emission spectrum is more than sufficient to render the production of such short pulses possible.

For further details concerning dye lasers the reader is referred to the excellent monograph edited by F. P. Schäfer (1973).

At this point it must be said that at present wavelength-tunable excitation from laser systems has its limitations. Optical parametric oscillators are usually temperature-tuned and hence a stable change in wavelength cannot be achieved quickly. Moreover, tunable ultraviolet cannot be produced by such means using presently available crystals and ultraviolet laser pumps. Dye lasers, on the other hand, can be rapidly tuned but usually only over restricted wavelength ranges. The scanning of excitation wavelength over a large spectral range may require several changes of dye, and the production of tunable ultraviolet below about 340 nm at present requires SHG or sum frequency techniques with the necessity for angle-tuning of the crystal. Diode lasers have similar limitations and even with SHG cannot at present provide ultraviolet below 300 nm.

However, in many circumstances the advantages of lasers far outweigh their disadvantages and enable new techniques to be exploited. The application of lasers to luminescence excitation has been evident in new fields of research, particularly those associated with the effects of high excitation densities, such as the study of electron–hole droplets (EHDs) in Si and Ge (Benoît à la Guillaume, 1976; Thomas, 1976). Furthermore, the availability of picosecond pulses has made spectroscopic studies on the time scale of vibrational relaxation processes possible.

3.2.4. Synchrotron Radiation Sources

A full understanding of the properties and behaviour of organic and inorganic materials necessitates spectroscopic studies at least up to and beyond the ionization potential of those materials. This fact, and the interest in studying the behaviour of materials when exposed to high energy radiation (both electromagnetic radiation and charged particles), has been the cause of greatly increased research activity in the vacuum ultraviolet (VUV) and X-ray regions. A major problem in the VUV region is to obtain sufficiently intense continuum sources covering a wide wavelength range and which, of course, do not require use of window materials.

A recent (post-1960) development has been the use of synchrotron radiation for studying inorganic materials such as alkali halides, chalcogenides, inorganic crystals, simple molecular gases, and for observing the absorption and emission spectra of a wide range of organic solids (Marr et al., 1973, 1975).

Synchrotron radiation is emitted as a smooth continuum spectrum extending from the X-ray region to the far infrared region whenever electrons or positrons are accelerated in a circular accelerator. At quite low energies (more than 150 MeV), electrons already are travelling at speeds very close to the speed of light. Under these circumstances, the accelerating electron radiates strongly only into a narrow cone (of angle about 10^{-3} rad) in the direction of electron travel. Because of this small cone angle, the radiation is observed only in a direction tangential to the circular electron orbit, when the observer sees the radiating electrons moving towards him with almost the speed of light. The resulting Doppler shift (and the very short time interval for observation) cause the radiation detected by the observer to be blue shifted from the radio frequency region and to lie predominantly in the far VUV and X-ray regions. For wavelengths in the VUV and X-ray regions, the divergence of synchrotron radiation is usually less than that from a laser (Fig. 3.6).

The detailed properties of synchrotron radiation can be calculated exactly from a knowledge of the accelerator parameters. This has been done at a number of different accelerators and complete agreement is reached when theory is compared with experiment. Figure 3.7 shows the synchrotron radiation emission spectra from a number of different synchrotron radiation sources. Because the intensity and geometry of the radiation at all wavelengths are calculable in absolute terms, the synchrotron source can be used as a primary standard for source or detector calibration purposes and relative and absolute quantum efficiency measurements.

Synchrotron radiation sources have several important properties which make them unique in their applications in spectroscopy particularly in the VUV region. The source size is determined by the cross-sectional dimensions of the orbiting electron beam and this may be very small indeed. In a storage ring the beam size may be of the order of 1 mm horizontal width by 0·1 mm vertical height. Since the total power (emitted around an entire machine) may be as much as 255 kW it is obvious that the source is exceedingly bright. The low divergence of the radiation combined with the small source size provide a spectroscopic source which is sufficiently intense when used

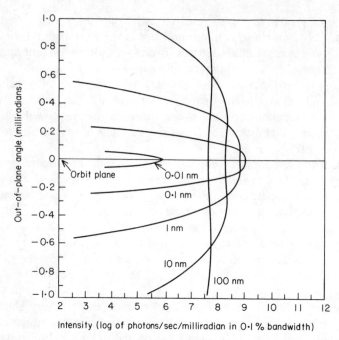

Fig. 3.6. The vertical angular distribution of synchrotron radiation at different wavelengths from the Daresbury synchrotron NINA at 5 GeV and 20 mA.

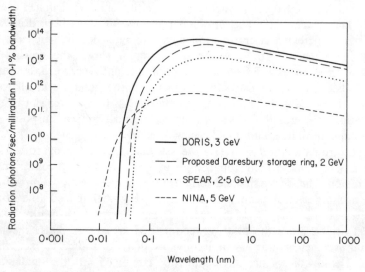

Fig. 3.7. Synchrotron radiation spectra from various accelerators.

with a monochromator to be effective for excitation measurements from about 0·01 nm into the far infrared region. The radiation emitted from a synchrotron source in the plane of the accelerator is completely (100%) linearly polarized at all wavelengths with the electric vector parallel to the plane of the machine. Although the radiation is elliptically polarized above and below the orbit plane the average degree of polarization even for visible light collected over all vertical angles remains about 80% or more. This property has been exploited in measurements of polarized absorption spectra of organic crystals. Electrons in an accelerator can be contained only within a high vacuum system ($\sim 10^{-9}$ Torr) and consequently the photon source is intrinsically clean and windowless. Any contamination will be associated with the sample itself or with its containment vessel.

A striking property of synchrotron sources is the way in which the emitted light is time modulated. The emitted light comes from electrons passing through the tangent point viewed by the observer. As a consequence of the radio-frequency acceleration mechanisms of the electrons necessary to compensate for the energy lost by synchrotron radiation, the electrons do not occupy the orbit uniformly but are bunched in space. The bunch length and therefore the time width of the observed light pulse depend only on the frequency of the accelerating radio-frequency field. For frequencies of 400–500 MHz the light pulse width is less than a nanosecond. The light emission is totally modulated and each pulse has an accurately measurable profile which is close to a Gaussian. For measurements of fluorescence emission lifetime, many storage rings can be used in the "single bunch" mode. This can produce something like 10^6 photons per pulse within a 0·1 nm band at 100 nm where the pulse width is approximately 0·2 ns and the pulse repetition frequency is approximately 1 µs and is defined by the ring frequency. Already synchrotron sources have been used to measure fluorescence time decay, time-resolved spectra, and the lifetimes of individual vibronic states in organic materials (see Section 3.5.6) (Marr and Munro, 1973; Lopez-Delgado et al., 1974).

3.2.5. Electron Excitation

The mechanisms involved in particle-excited luminescence are complex, involving many possible stages (Garlick, 1966). However, often the cathodoluminescence emission spectrum is identical, or very nearly so, to the photoluminescence spectrum and, although the absolute energy efficiency of the photoluminescence process may be greater than that of the cathodoluminescence process, high excitation densities are much easier to obtain with electron-beam excitation.

Moreover, many inorganic materials in particular are cathodoluminescent but not photoluminescent since ultraviolet is not absorbed to any appreciable extent. Excitation of centres often occurs predominantly via conduction electrons or excitons rather than directly, hence if such materials have a large band gap they will not be photoluminescent.

Electron-beam excitation has several advantages over optical excitation for emission spectroscopy; there is no background light from the excitation source to confuse the emission spectrum, and electron beams of keV energy can easily be focused, modulated, and rastered (as in a cathode ray tube, CRT).

The intensity of excitation can be varied by adjustments in beam energy or current density. A higher energy will increase the penetration depth and hence the volume of excitation, whereas a higher current density will increase the density of excited centres. There are, however, obvious limitations on both these parameters. The limitation on higher energies is determined by high voltage insulation problems and the necessity of radiation shielding on account of X-ray emission. Moreover, there is no point in exciting centres which are at a depth such that only a small proportion of the emission will escape from the sample to be measured. The limitation on beam current density is more fundamental; the number of centres in a given volume is finite and therefore a saturation level of excitation will be reached.

The work of Ehrenberg and Franks (1953) demonstrated that the excitation volume is approximately spherical for low energy electrons. For higher energies (~ 1 MeV) the excitation volume is a cylindrical channel ending in a nearly spherical volume of diffused electrons, both primary and secondary.

Besides the saturation effect, there are other effects of high current densities which limit the light output often well before the saturation level is reached. The low electrical conductivity of the sample may result in charge build-up which may tend to repel or deflect the incident beam. If the sample is in the form of a powder, then such charge build-up also causes electrostatic scattering of the powder grains. Highly focused electron beams can also produce very high localized temperatures, and it is not difficult to burn a hole in a thin sample. In a CRT these effects are overcome by back-aluminizing of the screen (which, incidentally, has many other advantageous effects). In electron microscopy and electron microprobe work it is usual to coat the sample with a thin evaporated conducting layer of carbon.

Modulation of an electron beam can be achieved by simple magnetic or electrostatic deflection of the beam. However, if a triode electron gun is used, any modulation waveform can be applied to the grid which is suitably dc biased. Such electrical "chopping" of the electron beam, using, say, a square wave, is particularly suitable if lock-in amplification techniques are to be used to recover the luminescence signal. Similarly, the decay curve of the luminescence can be determined by using a boxcar detector instead of a lock-in amplifier (see Section 3.4.4).

Highly focused electron beams have been used for many purposes in luminescence research; e.g. electron microprobes have been used in scanning or raster mode to identify luminescent minerals in thin rock sections (Geake et al., 1973), and the correlation of crystal defects with areas of low luminescence efficiency in GaP crystals has been studied using a scanning electron microscope (Davidson and Rasul, 1977).

Radioactive β-emitters, such as ^{32}P sources, are convenient and compact sources of high energy (MeV) electrons. However, such sources emit electrons with a range of energies, making focusing difficult, and do not usually give high excitation densities. Nevertheless, β-emitters are very useful for the excitation of luminescence in organic materials where high excitation densities caused by keV electron beams may cause decomposition or vaporization. Many organic materials will, moreover, quickly evaporate or sublime in a vacuum unless cooled to very low temperatures. β-Excitation, on the other hand, can be achieved in air at room temperature.

3.3. FILTERS AND SPECTROMETERS

3.3.1. Filters

The simplest and most efficient method of producing monochromatic radiation or radiation of a limited wavelength bandwidth in the visible or near ultraviolet regions is by the use of a filter. Filters are usually robust, can be of large area, and selected to have a very narrow (< 1 nm) or very large (~ 100 nm) bandpass with peak transmissions up to 90%. In luminescence studies filters are commonly used where fluorescence intensities are low as in fluorescence lifetime measurements. Filters are also frequently used to eliminate or reduce the stray light background from grating mono-chromators (Section 3.3.5).

A wide range of filter types are available; absorption filters (using glasses, gelatin films, chemicals in solution or gas form, and pure liquids), reflection filters, interference filters (multilayer dielectric or metal-dielectric), bi-refringent filters (such as the Lyot filter), and scattering filters (Christiansen type) are all widely used.

Gelatin-based and doped-glass absorption filters, which can be either broad bandpass or short wavelength cut-off (i.e. long pass) filters, are cheap, possess good transmission properties, and are readily obtainable com-mercially (Fig. 3.8). However, care must be exercised when using such filters since many are luminescent. As gelatin and normal silicate glass are opaque below 300 nm, fused silica, sapphire or alkali-halide-based materials are necessary in this spectral region. Chemical filters with suitable mixtures of materials having convenient transmission properties have been used in solid, solution, and vapour phase (Bowen, 1946; Kasha, 1948; Braga and Lumb, 1966; Pellicori *et al.*, 1966). Solution filters are, of course, restricted to wavelengths where the solvent is transparent, but the onset of air (oxygen) absorption is usually the limitation if the solvent is pure (Fig. 3.8).

Gelatin and chemical filters in general have the disadvantage that they may not withstand high temperatures and also often deteriorate as a result of photo-induced processes during prolonged irradiation particularly from ultraviolet sources.

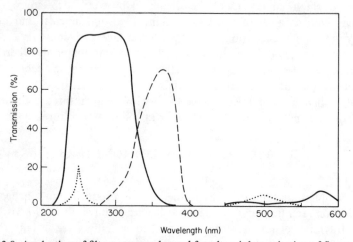

Fig. 3.8. A selection of filters commonly used for ultraviolet excitation of fluorescence: ——, liquid filter [$NiSO_4$ (500 g l^{-1}) + $CoSO_4$ (140 g l^{-1})], 10 mm thick; – – –, commercial ultraviolet filters (Chance OX1, Corning 7–39, Wratten 18A, Schott UG–2); · · · ·, narrow band interference filter (Barr and Stroud U1).

Interference filters provide a means of selecting a narrow bandpass with high peak transmission. An interference filter is essentially a Fabry–Pérot etalon with a very small spacing (of the order of a wavelength) filled with dielectric between the reflecting surfaces. If the spacing or fringe order is small, the free spectral range (i.e. the range of wavelength for which only one wavelength band is transmitted) is large. However, the width of the passband is also inversely proportional to the spacing or order but a narrow passband and large free spectral range can be simultaneously achieved by using several units in series with multiple spacings.

Narrow bandwidth also requires a high surface reflectivity and high transmission requires a low absorptivity of the reflecting surfaces. On account of these requirements multi-layer dielectric stacks are often used as reflectors instead of metal films particularly for the visible region (Longhurst, 1973). Such multi-layers consist of alternate high and low-refractive-index dielectrics of quarter-wavelength thickness; layers of zinc sulphide and cryolite are commonly used for this purpose.

The quality of an interference filter is judged by the ratio of the transmission at the desired wavelength to that at all other wavelengths within the range used or detected. A good filter will transmit less than 0·01% of light of any particular wavelength outside the passband. Sometimes the integrated intensity of light of all wavelengths outside the passband can be a significant fraction of the total transmitted intensity and an auxiliary broad-band absorption filter may be necessary. In fact, a glass or gelatin filter is often incorporated in the mounting of an interference filter in order to block the sidebands which would otherwise be transmitted.

Peak transmission in the visible region can be high (> 60%) particularly for larger bandwidths. However, in the ultraviolet peak transmission is usually less than 30% and can be as low as 10% for a bandwidth of about 20 nm at a wavelength of 200 nm. The light beam traversing the filter should be parallel if the passband is not to be broadened. Moreover, an inclination of the filter to the beam direction other than normal will shift the peak transmission to shorter wavelength.

Recently, interference-wedge filters have become commercially available for the visible region. Such filters transmit shorter wavelengths near the wedge apex and progressively longer wavelengths further away from the apex. By using a slit parallel to the apex it is possible to scan the wavelength of the transmitted light by moving the wedge or slit in a direction perpendicular to the slit length. Thus it is possible to construct a compact, high-transmission, scanning, filter spectrometer simply and cheaply.

Birefringent filters have also been devised which possess passbands of similar width to interference filters. Such filters are constructed using polarizers and retardation plates; wavelength selection depends on the dispersion of the retardation plates. Most types are based on the Lyot or Solc designs (Clarke and Grainger, 1971a) and some can be tuned by rotation about the filter axis.

Although in general filters are cheaper, more convenient, and usually more efficient in terms of light throughput compared with spectrometers, spectrometers are more versatile in terms of choice of wavelength and passband and usually have a lower stray light background.

3.3.2. Spectrometers

Spectrometers are used either to produce monochromatic light from a polychromatic source for sample excitation or absorption, or to disperse light emitted by the sample enabling the emission spectrum to be observed.

The dispersive element of a spectrometer, which may be a prism, diffraction grating, or interferometer, separates light of different wavelengths spatially. In the case of a prism or diffraction grating, the use of slits enables a particular wavelength or range of wavelengths to be isolated. In most spectrometers light from the entrance slit is collimated by a lens or mirror so that parallel light is incident on the dispersive element. After dispersion, light of a particular wavelength is focused on the exit slit by another lens or mirror and the wavelength of the light emerging can be changed by rotating the dispersive element.

For luminescence spectroscopy, the most important parameter which defines the merit or usefulness of a spectrometer is the light flux throughput or luminosity (defined as the output flux using a source of unit luminance), since this factor will limit the signal-to-noise ratio attainable. For any particular spectrometer, there is, however, always a trade-off between luminosity and resolving power and therefore comparisons of the luminosity for different instruments must always be made at a given resolution or wavelength bandwidth which is determined by the slit width and the dispersion. (The slit length is also important in this comparison, see Section 3.3.4.) Obviously the greater the dispersion available, the larger is the slit width for a given wavelength bandwidth. Dispersion, resolving power, and luminosity are therefore strongly inter-related and these inter-relationships will be discussed in the following sections. Although it can be argued that interferometers are superior to prisms and gratings from the point of view both of high luminosity and high resolving power, they are not suited to the analysis of broad-band spectra. Moreover, the multiplex advantage of Fourier transform spectroscopy is lost when the limitation on signal-to-noise ratio is photon shot noise rather than signal-independent thermal noise (see Section 3.4.1).

Good prism instruments usually have lower stray light (see Section 3.3.5) than grating instruments and have the advantage of freedom from overlapping orders inherent in the latter. However, for gratings and prisms of comparable physical dimensions, the grating gives a superior luminosity at all wavelengths for a given wavelength bandwidth and angular slit height (see Section 3.3.4).

Figure 3.9 shows common mountings for prism and grating instruments. The Littrow arrangement is a compact design which uses the same lens (or mirror) for the collimation of undispersed light and the focusing of dispersed light. In such an arrangement the light almost retraces its path, i.e. apart from a slight sideways deflection which is necessary in order that the exit slit can be spatially separated from the entrance slit; the dispersed light beam is also usually reflected through 90° to separate entrance and exit slits. The Ebert mounting for grating instruments also uses the same mirror for collimating and focusing, but here the entrance and exit slits are symmetrically placed on opposite sides of grating. The symmetry of this arrangement gives it relative freedom from optical aberrations. The Czerny–Turner arrangement, which has in recent years achieved universal popularity, is essentially an Ebert mounting with separate mirrors for collimation and focusing. The diameter of the mirrors is, of course, determined by the size of the grating but the total mirror area is usually smaller than in an Ebert thereby reducing stray light. There is a limitation on slit length if straight

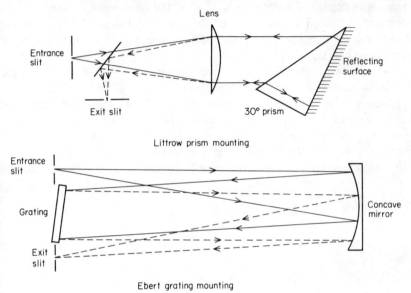

Fig. 3.9. Common mountings for prism and grating instruments.

slits are used but this limitation can be overcome by using curved slits which will not degrade the resolution. The spectral range of an Ebert or Czerny–Turner spectrometer is very large, and a range from the VUV up to about 20 μm in the infrared can be accommodated by interchanging gratings. For luminescence purposes one grating usually suffices covering the ultraviolet and visible regions. Although it is possible to use such an instrument, suitably evacuated in the VUV, concave grating spectrometers which have only one reflecting surface are usually employed, since the reflectivity of aluminium begins to fall significantly in this region. However, in Littrow, Ebert or Czerny–Turner mounts, plane reflection gratings permit the wavelength to be scanned simply by rotating the grating on its axis. If the grating is motor driven (at a constant motor speed) via a sine drive, then the spectrometer will scan equal wavelength intervals in equal times. Alternatively, a linear wavenumber scale can be obtained using a cosecant drive.

It is essential that a high quality grating is used which is free from "ghosts", particularly if the spectrometer is to be used for emission spectroscopy.

"Ghosts" are less intense replicas of a principal maximum which are displaced in diffraction angles from such a maximum and are caused by periodic errors in the grating rulings. If a single periodic error is present then Rowland ghosts are produced which are symmetrically spaced about the principal maximum and of highest intensity close to the principal maximum. Lyman ghosts, on the other hand, can occur with appreciable intensity at diffraction angles significantly different from that of the principal maximum. These are caused by a periodic error of very small period or by the presence of two commensurate periodic errors. Such defects are, however, much less likely in modern holographic gratings.

Ghost intensities which are less than 10^{-3} of the principal maximum can still be troublesome in emission spectroscopy (see also Section 3.3.5).

Modern diffraction gratings are invariably of the blazed reflection type, i.e. the grooves are aluminized and have a precise shape such as to concentrate the maximum amount of light into a particular diffraction angle on a particular side of the grating (see Fig. 3.10). Effectively, the single-slit diffraction envelope (i.e. the Fraunhofer pattern due to a single groove),

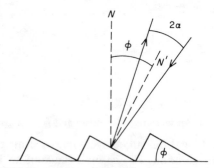

Fig. 3.10. Blazed diffraction grating. Reflected ray is drawn for blazed wavelength.

which for unblazed gratings has its principal maximum coincident with the zero order, is shifted so that it is coincident with, say, the first order for some particular wavelength λ_b determined by the blaze angle ϕ. In fact,

$$\lambda_b = 2d \sin \phi \cos \alpha \tag{3.13}$$

where 2α is the angle between entrance and exit beams and d is the grating periodicity. 2α is usually constant and ideally zero for a Littrow mount although it is usually fairly small even for an Ebert mount. Thus to a good approximation

$$\lambda_b = 2d \sin \phi \tag{3.14}$$

Obviously, second order for $\lambda_b/2$ is also maximized, but overlapping orders can usually be eliminated by an appropriate filter.

In general, the problem of overlapping orders in grating instruments is easily overcome using filters. For the ultraviolet such filters are unnecessary since the atmosphere absorbs wavelengths shorter than about 200 nm. In the range 300–600 nm a glass plate suffices as a filter.

3.3.3. Dispersion and Resolving Power

The spatial separation of light of different frequencies or wavelengths, caused by a dispersive element, is quantified by the angular dispersion defined as $d\theta/d\lambda$, i.e. the rate of change of the angle of deviation or diffraction with wavelength. However, since the actual linear dispersion available at the exit slit of a spectrometer depends on the focal length f of the focusing lens or mirror, the linear dispersion or, more usually, the reciprocal linear dispersion of the instrument is quoted:

$$\text{Reciprocal linear dispersion (usual unit: nm/mm)} \quad \frac{d\lambda}{dl} = \frac{1}{f}\frac{d\lambda}{d\theta}$$

dl being distance measured across the slit. Its convenience is apparent since the static wavelength bandwidth is obtained by multiplying by the slit width used.

The angular dispersion available from a prism depends on the prism angle, the dispersive power $d\mu/d\lambda$ of the prism material and also on its refractive index μ. For a grating, however, differentiation of the grating equation yields $d\theta/d\lambda = n/(d\cos\theta)$ when n is the order and d the spatial period of the grating rulings. For first order, the diffraction angle θ is usually such that $\cos\theta$ does not differ much from unity, hence $d\theta/d\lambda \approx 1/d$, i.e. the angular dispersion (in rad/mm) is approximately equal to the number of lines/mm. For example, a grating with 1200 lines/mm would therefore have an angular dispersion of $1\cdot2 \times 10^{-3}$ rad/nm in first order which would give a reciprocal linear dispersion of about $0\cdot8$ nm/mm in a 1 m spectrometer. (In second order the reciprocal linear dispersion would be $0\cdot4$ nm/mm). It is also noteworthy that the angular dispersion is almost independent of wavelength for a grating since, for a particular order, $\cos\theta$ is a slowly varying parameter. On the other hand, the angular dispersion of a prism is very dependent on wavelength differing by an order of magnitude or more across its usable wavelength range and competing with the grating only near the absorption edge of the prism material. Note that angular dispersion is a property only of the dispersive element and is independent of the overall physical size of the element.

Chromatic resolving power is clearly related to the angular dispersion; the ability to separate images of the entrance slit in two particular wavelengths obviously depends on their spatial separation. However, it will also depend on the width of such images. In the absence of optical aberrations, the width of the images will usually be governed by the width of the entrance slit. Nevertheless, narrowing the entrance slit will not always produce a narrowing of the image; there will come a point where the width of the image is limited by the width of the diffracting aperture or aperture stop of the optical system, since the angular width of the principal maximum in a Fraunhofer diffraction pattern is inversely proportional to the width of the aperture stop. In the case of a prism spectrometer, the height of the prism (provided it is filled with light) is clearly the limiting parameter. For a diffraction grating, where the number of grooves N is large, the profile of a diffracted image of the entrance slit (i.e. a principal maximum or order) can be shown to be identical to a Fraunhofer pattern of a rectangular aperture (i.e. a sinc function) equal in width to the total width of the grating (or to the width of the incident beam if all the grating is not filled with light). The maximum resolving power is therefore attained when the image is limited only by the width of the grating or prism.

Chromatic resolving power is defined as $\lambda/\Delta\lambda$, where $\Delta\lambda$ is the minimum wavelength separation which will produce resolving images as defined by Rayleigh's criterion, i.e. the images of the entrance slit in wavelengths λ and $\lambda+\Delta\lambda$ are theoretically defined as just resolved if the principal maximum in the Fraunhofer profile of one image falls on the first minimum in the profile of the other. In other words, the limit of resolution is defined to occur when the centres of the images for the two wavelengths have an angular separation $\Delta\theta$ equal to the angular half-width of the principal maximum in the diffraction intensity profile, i.e. $\Delta\theta = \lambda/W$, where W is the width of the grating or prism. Or in linear terms, the corresponding linear separation of the images is $f\lambda/W$ at the resolution limit. There is, therefore, nothing to be gained by using an exit slit narrower than this. Now resolving power is:

$$\frac{\lambda}{\Delta\lambda} = W\frac{\Delta\theta}{\Delta\lambda} = W\frac{d\theta}{d\lambda} \qquad (3.15)$$

i.e. the maximum resolving power attainable is equal to the width of the grating or prism (normal to the incident light beam) multiplied by the angular dispersion. For a 1 m spectrometer with a 100 mm wide grating, the optimum slit width $f\lambda/W$ is 5 μm at a wavelength of 500 nm. However, usually slit widths considerably larger than this are used in order to increase the light flux through the instrument at the expense of a considerable reduction in resolving power. In broad-band luminescence work, slit widths of the order of 1 mm are typical in such a spectrometer giving a static wavelength bandwidth of about 1 nm. However, when a spectrometer is used in scanning mode, the bandwidth is the sum of the static and dynamic bandwidths the latter is defined as the scanning rate (in nm/s) multiplied by the time constant of the associated detector instrumentation.

A better signal-to-noise ratio will obviously be obtained with a larger time constant (i.e. a smaller Δf, see Section 3.4.2) and therefore a slow scanning rate is required if the dynamic wavelength bandwidth is not to become unduly large. In fact, for a given overall wavelength bandwidth the scanning rate must be as slow as practicable for the best signal-to-noise ratio. Time constants are in practice restricted by the stability of the instrumentation and, to a lesser degree, by the inconvenience of long measurement times. Again the trade-off between resolution on the one hand and signal-to-noise ratio (and hence time of measurement) on the other is emphasized.

3.3.4. Light Flux Throughput

The flux throughput of any optical instrument is given by $\phi = BS\omega$, where S is the area of the source, B is the brightness or luminance of the source (defined as the flux per unit area per unit solid angle), and ω is the solid angle of the cone of acceptance or angular aperture of the instrument; ω is determined by the aperture stop of the system. Applied to a spectrometer, S is the area of the slit and ω is determined by the usable area of the collimator lens or mirror which is in turn determined by the effective area A (i.e. area projected perpendicular to the light beam) of the dispersive element. The luminosity of an optical instrument is defined as ϕ/B (Jacquinot, 1954).

Now if the angular aperture ω of a spectrometer is filled with light and the whole of the entrance slit is illuminated, then ϕ depends only on the luminance B of the slit (which is the same as the luminance of the light source illuminating it). Thus if the source of light is correctly aligned with the entrance slit-collimator system and is of a size and at a distance such as to fill, or more than fill (but see Section 3.3.5), the angular aperture of the instrument, there is no necessity for any condensing lens or mirror. However, if the source is too small or too far away to fill the angular aperture, then a lens is usually employed between the source and the entrance slit to ensure that the whole of the slit is illuminated by a light beam with an f-number to match that of the instrument. Let us suppose that the image of the source is formed in the plane of the entrance slit (although this is not always desirable particularly if even illumination of the slit is required), then it would appear that moving the lens nearer the source and, if necessary, correspondingly increasing its diameter to match the f-number, would increase the solid angle of collection and hence increase the flux collected. However, such action would also increase the image magnification and the net result would be to collect more light from a smaller area of the source. In fact, the flux collected in the angular aperture of the instrument is independent of the position and focal length of such a condensing lens provided the angular aperture and slit area are filled with light, since the luminance of the image cannot be greater than the luminance of the source.

Now the angular aperture $\omega = A/f^2$, where f is the focal length of the collimator or focusing lens or mirror (usually equal). Therefore $\phi = SAB/f^2$. However, if ϕ is the flux leaving the exit slit then we have to allow for the fact that the reflectivity of the mirrors or transmission of the lenses and the efficiency of the dispersive element are not 100%; thus $\phi = \tau(SAB)/f^2$, where τ is the transmission coefficient which may have a value of about 0·6.†

Now $S/f^2 = \alpha\beta$ where α and β are respectively the angular slit width and height. But $\alpha = D\Delta\lambda$, where D is the angular dispersion and $\Delta\lambda$ the wavelength bandwidth determined by the width of the slit. Hence

$$\phi = \tau AB\beta D\Delta\lambda. \qquad (3.16)$$

However, if we are dealing with broad-band spectra the luminance $B = B_\lambda \Delta\lambda$, if we assume that $\Delta\lambda$ is small enough for the luminance at a particular wavelength λ to be essentially constant over such a wavelength range. The output flux ϕ is then given by

$$\phi = \tau B_\lambda A\beta D(\Delta\lambda)^2$$

or

$$\left.\phi = \frac{A\tau B_\lambda \beta D\lambda^2}{R^2}\right\} \qquad (3.17)$$

in terms of the effective slit width limited resolving power R. Therefore, for a given bandwidth $\Delta\lambda$ (or effective resolving power R) and angular slit height β, ϕ depends only on the properties of the dispersive element, namely, its angular dispersion and effective area. For a grating instrument, ϕ therefore depends only on the order, number of grooves/mm, and area of the grating. *Note that, provided the ratio β of slit height to focal length is kept constant, ϕ is independent of the angular aperture or f-number of the instrument.* The f-number does not therefore affect the signal-to-noise ratio attainable when photoelectric or photoconductive detectors (which measure flux) are used.

If photographic detection is employed, however, it is the illumination (flux per unit area) in the plane of the exit slit, rather than the total flux, which determines the plate blackening. Now illumination $E = B\omega = BA/f^2$, i.e. the illumination is inversely proportional to the square of the f-number (as for a camera) and for monochromatic radiation (i.e. natural linewidth less than instrumental linewidth) is independent of the slit width. However, for a continuous or broad-band spectrum, increasing the slit width would increase the illumination since images of slightly different wavelength will overlap, but, of course, only at the expense of resolution.

However, for a given resolution on the film f is effectively fixed since the separation of images of different wavelength is governed by the linear

---

† The quantity $\phi/B\tau$ is sometimes used and is called the étendue.

dispersion fD. A larger slit width and hence higher illumination would therefore mean a worse resolution for a given dispersive element.

3.3.5. Stray Light

Stray light is that light which emerges from the exit slit which is not of the desired wavelength, i.e. the detected resultant flux of light of all wavelengths outside the wavelength bandwidth determined by the slit width which is centred on the desired wavelength at which the spectrometer is set. The stray light ratio is the ratio of the flux of such spurious radiation (as detected) to that of the desired or primary radiation (as detected). These definitions are, however, far from satisfactory, and a manufacturer's statement that the stray light ratio is, say, 10^{-5} can be misleading unless more precise details of the circumstances of measurement are given. Obviously, the stray light ratio measured at a particular wavelength setting will depend on the efficiency of the spectrometer at that wavelength, the spectral distribution of the source used, the slit width, and the spectral sensitivity of the detector used. It is common experience that the stray light ratio becomes much worse near the extremities of the spectral range where the grating efficiency is low, the source output is low, and the detector response is low. The stray light ratio is also increased by overfilling the angular aperture of the instrument.

The main source of stray light is light scattering by the optical components, particularly at gratings. Grating "ghosts" will obviously contribute to the background light but so also will any ruling error which is random rather than periodic, although random ruling errors will produce a background level which is much less wavelength dependent. However, even a grating with no ruling errors would scatter light from groove edges, etc. In addition, matt black surfaces are not perfectly absorbing and reflected light levels are measurable. Usually, the largest contribution to stray light comes from wavelengths which are relatively close to the nominal wavelength setting, since only small-angle scattering is necessary for such light to pass through the exit slit. Such small-angle scattering gives rise to "tails" in the slit function (see Fig. 3.11). Ideally, the slit function (i.e. the shape of the spectral profile when a monochromatic line is scanned) should be triangular, but diffraction effects and small-angle scattering tend to round off the profile.

In any particular case, stray light background can be estimated using a filter which will absorb the primary radiation but transmit other wavelengths which contribute to such a background. Poulson (1964) has suggested some aqueous solutions of alkali halides for use in the ultraviolet region below 300 nm. Using a monochromatic source such as a laser, it is possible to determine the stray light at any other wavelength by setting the spectrometer to that wavelength. Although this gives unambiguous measurements of the stray light background when the particular laser source is used, it does not, of course, give any meaningful indication of what stray light levels would be with, say, a continuous spectral source.

Where extremely low stray light levels are necessary, double spectrometers are used.

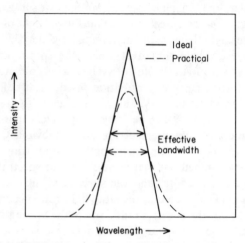

Fig. 3.11. Effect of diffraction and small-angle scattering on the slit function.

3.3.6. Double Spectrometers

In practice, the stray light ratio, which should be ideally zero, can be very significantly reduced by using a double spectrometer system; the stray light ratio of a double monochromator is the square of that of a single instrument of the same type used under similar conditions.

It is possible to use two separate monochromators in series such that the output from the exit slit of one is focused on to the entrance slit of the second. However, there are advantages to be gained by arranging that the exit slit of one is the entrance slit of the other and by having the rotation of the dispersive elements controlled by the same lead screw. In both arrangements it is possible to have either additive (usually double) or subtractive dispersion depending on the relative optical arrangement of the two parts.

In the double Czerny–Turner system shown in Fig. 3.12(a), double dispersion is achieved with the two gratings orientated in the same way and rotating in the same direction. If the second grating is mounted such that it has been rotated through 180° about the axis with respect to the first, subtractive dispersion will occur and the gratings will have to rotate in opposite directions for scanning purposes (Fig. 3.12(b)).

From Eq. (3.17) of Section 3.3.4, it appears that double dispersion would yield twice the flux throughput for a given wavelength passband compared with a single monochromator. However, the transmission factor τ for the double monochromator would be the square of that for the single instrument of the same type. Hence if, for example, $\tau = 0.5$ for the single instrument at a particular wavelength, the flux throughput of the double monochromator with double dispersion would be identical to that of a similar single instrument. (The slit width of the double monochromator would, of course, be twice that of the single instrument passing the same bandwidth.)

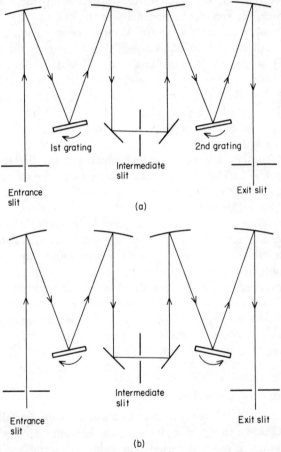

Entrance slit

1st grating

Intermediate slit

2nd grating

Exit slit

(a)

Entrance slit

Intermediate slit

Exit slit

(b)

Fig. 3.12.(a) Double Czerny–Turner spectrometer with additive dispersion. (b) Double Czerny–Turner spectrometer with subtractive dispersion.

When the optical arrangement is such that subtractive dispersion occurs n a double monochromator (as in Fig. 3.12(b)) then, provided the dispersion n the two halves is the same in magnitude, there is no dispersion at the exit slit; the light emerging is homogeneous in wavelength but nevertheless has a wavelength bandwidth determined by the dispersion of the first element and the width of the intermediate slit. At first, it may seem rather strange to employ two dispersive elements to produce no dispersion! However, such a system is in effect a variable passband filter with an exceptionally low background transmission outside the passband. But then so is a double-dispersion instrument and, moreover, the flux throughput of the latter is twice that of a comparable subtractive dispersion instrument for the same wavelength passband. Nevertheless, the subtractive dispersion instrument has some useful properties; the spectral homogeneity of the output beam ensures that different parts of a sample under excitation or of the photocathode of a detector receive light of identical wavelengths. Moreover, the passband is almost entirely determined by the intermediate slit width rather than entrance and exit slit widths. If the intermediate slit is much larger in width than the entrance and exit slits, the result is a large "square" passband and a concomitant large flux throughput (Lipsett et al., 1973).

If two separate, identical, single monochromators are used in series then the criteria for additive and subtractive dispersion are usually reversed compared with a combined double monochromator using the same internal optical arrangements. The reason is that the two separate monochromators are usually linked optically by a single lens which focuses the exit slit image from the first monochromator onto the entrance slit of the second (without magnification) with consequent image inversion. Hence the wavelength order or dispersion across the second entrance slit is reversed compared with that across the exit slit of the first monochromator. Often, however, there are alignment problems when this arrangement is used.

Double-pass monochromators are also available in which a single dispersive element is used twice by suitable optical design. Again double or subtractive dispersion may result depending on the optical arrangement. The stray light ratio of such a system is, however, usually no better than that of an ordinary single monochromator.

3.3.7. Calibration of Spectral Response

The wavelength or wavenumber calibration of commercially available spectrometers is normally displayed by a counter. However, such a calibration should be checked periodically using a spectral line source such as a mercury lamp. Although gratings are usually accurately kinematically mounted, the removal and subsequent replacement of such a grating often results in a small shift in the calibration; an interchange of gratings, even of the same type, will necessitate recalibration, but this is a relatively simple matter.

Spectral response calibration of spectrometers, sources, and detectors on the other hand, is neither as simple nor as accurate as wavelength calibration, and indeed in certain circumstances there are ways of avoiding the necessity for such calibration altogether.

The variation of intensity with wavelength of light emerging from the exit slit of a spectrometer depends on the output of the source and the efficiency of the spectrometer at different wavelengths. Both these parameters can vary over a wide range with variations in wavelength, and it is therefore necessary to correct for such variations when measuring excitation or absorption spectra. Similarly, for emission spectra, corrections must be made for variations in the efficiencies of the spectrometer and detector with wavelength. Methods for determination of these instrumental correction functions have been discussed at length in the literature (Christensen and Ames, 1961; Melhuish, 1962, 1973; Allison et al., 1964; Argauer and White, 1964; Eastman, 1966; Parker, 1968; Yguerabide, 1968; Budde and Dodd, 1971).

In absorption spectroscopy the necessity for spectral calibration is usually overcome by using a dual-beam technique. The monochromatic light is split into two beams, one passing through the sample and the other bypassing it or passing through a "blank" cell; the detector then alternately samples each beam. In excitation spectroscopy part of the light used for excitation may be viewed by a "non-selective" detector which gives an output proportional to the light intensity but independent of wavelength. The output from such a reference detector can then be used to correct for the incident intensity variations. Such a correction can be made automatically, using a dividing circuit or a ratio-taking recorder. It should be remembered that the time constants of the circuits processing the two signals prior to dividing should be well matched otherwise the resulting spectral profile may be distorted. Accurate zeroing of both signals is also essential when dividing.

The "non-selective" detectors most often used are the thermopile or some type of quantum counter. There is evidence, however, that thermopiles may not always have the ideal response in practice (Betts, 1965); sensitivity is also poor. Quantum counters, on the other hand, have a limited wavelength range and may deteriorate with time. A quantum counter can be any fluorescent sample that absorbs almost all the incident radiation falling upon it in a thin layer of the sample, and has a fluorescence spectrum and quantum efficiency (preferably high) which are independent of excitation wavelength. Integrating solutions of organic dyes or scintillators are often used for this purpose (e.g. 10^{-2} M Rhodamine B in ethylene glycol or 10^{-2} M PPO in cyclohexane). A microcrystalline deposited layer of sodium salicylate has been widely used as a quantum counter in the ultraviolet and VUV. Such a layer deposited on a phototube window renders the tube sensitive to short wavelength radiation which would otherwise be absorbed by the window Allison et al., 1964; Knapp and Smith, 1964). However, if a quantum counter is to be used in the visible region, it must of course absorb most visible radiation and consequentially emit in the far red or infrared, where even photomultipliers with S20 photocathodes have a much reduced sensitivity. (Since the photomultiplier is always viewing the same spectral distribution, i.e. the fluorescence spectrum of the quantum counter, the photomultiplier output is a wavelength-independent measure of the photon flux falling on the quantum counter.)

Silicon p–i–n photodiodes usually have a spectral sensitivity which is reasonably wavelength-independent except near the extremities of the sensitive range, and can often be used as a reference monitor in the visible and near infrared. A recent development has been the possible use of pyroelectric detectors, which are now commonly used as infrared detectors, for calibration purposes in the visible and ultraviolet regions (Phelan and Cook, 1973). Pyroelectric material is used to form a thin parallel plate capacitor by evaporation of thin metallic electrodes on either side; any change in temperature caused by absorption of incident radiation results in an electric charge on the plates of the capacitor. Since leakage counteracts any standing polarization change, modulation of the incident radiation is necessary. However, lock-in amplification can be used with a suitable very high impedance FET preamplifier. Usually, only part of the incident radiation is absorbed by the device, and therefore it is necessary to determine the wavelength dependence of this absorption. Pyroelectric detectors can be calibrated absolutely by electrical means, and details of such calibration are given in the reference quoted. Like the thermopile the pyroelectric detector is "non-selective" in its response to differing wavelengths of the incident radiation and measures radiant power.

In emission spectroscopy the photomultiplier current is given by $i_\lambda = S_\lambda I_\lambda$ where I_λ is the true "intensity"† of emission and S_λ is the overall spectral response factor of the spectrometer and detector. We shall define I_λ to be in units of photons/s/unit wavelength interval although it is seldom necessary to know its absolute values but rather its relative values. Spectral response calibration determines S_λ or its reciprocal. Such a calibration of the analyzing spectrometer and detector is usually carried out using a calibrated source. Such a source may be a continuous-spectrum light source plus monochromator calibrated by one (or more) of the methods just described. Alternatively, fluorescence spectra of well-known fluorescence standards may be measured and the necessary correction factors thereby determined (Crosby et al. 1973; Melhuish, 1973; Velapoldi, 1973).

When any calibration source is used it is well to ensure that the light is unpolarized; a "depolarizer" may be used (see Section 3.3.8), but the simplest method is to scatter the light from a freshly prepared MgO screen

---

† Strictly speaking, I_λ is photon flux per unit wavelength interval, whereas intensity is photon flux per unit solid angle.

Aluminized mirrors should of course not be used. In the wavelength range 350–900 nm, a tungsten filament lamp of known colour temperature is often used as a calibrated source. Such standardized lamps are commercially available and are usually operated at a colour temperature in the vicinity of 3000 K (usually 2854 K) when a prescribed calibrated current is passed through the filament. The radiant energy distribution with wavelength is calculated from Wien's formula. However, since colour temperature is critically dependent on the current i (approximately $\propto i^2$), the latter must be both accurately measurable and highly stable. Moreover, the lamp emission is, of course, calibrated in a particular direction.

Spectra are usually presented as plots of "intensity" against wavelength or wavenumber, but the actual shape of the spectral profile and relative intensities of spectral features will depend on what units the spectra are plotted in. The "intensity" may be in relative (or absolute) units of radiant power or number of photons/s, and although the latter units are more commonly used the former are equally correct. If E_λ is the radiant power/unit wavelength interval and I_λ is the number of photons/s/unit wavelength interval, then clearly $I_\lambda = \lambda E_\lambda/hc$. Since h and c are the usual universal constants, to convert relative radiant power ordinates to photons/s ordinates it is necessary to multiply ordinates by λ for each point.

Most grating spectrometers scan linearly in wavelength rather than wavenumber, and therefore if spectra are to be plotted in terms of wavenumber or transition energy, the ordinates of photons/s/unit wavelength interval I_λ must be converted to photons/s/unit wavenumber or energy interval $I_{\bar{\nu}}$.

Now $I_\lambda \, d\lambda = I_{\bar{\nu}} \, d\bar{\nu}$, i.e. we do not change the integrated intensity simply by changing the units, and since $\bar{\nu} = 1/\lambda$, $d\bar{\nu} = -d\lambda/\lambda^2$, therefore

$$I_{\bar{\nu}} = \lambda^2 I_\lambda \qquad (3.18)$$

ignoring the negative sign which simply emphasizes the fact that λ increases as $\bar{\nu}$ decreases. Thus a conversion from I_λ to $I_{\bar{\nu}}$ requires ordinates to be multiplied by λ^2 for each point.

When calibration is carried out, it is therefore essential to ascertain whether the calibration is in terms of radiant power (thermopile, etc.) or photons/s (quantum counter). Black-body curves are calculated and usually presented in terms of radiant power. However, for convenience, Table I gives the spectral output of a standard tungsten lamp, operated at a colour temperature of 2854 K in terms of photons/s/unit wavelength interval.

Correction factors for emission spectra can be applied point by point manually or, more usually nowadays, by computer (Costa et al., 1969; Yuan and Wild, 1973). However, unless the detector output is on line to the computer, it is a great convenience to have the corrections applied automatically. A number of commercial systems now have this facility, although

TABLE I

Relative Spectral Energy Distribution Q_λ of a Black Body at 2854 K, in Quanta s^{-1} Unit Wavelength Interval^{-1}, against Wavelength λ

λ (nm)	Q_λ	λ (nm)	Q_λ	λ (nm)	Q_λ
340	55·00	460	770·5	580	2946
345	64·00	465	825·4	585	3060
350	74·50	470	883·6	590	3186
355	85·91	475	969·0	595	3308
360	99·00	480	1032	600	3432
365	113·9	485	1096	605	3557
370	131·1	490	1166	610	3697
375	148·5	495	1247	615	3831
380	167·2	500	1325	620	3956
385	187·9	505	1404	625	4094
390	210·2	510	1494	630	4227
395	233·4	515	1586	640	4486
400	260·0	520	1674	650	4758
405	288·8	525	1775	660	5036
410	320·6	530	1866	670	5313
415	354·0	535	1963	680	5596
420	390·6	540	2063	690	5879
425	427·6	545	2164	700	6160
430	468·7	550	2266	710	6440
435	509·0	555	2381	720	6721
440	558·8	560	2464	730	7001
445	607·4	565	2599	740	7282
450	659·3	570	2719	750	7553
455	716·6	575	2829	760	7828

at some considerable cost, and several systems have been described in the literature (Haugen and Marcus, 1964; Hamilton, 1966; Cundall and Evans, 1968; Witholt and Brand, 1968; Pailthorpe, 1975).

3.3.8. Grating Anomalies

The spectral transmission characteristic for a grating instrument is not always a smooth curve; sometimes there are one or more sharp cusps in the curve (Fig. 3.13). In fact, if the spectral transmission curve is determined for polarized light when the plane of polarization is parallel to the slits or grating grooves (P-polarization), it will probably be found that it is quite different from the curve determined when the light is polarized perpendicular to the grating grooves (S-polarization). In fact, the spectral transmission curves for P- and S-polarizations intersect at the grating blaze wavelength

The spectral shape of the transmission curve for unpolarized light is, not surprisingly, intermediate between these two cases (Fig. 3.13). Such variation of the spectral shape of the transmission curve can lead to large errors in spectral correction if the state of polarization of the light to be spectrally analyzed is not known. A small error in wavelength or in the determination of the zero intensity level can result in the cusp appearing on all "corrected" spectra.

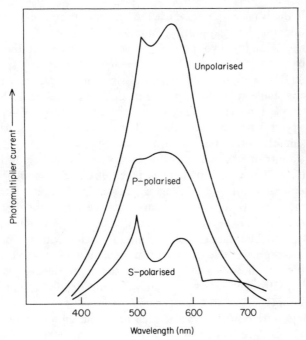

Fig. 3.13. Effect of polarization on the transmission characteristics of a typical grating spectrometer using a tungsten lamp as source.

These peaks or dips in the spectral transmission curve are called grating or Wood anomalies. Unlike ghosts they are a property of the grating which is not in any way associated with imperfections in the rulings. As long ago as 1907 Lord Rayleigh developed a theory which predicted at what wavelengths such anomalies would occur but, as yet, no prediction can be made concerning whether the anomaly will give rise to a sudden increase or decrease in intensity. Moreover, Rayleigh's theory applies only to S-anomalies (i.e. those which occur when the plane of polarization is perpendicular to the grooves). The basic idea is that when an order disappears (e.g. when for normal incidence the diffraction angle exceeds 90°), the light formerly in that order is redistributed among the remaining orders. Whether such a redistribution leads to an increase or decrease in intensity in one of the remaining orders is dependent on the phase differences involved. For further discussion and references to the subject of Wood anomalies see, for example, Stewart and Galloway (1962).

If a Wood anomaly is suspected then its presence can be confirmed using Rayleigh's formula:

$$n\lambda_R = d(1 \pm \sin i)\dagger \qquad (3.19)$$

where λ_R is the wavelength at which the anomaly occurs, i is the angle of incidence, and n is the disappearing order.

There are essentially two possible ways of overcoming the problems presented by such grating anomalies; the light entering the spectrometer should preferably be either fully plane-polarized or completely unpolarized. If the light is made fully plane-polarized then it is preferable to polarize parallel to the grating grooves (P-polarization) since this usually gives the smoothest spectral transmission curve (Fig. 3.13). In the visible region a sheet of suitably orientated polaroid suffices. However, this method necessarily results in a considerable fraction of the incident light being lost.

Alternatively, if the incident light is to be effectively unpolarized, some form of depolarizer or polarization scrambler is required. Such devices produce what is known as pseudodepolarization, i.e. a mixture of polarization forms which are, for the purposes in question, indistinguishable from unpolarized light. The state of polarization of a light beam can be varied in space, in time or in wavelength. Polarization scramblers that vary the state of polarization with wavelength, such as the Lyot depolarizer, are not suitable for spectrophotometry since they will not depolarize narrow bandwidth radiation (Clarke and Grainger, 1971b).

Polarization variations in time, if such are rapid enough, are acceptable in most circumstances and can be achieved by a rotating quarter-wave plate followed by a half-wave plate rotating at twice the speed (Billings, 1951). However, a simpler and more convenient polarization scrambler is the quartz-wedge type in which the state of polarization is varied spatially across the beam (Hughes, 1960).

3.3.9. Derivative Spectroscopy

Recording of the derivative of the signal, with respect to some physical variable, rather than the signal itself, has long been common practice in the field of magnetic resonance spectroscopy (e.g. ESR and NMR). However, there are advantages to be gained by recording optical derivative spectra,

---

† If the grating is coated with a dielectric of refractive index μ then the formula becomes $n\lambda_R = d(\mu \pm \sin i)$.

i.e. the derivative of intensity with respect to wavelength as a function of wavelength (Bonfiglioli and Brovetto, 1964). Obviously a derivative spectrum does not contain more information than a normal spectrum but fine structural features are often more readily apparent in the derivative spectrum; for example, a weak peak in the tail of a strong peak would be more easily discernible. Furthermore, the actual position of the centre of a spectral band is more easily determined from a derivative spectrum.

However, unlike ESR spectra, the determination of an optical derivative spectrum is experimentally much more difficult than that of a normal spectrum. In many other types of optical modulation spectroscopy (e.g. electroreflectance, piezoreflectance, etc.), the modulation is applied to the sample itself and therefore does not affect the incident intensity or detector sensitivity. In the case of wavelength modulation, however, the incident intensity, detector sensitivity, and other factors will also be modulated if such are wavelength-dependent; fine structure in the light source output will be particularly troublesome (Cardona, 1969).

The theory of wavelength modulation has been expounded several times elsewhere (e.g. Cardona, 1969; Welkowsky and Braunstein, 1972); only the briefest outline consistent with clarity will therefore be given here.

If the reflectivity, say, of a sample is to be measured as a function of wavelength, then the signal measured $S(\lambda)$ will be given by

$$S(\lambda) = R(\lambda) I_0(\lambda) \tag{3.20}$$

where R is the sample reflectivity and $I_0(\lambda)$ the incident intensity as measured (i.e. including spectrometer transmission, detector response, and other wavelength-dependent factors).

The derivative of $S(\lambda)$ with respect to wavelength is given by

$$S'(\lambda) = R'(\lambda) I_0(\lambda) + R(\lambda) I_0'(\lambda) \tag{3.21}$$

It is the second term on the right-hand side of this equation which is the source of experimental difficulties. If we divide Eq. (3.21) by Eq. (3.20) and rearrange, we obtain

$$\frac{R'(\lambda)}{R(\lambda)} = \frac{S'(\lambda)}{S(\lambda)} - \frac{I_0'(\lambda)}{I_0(\lambda)} \tag{3.22}$$

If we can in principle measure both terms on the right-hand side, then $R'(\lambda)/R(\lambda)$ can be determined by subtraction. In practice, fine structure in I_0 will appear in R' unless almost perfect compensation is achieved.

The actual wavelength modulation can be done in a number of ways; the most obvious is to vibrate the spectrometer slit in a direction parallel to the slit width. Oscillating or rotating plates inclined at an angle to produce a variable lateral shift of the light beam have also been used. However, the most common method is to use a vibrating mirror with a modulation frequency of about 1 kHz to vibrate the light beam across the spectrometer slit. At a particular wavelength setting of the spectrometer, λ_0, the wavelength output will be $\lambda_0 + \Delta\lambda \cos \omega t$, where $\Delta\lambda$ is the wavelength modulation amplitude and ω the angular frequency of the modulation. If we substitute $\lambda_0 + \Delta\lambda \cos \omega t$ for λ in Eq. (3.20) and expand as a Taylor series, we obtain

$$S(\lambda_0, t) = R(\lambda_0) I_0(\lambda_0) + \Delta\lambda \cos \omega t (R I_0' + R' I_0) \tag{3.23}$$

where terms in higher orders of $\Delta\lambda$ have been ignored on the assumption that $\Delta\lambda$ is small.† This equation can be rewritten

$$S(\lambda_0, t) = S(\lambda_0) + \Delta\lambda \cos \omega t S'(\lambda_0) \tag{3.24}$$

The measured signal therefore consists of a dc term and an ac term oscillating at the modulation frequency. The amplitude of the latter, which can be measured using a lock-in amplifier, is a measure of $S'(\lambda_0)$. The dc term

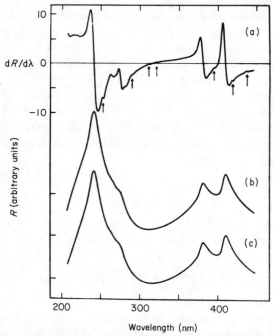

Fig. 3.14. Ultraviolet reflectance spectra of GaAs at 4 K: (a) derivative spectrum: arrows indicate residual Xe lines; (b) conventional spectrum as measured; (c) conventional spectrum obtained by numerical integration of (a). (From Zucca and Shen, 1973.)

† The series converges rapidly since it can be shown that higher-order derivatives of the intensity decrease dramatically with increasing order.

$S(\lambda_0)$ can also be measured and used to normalize S' (Welkowsky and Braunstein, 1972). A double-beam system is essential in order to compensate for the I_0'/I_0 term. The reference beam effectively has $R = 1$ and $R' = 0$ in Eq. (3.22), thus enabling I_0'/I_0 to be measured and simultaneously subtracted from S'/S. The validity of neglecting higher order terms in Eq. (3.23) can be checked experimentally by observing whether the measured amplitude of the ac term is a linear function of $\Delta\lambda$ in the vicinity of the value of $\Delta\lambda$ used.

Recently Zucca and Shen (1973) have described a wavelength modulation spectrometer which uses a feedback loop instead of subtracting the two normalized signals. It is claimed that this method leads to better compensation. Figure 3.14 shows the derivative and normal reflection spectra of GaAs in the ultraviolet measured with this spectrometer.

3.4. DETECTORS AND SIGNAL RECOVERY

3.4.1. Detectors

In most circumstances the detector is the key factor in luminescence measurements. Depending on the measurement the detector's wavelength or time response or its signal-to-noise ratio will be the limiting factor. Though there are a number of different types of photodetector, the photomultiplier is almost universally used in luminescence work because of its outstanding performance in most respects. However, in the last few years semiconductor photodiodes have been considerably improved and now also offer good performance, with some advantages in wavelength response, size, robustness, and power supply. General discussions of photodetectors are given in Engstrom (1963), Smith (1965), Anderson and McMurty (1966), Ambroziak (1968), Dance (1969), Melchior et al. (1970), Seib and Aukerman (1973) and Hamilton (1977). Photomultipliers are discussed more particularly in RCA (1970) and EMI (1970) and photodiodes in EG and G (1967, 1973) and Hamstra and Wendland (1972). Only photomultipliers and photodiodes will be considered here.

The wavelength response of a photomultiplier depends primarily on the material of the photocathode, but at short wavelengths this is limited by the optical transmission of the envelope material. There are many different types of photocathode (EMI, 1970; RCA, 1970), but a considerable number of these differ in only minor respects. The response of some representative types is shown in Fig. 3.15, where the scales show sensitivity and quantum efficiency. Figure 3.16 shows the relative response of silicon and germanium photodiodes. It is clear that any broad spectrum will be considerably distorted by these variations in response and other than for simple comparative work corrections must be applied. Techniques for carrying this out are

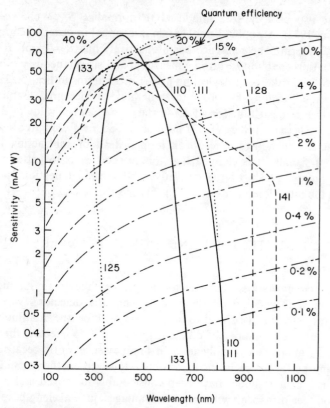

Fig. 3.15. Response of some photocathodes as a function of wavelength. The type number designations are: 110 = multi-alkali (S20); 111 = multi-alkali; 125 = CsTe; 128 = GaAs; 133 = bi-alkali; 141 = GaInAs. (Courtesy RCA Solid State Division.)

discussed in Section 3.3.7, but determination of the correction factors presents some difficulty.

Photomultipliers have two common configurations, the side-window type with solid photocathode and the end-window with semi-transparent photocathode. The former are generally smaller, cheaper, operate at lower overall voltage, and are the ones usually used in commercial fluorimeters. The end-window types are more expensive and operate at higher voltages, but have higher gains and quantum efficiency and a greater range of photocathodes. Figure 3.17 shows the construction and voltage divider network for such tubes. The higher gain and quantum efficiency are important factors when weak signals are encountered, and also allow the detection of individual photons. This means that the output signal can be measured in two ways—either counting the individual pulses or integrating them into an average current. The latter is considerably simpler and in many cases quite satis-factory, but pulse counting has an advantage if the data are to be digitally processed since it is directly available in digital form. These techniques are considered further in Section 3.4.5.

The time response of detectors is important when measuring lifetimes, time-dependent spectra, or when using pulse counting. In photomultipliers the limiting factor is the width of the output pulse originating from a single

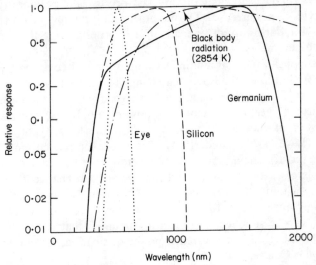

Fig. 3.16. Relative spectral response of silicon and germanium photodiodes. (After Hamilton, 1977.)

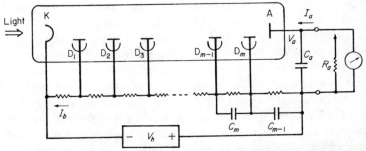

Fig. 3.17. Schematic diagram of a photomultiplier. The photocathode K and dynodes D_n are maintained at the desired voltages by means of the resistor chain connected to the high voltage power suppl·. For pulse operation the lower resistors may be bypassed by capacitors C_D. Output current flows from the anode A to a current meter M or, alternatively, through a local resistance R_L in which case M will be a voltmeter. The stray capacity C_A at the anode will limit the frequency response of the system. (Adapted from Hamilton, T. D. S. (1977). "Handbook of Linear Integrated Electronics for Research." McGraw-Hill, New York.)

photoelectron leaving the photocathode. The width depends on the detailed structure of the multiplying stages, which are conveniently divided into two types—focused and unfocused. In the focused structures electrons follow well-defined paths with minimum variation of path length and hence a minimum transit-time spread (≈ 1 ns). In the unfocused types (commonly having venetian-blind dynodes) there can be substantially greater variation in path and hence a greater spread of times (≈ 20 ns) and wider output pulse. Thus for measurement of very fast processes, e.g. fluorescence lifetimes, focused types are used. On the other hand, the unfocused types tend to have lower dark current and are thus appropriate for the more usual spectral measurements.

Power supplies for photomultipliers require careful consideration. For a typical photomultiplier the fractional change in gain will be about 10 times the fractional change in overall applied voltage. Thus for good measurement repeatability a very well-stabilized power supply is necessary. The characteristics of the supply may also affect the linearity of the photomultiplier (Lush, 1965; Land, 1971). The anode output current I_a affects the dynode resistor chain current I_b. If I_a is too large I_b, and hence the interdynode voltages, will be altered and the overall gain changed. As a guide, for linearity within 1%, I_b should be $> 100\ I_a$. For a given maximum I_a the power supply must be able to provide the necessary I_b. However, even if the required I_b can be supplied, there is a maximum I_a that should be used set by allowable photocathode currents, power dissipation, and fatigue effects (Keene, 1963; Coates, 1975).

In complete darkness there will still be some anode current. This "dark" current arises from several sources, such as field emission from sharp points on electrodes, radioactivity, cosmic rays, and most importantly from thermionic emission from the photocathode (Young, 1966; Jerde et al., 1967; Pettifer and Healey, 1974). The latter has an approximately exponential dependence on temperature so that cooling can reduce this contribution substantially. Cooling below about $-20\ ^\circ$C produces little further improvement for most photocathodes as other sources then become predominant (Oliver and Pike, 1968; Foord et al., 1969, 1971; Davies, 1972). Cooling should be undertaken with care so as not to cause thermal stresses and possible fracture, and to avoid condensation on the tube and pins which can greatly increase noise owing to leakage currents. If the optical image is significantly smaller than the photocathode area, then an alternative technique may be used to reduce the thermionic contribution. By placing a pancake coil adjacent to the photocathode the effective area of the photocathode can be varied by controlling the magnetic field produced by the coil (Knight et al., 1968; Topp et al., 1969). Dark current is in itself not a limitation when measuring small signals as in principle it can be backed-off, but fluctuations in this, and other sources of noise, cannot be compensated. In practice there will be an optimum overall voltage for best signal-to-noise ratio (SNR), above which the noise will increase faster than the signal. This is not necessarily the best operating voltage however, since the noise performance of any subsequent circuits must also be considered. Such considerations are examined in Faulkner (1975). Nevertheless, the photo-multiplier is a very good low-noise wide-band detector indeed and its gain should be used if possible rather than adding an external amplifier.

There are two fundamental sources of noise in the photomultiplier—the statistical probability of emission of a photoelectron from the photo-cathode and the fluctuation of secondary emission at the dynodes. There is a further noise component due to the random arrival of photons at the photocathode; what may be called noise-in-signal and which we cannot do anything about. This noise-in-signal is not significant when detecting low frequency electromagnetic signals, as detectors in this region cannot remotely approach the detection of single photons. The effects of these noise sources are considered in Eberhardt (1967), Oliver and Pike (1968), Foord et $al.$ (1969), RCA (1970), and Ingle and Crouch (1972). Some results are quoted here. If $\langle n_p \rangle$ is the average number of photons arriving in the observing time then the SNR for the photon beam is

$$\text{SNR}_p = \langle n_p \rangle^{\frac{1}{2}} \tag{3.25}$$

After detection by a photocathode of quantum efficiency η the SNR for the photoelectrons is

$$\text{SNR}_{pe} = (\eta \langle n_p \rangle)^{\frac{1}{2}} \tag{3.26}$$

so that for a good efficiency $\eta = 0.3$ the input ratio is reduced to about 55%. After m stages of multiplication with a dynode secondary emission factor δ the SNR at the anode is

$$\text{SNR}_a = \left(\eta \langle n_p \rangle \frac{\delta - 1}{\delta} \right)^{\frac{1}{2}} \tag{3.27}$$

For normal operation $\delta \doteq 4$ which degrades SNR_a to about 75% of SNR_{pe}. Recently developed semiconductor dynodes (Morton et $al.$, 1969; RCA, 1970) with $\delta \doteq 40$ would make $\text{SNR}_a \doteq \text{SNR}_{pe}$ in which case the multiplication process is essentially noiseless. It is this low noise contribution that makes the photomultiplier such an attractive detector.

The same results are obtained by considering the shot noise of the various currents. The output current will flow in some anode load resistor R_a with shunt capacity C_a. To maintain the SNR the Johnson noise from R_a should be negligible, which requires:

$$\delta^m \gg 0.1/(2I_a R_a) \tag{3.28}$$

If this is satisfied then the maximum ratio for a photon arrival rate I_p is

$$\text{SNR}_a(\text{max}) = (2\eta I_p R_a C_a)^{\frac{1}{2}} \qquad (3.29)$$

This allows the selection of operating conditions. For a desired SNR_a the compromise is between minimum intensity I_p and maximum time constant $R_a C_a$—increasing the latter requires increasing the recording or scan time.

Although photomultipliers still offer superior performance to the presently available solid-state photodetectors, considerable progress has been made in recent years in the development of photodiodes. The most commonly used type is the silicon p–i–n photodiode which has a sensitive range extending from the ultraviolet well into the infrared (1·1 μm) (Fig. 3.16). Provided the p-layer, through which incident light must pass, is made thin enough, a quantum efficiency in excess of 30% can be maintained into the ultraviolet down to 250 nm. Quantum efficiencies of 70% or more can be achieved in the visible and near infrared. Photons are absorbed in the depletion layer of the junction, producing electron–hole pairs. If the junction is reverse biased the electric field sweeps the electrons to the n-side and the holes to the p-side. The photocurrent produced will be proportional to the optical generation rate of electron–hole pairs. Although the width of the depletion layer in a p–n junction can be controlled by doping levels, it can be made effectively much wider by having an intrinsic silicon layer between the p- and n-layers. A wide effective depletion layer results in a high quantum efficiency and in a smaller junction capacitance. This latter parameter together with the parasitic resistance of the p-layer (which is higher for a thinner p-layer) determine the minimum time constant of the detector circuit. There is also an upper limit to the width of the depletion layer governed by the time required for the carriers to drift to the p or n side. There are therefore trade-offs between speed of response, ultraviolet sensitivity, and quantum efficiency. Silicon p–i–n photodiodes have been made with time constants as short as 10 ps but this is not compatible with good ultraviolet response.

The p–i–n photodiode has several advantages over a photomultiplier tube. It is very much smaller and lighter, requires only a very modest low voltage supply and is not appreciably affected by magnetic fields, over-loading, or variations in temperature. On the other hand, it suffers from the same problems as a simple photo-emissive cell since it requires a high gain, high input resistance amplifier for sensitive work, which precludes fast response. Moreover, although p–i–n photodiodes can be made with very low dark currents, such are very small-area devices (~ 1 mm²). Devices with a large sensitive area of ~ 100 mm² often have dark currents approaching 1 μA. Dark noise currents from large area devices will therefore considerably limit the lower light levels at which these devices are useful.

Perhaps the most important challenge to the photomultiplier comes from the development of avalanche photodiodes (APD) where the multiplication of carriers is analogous to the electron multiplier in a photomultiplier. These are operated at high enough reverse bias to cause avalanche multiplication of the photogenerated carriers to give high gain with fast and linear response. However, the gain is much lower than a photomultiplier so that the performance of the following amplifier is of much greater importance. If M is the multiplication factor for the APD the noise current i_N^2 is found to increase as M^d, where d is 2·3 for silicon and 3 for germanium. This means that the noise current increases faster than the signal current, $i_S^2 \propto M^2$, but increasing M by increasing applied voltage can still give an improvement of SNR since the noise of the following amplifier will dominate at low M. Gains of several hundred can be achieved and an optimum value of M derived (Barelli, 1973). Such values of M imply working near breakdown voltage where M varies rapidly with voltage and depends significantly on temperature (Nishida and Nakajima, 1972). Stabilization is most effectively achieved by using matched dual devices with one diode masked from the light and operated at constant current. The changes in voltage with temperature across this diode are monitored and used to control the bias voltage applied to the active diode (EMI).

Both photomultipliers and APDs operate as current sources, i.e. have a high output impedance. Most simply the current can be fed directly to a galvanometer, or more commonly, when chart recording is required, through a high valve resistor across which the voltage is measured by means of a high input resistance amplifier. However, this is not the optimum technique to use, a transimpedance amplifier being capable of better performance (Faulkner, 1975).

Beyond 1 μm, the most common type of detector used is the photoconductive cell. In the range 1–3 μm lead sulphide cells are commonly used although their relatively slow response (~ 100 μs) can be a serious disadvantage. Indium antimonide, on the other hand, has a much faster response ($\sim 0\cdot1$ μs) and can be used to beyond 6 μm but is generally much less sensitive than PbS unless cooled to very low temperatures. PbS cells are usually current-noise limited ($1/f$-noise) and therefore the radiation is chopped at a frequency of about 1 kHz; InSb cells are usually Johnson (thermal) noise limited (Conn and Avery, 1960; Smith et al., 1968).

3.4.2. Signal Recovery

When signals are large there is little difficulty with noise. However, in many of the more interesting situations the signal is comparable with or less than the noise, so that special techniques must be used to extract the signal. In order to be able to decide on the most appropriate technique, some understanding of the characteristics and sources of noise is necessary (Bell, 1960; St. John et al., 1966; Baxendall, 1968; Betts, 1970; Inst. of Physics, 1973; Usher, 1974; Hamilton, 1977). Fundamental noise arises from two sources, referred to as Johnson and shot noise. Any system will exhibit random fluctuations in the variables or parameters which specify its state, for example, energy, voltage, or current. For a system in thermal equilibrium then the thermal fluctuations can be derived from thermodynamic arguments without in any way considering the actual nature of electricity. This thermal or Johnson noise manifests itself as a noise voltage across any resistance R at temperature T given by its mean squared value

$$\langle v_{\mathrm{NR}}^2 \rangle = 4kTR\,\Delta f$$

or

$$v_{\mathrm{NR}} = 0{\cdot}13[R\,(\mathrm{k}\Omega) \times \Delta f\,(\mathrm{kHz})]^{\frac{1}{2}}\ \mu\mathrm{V\ r.m.s.}\quad \text{at 300 K}$$

$$\left.\right\} \qquad (3.30)$$

where k is Boltzmann's constant and Δf is the frequency interval over which the noise is measured. Thus, for a given resistance R, the noise can only be reduced by decreasing the temperature T or the frequency bandwidth Δf.

The fact that an electrical current does consist of a stream of individual charges gives rise to a second source of random fluctuation called shot noise. For an average current $\langle I \rangle$ the mean squared noise current in a frequency interval Δf is

$$\langle i_{\mathrm{N}}^2 \rangle = 2q\langle I \rangle\,\Delta f$$

$$i_{\mathrm{N}} = 5{\cdot}7 \times 10^{-4}[I\,(\mathrm{mA}) \times \Delta f\,(\mathrm{kHz})]^{\frac{1}{2}}\ \mu\mathrm{A\ r.m.s.}$$

$$\left.\right\} \qquad (3.31)$$

where q is the electronic charge. For a given current I this noise can only be reduced by decreasing Δf.

There is a further form of noise usually referred to as flicker or $1/f$ noise, which has various not well-understood origins probably of a technical rather than fundamental nature. In this category may also be included low frequency noise commonly referred to as drift and such effects as burst or popcorn noise sometimes found in semiconductors and integrated circuits. The important feature is its frequency spectrum, the noise per unit bandwidth increasing approximately as $1/f$. Johnson and shot noise, on the other hand, have a white spectrum in that there is the same noise power in an interval Δf at any frequency f, so that as far as these are concerned there is no advantage to be gained from working at any particular frequency; $1/f$ noise, however, can be discriminated against by operating at high enough frequency that this noise is insignificant relative to the white noise. As a guide, the crossover usually occurs in the region 100 Hz to 1 kHz.

As it is difficult and inconvenient to reduce the temperature by a significant factor this is seldom done. (Note, however, that this does not apply when thermionic emission is involved (see Section 3.4.1) since in this case the emission is an exponential function of the temperature.) The prime factor for reducing noise is then reduction of bandwidth Δf by some form of narrow band filter. In principle Δf can be made as small as necessary but in practice there are two limitations. First, as the filter bandwidth becomes narrower the requirements on the frequency stability of the signal are correspondingly increased. Secondly, the time it takes a filter to respond to an input increases as the bandwidth decreases. Thus the narrower the filter the slower the signal may be allowed to change if distortion is to be avoided, i.e. improvement in signal-to-noise ratio (SNR) is bought with time.

Any detection system will add noise to the input, which will itself have a noise component. Consideration must therefore be given to the degree to which the input SNR is degraded by the detection system. Optimum operation of the optical detector has been considered in Section 3.4.1 and there may be an additional contribution from any following amplifier. The degradation of SNR due to a system is commonly stated in terms of the noise figure F of the system. This is defined as the ratio of the input SNR to the output SNR and is usually given in dB. For an amplifier of gain A with effective signal source resistance R_S the noise figure is given by

$$F = 1 + \frac{\langle v_{NO}^2 \rangle}{4A^2 kTR_S \Delta f} \qquad (3.32)$$

where v_{NO} is the output noise voltage; $F = 1$ (0 dB) represents a noiseless amplifier and a value less than 2 (3 dB) is regarded as a low noise. A noisy amplifier is often characterized by input voltage and current noise generators or their equivalent noise resistances R_{NV} and R_{NI} (which will be functions of frequency) using Eq. (3.30) and Eq. (3.31) (Faulkner, 1968, 1975; Hamilton, 1977). The noise figure is then given by

$$F = 1 + \frac{R_S}{R_{NI}} + \frac{R_{NV}}{R_S} \qquad (3.33)$$

Thus resistance in series with the input (R_{NV}) should be small and that in parallel (R_{NI}) large relative to the source resistance R_S to achieve a low noise figure. From Eq. (3.33) the optimum source resistance for a given amplifier is readily found to be

$$R_{S \, opt} = (R_{NV} R_{NI})^{\frac{1}{2}} \qquad (3.34)$$

with

$$F_{min} = 1 + 2 \left(\frac{R_{NV}}{R_{NI}} \right)^{\frac{1}{2}} \qquad (3.35)$$

and

$$F - 1 = \tfrac{1}{2}(F_{min} - 1) \left[\frac{R_{S \, opt}}{R_S} + \frac{R_S}{R_{S \, opt}} \right] \qquad (3.36)$$

The low noise capability of an amplifier should be assessed in terms of $F_{\min}$, and if R_S is not open to choice F may be found from Eq. (3.35). Matching techniques can be used to improve F in the latter circumstance but transformers should be avoided unless there is some special requirements. An alternative is to use a number of amplifiers in parallel though this is only practicable for a small transformation ratio. In most laboratory experiments the source resistance, e.g. a photomultiplier anode load, is open to choice so the problem does not arise.

Many amplifiers make use of negative feedback because of the considerable advantage this provides such as gain stability and low output impedance. Simple application of negative feedback does not improve the noise figure and may in fact degrade it. However, this does not mean that a lower noise figure cannot be obtained using feedback than without it. On the contrary, when all the performance parameters are taken into account it is possible to achieve a considerably better performance with negative feedback than with a straight amplifier that otherwise gives the same performance (Faulkner, 1968). In the present context it may be noted that the photomultiplier is a very good low noise amplifier indeed (see Section 3.4.1) and it is often unnecessary to use any further additional amplifier. If it is, then the use of a high resistance anode load is undesirable, a feedback current, or transimpedance, amplifier being capable of substantially better performance (Faulkner, 1975).

A minor but sometimes confusing point is that of noise bandwidth relative to signal bandwidth. Taking a simple RC low pass filter characteristic the signal bandwidth refers to the -3 dB point which occurs at a frequency:

$$f_l = \frac{1}{2\pi RC} = \frac{1}{6 \cdot 3 RC} \tag{3.37}$$

Noise bandwidth, on the other hand, refers to the "square" passband which gives the same noise contribution as the actual response curve (Fig. 3.18). The difference arises due to the noise contribution in the cut-off region which extends nominally to infinite frequency. The equivalent cut-off frequency is found to be (Usher, 1974):

$$f_N = \frac{1}{4RC} \tag{3.38}$$

which is a little greater than f_l and is the value to be used in Eq. (3.30) and Eq. (3.31).

The design of low noise amplifiers is discussed in Faulkner (1968), (1975), Motchenbacher and Fitchen (1973) and Hamilton (1977). For bipolar transistors $R_{S\,\text{opt}}$ and $F_{\min}$ are given approximately by

$$R_{S\,\text{opt}} \doteq r_e(\beta_0)^{\frac{1}{2}}\ \Omega, \quad F_{\min} \doteq 1 + \frac{1}{\beta_0^{\frac{1}{2}}} \tag{3.39}$$

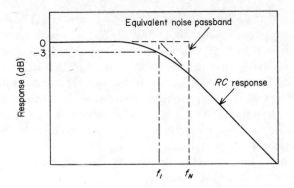

Fig. 3.18. Relation of noise to signal bandwidth.

where $r_e = 26/I_E$ is the emitter resistance for a given current I_E (in mA) and β_0 is the dc current gain. For typical values the optimum source resistance is in the range up to a few hundred $k\Omega$ and the transistor will be operating as a voltage amplifier (i.e. $R_{in} \gg R_S$) rather than the more usual current amplifier.

For junction field-effect transistors it is found that

$$R_{S\,opt} = 0\cdot185/(g_m I_G)^{\frac{1}{2}}\ \Omega, \quad F_{min} = 1 + 7\cdot2(I_G/g_m)^{\frac{1}{2}} \qquad (3.40)$$

where g_m is the transconductance (A/V) and I_G is the reverse gate leakage current in amperes. (IGFETs are rather poor for low-noise use.) For typical values the values of $R_{S\,opt}$ come out to be larger than a few hundred kohms though lower values are becoming possible with newer JFETs. The values of F_{min} turn out to be substantially better than for bipolar transistors so JFETs should be chosen when the source resistance allows.

Assuming detector and amplifier have been designed properly and are operating at optimum conditions there are several techniques available for improving the SNR. Two of these, lock-in amplification and signal averaging, are examples of correlation (Lee, 1960) while the third, photon counting, depends on the photon nature of light and the outstanding characteristics of photomultipliers. These three approaches will be discussed in turn (Hamilton, 1977) so that a better informed choice between them can be made.

3.4.3. Lock-in Amplification

The technique for escaping from low frequency noise and from the relative drift with very narrow bandwidth filters is known variously as lock-in, coherent, or phase-sensitive detection (Danby, 1970; Tekelec/Airtronic, 1970; Smith, 1972; Abernethy, 1973). The signal sought is modulated at a suitable frequency ω_c selected to make $1/f$ noise insignificant. This is commonly achieved by chopping the light beam, by wavelength modulation (Bonfiglioli and Brovetto, 1964; Evans and Thompson, 1969; Hager and Anderson, 1970; Zucca and Shen, 1973) or by modulating the sample directly (Cardona, 1969). A reference signal obtained from the modulating oscillator operates a synchronous switching detector which acts as a rectifying bandpass filter locked to the signal frequency since the same oscillator drives both. The problem of relative frequency drift is thus eliminated. The output of the detector is smoothed by means of a low pass RC filter whose bandwidth determines the overall bandwidth and hence the SNR improvement of the system.

The action of the switching detector is that of multiplication. For a signal e_s and a reference (or carrier) e_c:

$$e_s = E_s \cos(\omega_c t), \quad e_c = E_c \cos(\omega_c t + \phi) \tag{3.41}$$

then the output from a multiplier will be

$$e_m = \tfrac{1}{2} K E_s E_c [\cos(2\omega_c t + \phi) + \cos \phi] \tag{3.42}$$

The output filter eliminates the first term at frequency $2\omega_c$ to give a final output:

$$E_{out} = \tfrac{1}{2} K E_s E_c \cos \phi \tag{3.43}$$

and to get a signal proportional to E_s both E_c and $\cos \phi$ must be kept constant. The dependence on $\cos \phi$ accounts for the name phase-sensitive detection and for the need to include in the system some means of varying the phase to compensate for incidental phase shifts. Of course the phase itself may be the quantity of interest in which case E_s is kept constant. If e_s is not a simple sinusoid but contains harmonics (as it usually will) then their contribution to the output must be considered. Odd harmonics ($3\omega_c$, $5\omega_c$, etc.) will produce a dc output but even ($2\omega_c$, $4\omega_c$, etc.) will not. The contribution of these higher harmonics is often reduced by passing e_s through a band-pass pre-filter, but this must not be too sharply tuned as small variations in ω_c will give changes in e_{out} via the $\cos \phi$ term. This filter also performs another important function as will be described below.

To consider the discrimination of the lock-in against noise it is simplest to take one component, say at frequency ω_n. Then, as before, the output is:

$$e_{out} = \tfrac{1}{2} K E_n E_c [\cos(\omega_n + \omega_c) t + \cos(\omega_n - \omega_c) t] \tag{3.44}$$

The first term will be at least at frequency ω_c and hence will be rejected while the second will only give an output provided $\omega_n - \omega_c$ lies within the passband of the smoothing filter. The RC filter is often said to be performing an integrating function, but this is not quite true (Tavares, 1966). Though a true integrator is better by a factor of two (Rolfe and Moore, 1970), it is seldom used owing to the dc drift it may contribute.

What improvement in SNR may be achieved? In principle this can be as great as required if sufficient time is available. (An RC filter requires about 5 time constants for the output to get to within 1% of the final value.) In practice there are always limiting factors such as drifts in other parts of the system, e.g. light sources, detector, sample, etc. The equivalent noise bandwidth of the filter is from Eq. (3.38):

$$\delta f = f_N - 0 = \tfrac{1}{4}RC \tag{3.45}$$

so that if a signal is being extracted from white noise of bandwidth Δf the ratio of noise powers is $\Delta f/\delta f$, or of noise voltages $(\Delta f/\delta f)^{\frac{1}{2}}$. For example, if $\Delta f = 10$ kHz and $\tau = 10$ s the improvement will be by a factor of about 640. With $f_c = 1$ kHz the effective $Q = f_c/\delta f = 4 \times 10^4$. With such a Q a relative frequency drift of as little as 1 part in 10^5 would reduce the signal by 20%.

It is most important that a lock-in amplifier is not overloaded, as this introduces non-linearity and extra intermodulation products and hence extra additional spurious output. The pre-filter referred to earlier serves to limit the probability of this occurring, particularly due to large noise bursts. If the signal is considerably below the noise level this pre-filter may have to be much narrower than system stability would allow, in which case this should be a tracking or coherent filter (Danby, 1970) which operates on the same principle as the switching detector. Such pre-filters do not improve the SNR (in the absence of overloading) as their bandwidth will be substantially greater than the smoothing filter.

The output from a lock-in amplifier must be interpreted with care: the result will depend on the form of the modulation applied to the system. Say the quantity F being measured is some function of a parameter H (Fig. 3.19). Then in the vicinity of a particular operating point H_1 the function may be expressed in terms of a Taylor series:

$$F(H) = F(H_1) + (H - H_1)\frac{dF(H_1)}{dH} + \frac{1}{2!}(H - H_1)^2 \frac{d^2 F(H_1)}{dH^2} + \ldots \tag{3.46}$$

where $F(H_1)$ is the value of F at H_1, $dF(H_1)/dH$ is the value of dF/dH at H_1, etc. If H is modulated about H_1,

$$H = H_1 + H_m \cos(\omega_c t) \tag{3.47}$$

the resulting signal will be

$$e_s = F(H_1) + \frac{dF(H_1)}{dH} H_m \cos(\omega_c t) + \frac{1}{2!}\frac{d^2 F(H_1)}{dH^2} H_m^2 \cos^2(\omega_c t) + \ldots \tag{3.48}$$

Using Eq. (3.48) in Eqs (3.41), (3.42), and (3.43) it is seen that operating the lock-in at ω_c gives an e_{out} dependent on the first derivative dF/dH rather

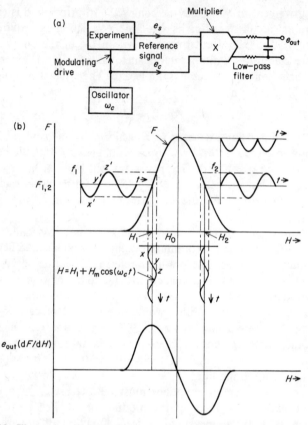

Fig. 3.19. Illustrating the signal output from a lock-in detector. F represents the variation of the experimental function and H the independent variable. The signals resulting from small amplitude modulations of H at various values $H_{0, 1, 2}$ are shown at $f_{0, 1, 2}$. Note the phase change arising from the opposite slopes of F and the frequency doubling effect at H_0. The signals $f_{0, 1, 2}$ are synchronously rectified and smoothed to give the output shown which can be seen to be the differential dF/dH of the original function F. (After Hamilton, 1977.)

than the signal F. If the reference e_c is at $2\omega_c$, the e_{out} is proportional to the second derivative $d^2 F/dH^2$ and so on. The expansion (3.46) is only good for small deviations $(H - H_1)$ so H_m should be kept small. However, e_{out} will be proportional to H_m so some compromise must be made (Russell and Torchia, 1962; Evans and Thompson, 1969; Hager, 1973). To obtain $F(H_1)$ directly, H is switched between a value where $F = 0$ and the value H_1. The resultant will be a square wave of amplitude $F(H_1)$, the first term in Eq. (3.48). This switching action would occur, for example, in using chopper wheels as in a phosphoroscope (Langouet, 1972), but again care must be exercised in the quantitative interpretation of the results (Parker, 1968; Gijzman and Razi-Naqvi, 1974).

3.4.4. Signal Averaging

An alternative approach to improving the SNR is to measure the spectrum many times and average the results. If the noise is random, then it is expected that it will eventually average to zero while the signal tends to its true value (Bonnet, 1965; Ernst, 1965; Trimble, 1968; Rex and Roberts, 1969; Beauchamp, 1973). This requires that the spectrum be repeatable and that there be some synchronizing trigger to which the signal is accurately related in

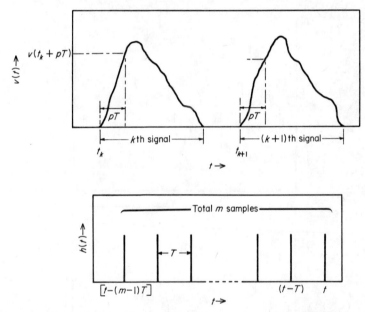

Fig. 3.20. Illustrating signal averaging, the symbols used are defined in the text. (After Hamilton, 1977.)

time. The signal must also be stored so that it may be averaged with subsequent scans, and for efficiency in time enough channels are required to store all points in the spectrum. Such multi-channel averagers use digital techniques as this allows long-time drift-free storage and the ready application of operations, such as dividing, over a wide dynamic range.

Consider a signal $v(t)$ containing the required signal $e(t)$ and random noise $n(t)$ (Fig. 3.20):

$$v(t) = e(t) + n(t) \tag{3.49}$$

Let the kth repetition of $v(t)$ start at t_k and be sampled at intervals of T. Then for a particular sample, say the pth, the amplitude will be

$$v(t_k+pT) = e(t_k+pT)+n(t_k+pT)$$
$$= e(pT)+n(t_k+pT) \qquad (3.50)$$

since the true signal e will be the same at every repetition, i.e. e is independent of t_k; n on the other hand, is not, and though it has zero mean value over a large number of samples it will have an r.m.s. value ρ for a single sample.

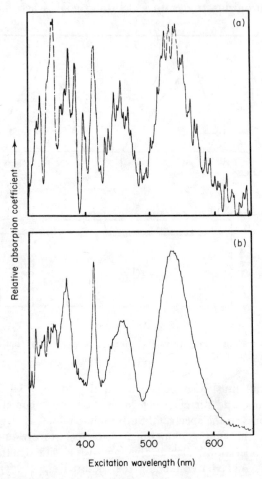

Fig. 3.21. The effect of signal averaging on the excitation spectrum of Mn^{2+} emission in natural forsterite (Green and Walker, unpublished work): (a) a single scan (time constant ~ 1 s); (b) after 64 scans.

For a single sample at pT the SNR is thus:

$$(\text{SNR})_1 = \frac{e(pT)}{\rho} \qquad (3.51)$$

After m repetitions the amplitude stored in the pth memory channel is the sum of m values given by (Eq. 3.50)

$$\sum_{k=1}^{m} v(t_k + pT) = \sum_{k=1}^{m} e(pT) + \sum_{k=1}^{m} n(t_k + pT)$$

$$= me(pT) + (m\rho^2)^{\frac{1}{2}} \qquad (3.52)$$

since $e(pT)$ is constant and for random noise the net r.m.s. value will be the square root of the sum of the squares. Thus after m repetitions, the SNR is

$$(\text{SNR})_m = \frac{me(pT)}{(m\rho^2)^{\frac{1}{2}}} = \frac{m^{\frac{1}{2}} e(pT)}{\rho} = m^{\frac{1}{2}}(\text{SNR})_1 \qquad (3.53)$$

and the SNR improves as the square root of the number of repetitions (Fig. 3.21). In practical instruments it is usual to divide by the number of repetitions after each one but this makes no difference to the above argument. For a substantial improvement a considerable number of scans are required which will take a corresponding time, so some consideration should be given to the best way to use a given time. Since there will generally be drifts in the experimental system, then for a given measurement time it is better to scan the spectrum many times quickly rather than a few times slowly (Ernst, 1965).

In the frequency domain signal averaging appears as a comb filter. As the number of scans is increased the width of the comb teeth decreases; this is illustrated in Fig. 3.22, where the amplitude of the peaks has been normalized for comparison. Thus the signal, whose harmonics occur at the frequencies of the peaks, is preferentially selected relative to the noise occurring at intermediate frequencies.

If very rapid signal changes are involved the standard signal averagers cannot cope and it is necessary to use a fast single-channel averager usually referred to as a gated integrator or boxcar detector (Clark and Kerlin, 1967; Amsel and Bosshard, 1970; Abernethy, 1971; Collins and Katchinoski, 1973). Here a fast gate is used to sample the signal during a short interval at some time delay after a reference trigger. By repeating the signal and taking a number of samples at the same delay and integrating or averaging, then the SNR is improved as in using a simple RC filter. The delay is then varied slowly to sweep over the whole time range of the signal, or the delay is kept constant and the wavelength scanned over the spectrum. With this technique it is possible to average signals of nanosecond duration at least.

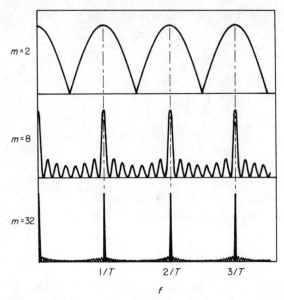

Fig. 3.22. Signal averaging as a comb filter in the frequency domain. The teeth of the comb occur at the harmonics of the signal and as the number of scans increases the bandwidth of the teeth decreases thus improving the selectivity. The peaks of the response have been normalized for comparison.

3.4.5. Photon Counting

The outstanding performance of photomultipliers makes it possible for single photons to produce output pulses large enough to be counted individually. This approach has been found to provide a means of improving the SNR by about an order of magnitude relative to lock-in amplification or signal averaging (Alfano and Ockman, 1968; Morton, 1968; Nakamura and Schwarz, 1968; Oliver and Pike, 1968; Tull, 1968; Foord et al., 1969; Franklin et al., 1969; Topp et al., 1969; Rolfe and Moore, 1970; Jones et al., 1971; Zatzick, 1970; Ingle and Crouch, 1972; Poultney, 1972).

There has been much discussion and some controversy regarding the advantage of photon counting over other methods but this is now well established. The advantage arises from the form of the pulse-height distribution of the output pulses and the digital measurement. Noise pulses arising from electrons not originating from the photocathode will probably be smaller than those that do, while pulses due to high energy sources, e.g. cosmic rays, radio activity, will probably be considerably larger. Thus by using a discriminator at the output a substantial number of these noise pulses can be eliminated from the count. The mid-range pulses that are accepted will still be subject to variation in height owing to statistical variations in gain, but provided the variation does not go outside the discrimination levels they are counted as 1 count irrespective of height. Both considerations improve the SNR but it should be noted that for some especially good photomultipliers the signal and noise pulse-height distributions are essentially the same in which case discrimination is unnecessary (Jones et al., 1971). The pulse-height distribution for each individual tube should be determined to ascertain the optimum operating conditions as even tubes of the same nominal type may differ significantly. As an example of the effectiveness of photon counting the measurement of a flux of 0·04 photons/s has been reported (Oliver and Pike, 1968).

Though photon counting is particularly advantageous at low light levels, the convenience of digital measurement and processing may encourage its use at much higher intensities. However, for good linearity there is an upper limit of about 10^6 pps set by the resolving time of discriminators and counters since the random arrival time of the photons must be allowed for. If two pulses arrive too close together the counter will see them as one so counts will be lost (Ingle and Crouch, 1972a). Above this limit normal current measurements must be made though a digital signal may be readily obtained by using a current integrator or current-to-frequency converter (Hamilton, 1977). Corrections for count rate losses can be made but the correction depends on the type of signal being measured (Coates, 1972; Davis and King, 1972; Ingle and Crouch, 1972a; Holzapfel, 1974; Klobuchar et al., 1974; Pace and Atkinson, 1974). If the observed counting rate is r_0 and the true count rate r_t, then for a resolution time τ the correction may be approximately found from

$$r_0 = r_t/(1 + \tau r_t) \qquad (3.54)$$

A variation of the photon counting technique is to measure alternately signal plus noise and noise alone, usually referred to as synchronous single photon counting (Arecchi et al., 1966). This allows the subtraction of the background from the measured value, but to obtain minimum relative variance the fraction of time spent counting signal plus noise and noise must be properly chosen. If the noise count rate is r_N and the signal plus noise rate $(r_S + r_N)$ then the latter should be counted for a fraction α:

$$\alpha = 1 + \frac{r_N}{r_S} - \frac{r_N}{r_S}\left(1 + \frac{r_S}{r_N}\right)^{\frac{1}{2}} \qquad (3.55)$$

the noise being counted for the remaining fraction $(1 - \alpha)$ (Oliver and Pike, 1968).

3.5. TIME-RESOLVED MEASUREMENTS

The most striking advances in luminescence instrumentation have occurred in the general area of time-resolved measurements (or fluorometry). During the past forty years there has been an extension of many measuring techniques into the very short time regions. Chronologically, and in order of increasing experimental complexity and cost, luminescence research has been directed towards phosphorescence measurements ($\sim 10^{-3}$ s), flash photolysis techniques ($\sim 10^{-6}$ s), measurement of fluorescence lifetimes ($\sim 10^{-9}$ s), and is now exploring the 10^{-12} s time domain using mode-locked lasers. The majority of experiments concerned with studying the time dependence of emitted luminescence rely on the relatively high level technology associated with nuclear modular electronics and fast photodetectors. The component parts necessary to measure fluorescence lifetimes can be assembled and used without too great difficulty and complete fluorescence lifetime apparatus is also available commercially. Nevertheless, measurements of a high accuracy are routinely carried out only in a very small number of research laboratories.

A study of the time dependence of fluorescence emission in the region from roughly 10^{-9}–10^{-6} s divides into two general categories: (i) periodic excitation where a measurement is made of the phase shift or modulation change introduced by any fluorescence process and (ii) pulsed excitation where the direct time decay of fluorescence is observed after a short initial excitation pulse. These two categories are different both in terms of the techniques employed and of their areas of applicability. Phase and modulation fluorometry requires knowledge of the actual decay function of the fluorescence before the results can be interpreted. Pulse fluorometry demands an excitation source whose duration is short compared with the fluorescence decay to be measured, otherwise a detailed correction has to be applied to the measured result to eliminate instrumental effects.

3.5.1. Phase and Modulation Fluorometry

The theory associated with this technique is analogous to ac circuit theory where a reactive circuit component (for example, a capacitor) will shift the phase of an ac voltage signal applied to it. In the general case, the measured fluorescence intensity as a function of time $R(t)$, will be given by the convolution of the source and excitation function used $I(t)$ with the actual sample fluorescence decay function $F(t)$.

$$R(t) = \int_0^\infty F(t')\,I(t-t')\,\mathrm{d}t' \tag{3.56}$$

Phase and modulation studies are generally carried out using an excitation source modulated at a single frequency ω. If the fluorescence decay is of a single exponential form then we can write

$$I(t) = i + j \cos \omega t \tag{3.57}$$

and

$$F(t) = F_0 \exp(-t/\tau) \tag{3.58}$$

This gives

$$R(t) = k\tau \left[i + \frac{j}{\sqrt{(1+\omega^2\tau^2)}} \cos(\omega t - \theta) \right] \tag{3.59}$$

where θ is the phase angle (phase shift) between $I(t)$ and $R(t)$.
The exponential lifetime of fluorescence can be derived in two different ways. One can measure the phase angle between the excitation and fluorescence light, when

$$\tan\theta = \omega\tau \tag{3.60}$$

or, alternatively, measure the modulation factor (m) where

$$m = \frac{1}{\sqrt{(1+\omega^2\tau^2)}} = \cos\theta \tag{3.61}$$

Here m is called the degree of modulation, or the ratio of the ac and dc amplitudes, of the fluorescence relative to that of the exciting light. In principle, the measurement of θ and m over a wide range of values of ω could yield the true functional form of $F(t)$ without the need for any assumptions. In practice, measurements are hardly ever made at more than one frequency.

An interesting application of a phase and modulation fluorometer has been to the analysis of the fluorescence decay function which occurs in systems exhibiting excimer formation (see Chapter 2 and Birks et al., 1963). In these systems, the time decay of emission from the excited dimer is of the form $I_D(t) = \exp(-\lambda_1 t) - \exp(-\lambda_2 t)$ and that from the excited monomer is $I_M(t) = \exp(-\lambda_1 t) + A\exp(-\lambda_2 t)$. A measurement of the phase shift and the degree of modulation both for the excimer and for the excited monomer fluorescence emission enables all three unknown parameters λ_1, λ_2, and A to be determined.

In circumstances when the decay function is unknown and is non-exponential, the phase and modulation method has the disadvantage that no precise significance can be given either to θ, the phase shift, or to m, the degree of modulation. This limitation is most apparent in the published literature where phase shift methods used with electron beam excitation yield information which may be very misleading.

3.5.2. Phase Shift Measurements

Until recently the most popular method to measure fluorescence lifetimes has been to use a phase shift technique. The techniques of light source modulation and phase (and modulation) comparison of the signals vary widely and the reader is directed to more detailed reviews for a fuller description of each apparatus (Birks and Munro, 1967; Ware, 1971).

The essential elements of phase shift apparatus are shown in Fig. 3.23. A source whose intensity can be periodically modulated at a fixed frequency is used to stimulate emission from the sample. A portion of the modulated excitation light is then compared in phase with the emitted fluorescence light. The phase comparison may be a direct comparison or a null method and the relative phases can be compared using either optical or electrical methods.

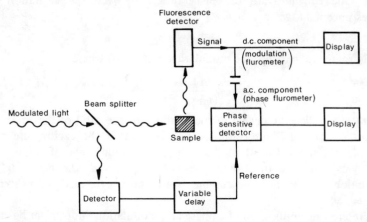

Fig. 3.23. Schematic arrangement of a phase and modulation fluorometer.

Phase shift measurements can be made to an accuracy of about 1° in phase difference. Since phase fluorometers normally operate only at a single frequency, a frequency range of from about 1 to 20 MHz is used for lifetimes in the range 10^{-7} s to 10^{-10} s. Intensity modulation at these frequencies can be achieved either by intrinsic r.f. modulation of the discharge in a source or by passing a steady light beam through a device with variable optical density. The upper frequency limit for radio-frequency source modulation is limited by the lifetimes of the atomic or molecular transitions yielding the emission. As the frequency is gradually increased the degree of source modulation will become smaller and in addition the modulation of the fluorescence signal itself may become inconveniently small.

The construction and operation of a radio-frequency-modulated lamp requires care since the coupling between the power source and the lamp is critical. In order to maintain a frequency stability in the power supply which will be compatible with the tuned circuit narrow band response of the detector, a quartz crystal is invariably used as a control. Typical lamps use tungsten electrodes separated by about 1 mm in hydrogen at about 100 Torr. The output from a 30 W lamp will be up to 70% modulated at 10 MHz.

For modulation of high source intensities such as from high pressure mercury or xenon arc lamps, the light is passed through a device with variable optical density using a Kerr cell, an ultrasonic diffraction grating (Debye–Sears method), or an electro-optical crystal (the Pockels effect). The development and use of these devices reveal a great deal of ingenuity. However, they are limited by the transmission properties of the modulating material or of its containment vessel. Kerr cell modulation with high efficiency is restricted to the visible and near ultraviolet and both ultrasonic gratings and the Pockels effect are restricted to sample excitation at wavelengths longer than approximately 200 nm.

The geometry of the sample is of considerable importance in phase shift measurements since the presence of even small amounts of scattered light mixed in with the phase-shifted fluorescence light can introduce large systematic errors. The signal detectors are almost invariably high gain, wide bandwidth photomultiplier tubes usually of the focused dynode type. The detector output then is fed into the phase or modulation comparator which may make the comparison either at the modulating frequency or after frequency conversion. A wide range of interesting methods have been devised for phase comparison (Birks and Munro, 1967; Ware, 1971) and Fig. 3.23 shows one such method where both phase and modulation changes can be measured (Birks and Dyson, 1961). In this case the reference signal is provided directly by the electrical power supply to the light modulator and comparison with the fluorescence signal is made using the centre tap

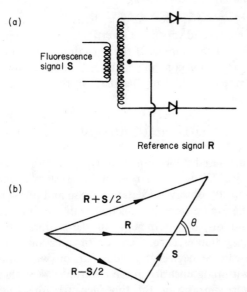

Fig. 3.24.(a) Simple radio-frequency phase sensitive detector. (b) Phase and amplitude relationship between reference signal R and fluorescence signal S, used to measure fluorescence lifetimes.

transformer. If the reference signal at the transformer is R (a vector representing the phase and amplitude of the voltage) and the signal voltage induced on the secondary is S, then the voltage at the end of the coil is given by $R+S/2$ and $R-S/2$. This will produce a steady signal at the output, after rectification, of $|R+S/2|-|R-S/2|$. When $|R|$ is selected to be much greater than $|S|$, the current at the output will vary approximately as $S\cos\theta$ where θ is the phase difference between R and S (see Fig. 3.24). When θ is changed by altering the variable delay, the output current oscillates sinusoidally about zero with an amplitude proportional to S and independent of R and any noise components in the signal. Modulation measurements can be made by combining photomultiplier anode current measurements with R_{max} measurements. The resolution of the instrument is about 10^{-10} s with an overall accuracy of $\pm 2\%$ when driven at 10·7 MHz.

In addition to scattered light and frequency stability problems, errors arise in phase fluorometry from non-linearity in the photodetector and sometimes from the different degrees of modulation associated with different emission lines in the source spectrum. The ways of identifying and correcting for systematic errors in phase fluorometry have been reviewed (Brewer *et al.*, 1963; Lawry *et al.*, 1965; Ware, 1971).

3.5.3. Pulsed Excitation Source Methods

Pulse fluorometry measures the real time decay of fluorescence following stimulation by an intense and very short duration light pulse giving effectively δ-function excitation of the sample. If the detection system which views the fluorescence emission is not constrained by a limited bandwidth, then the decay profile measured would truly be that of the sample under study. In the general situation

$$R(t) = \int_0^t F(t')\, I(t-t')\, dt' \qquad (3.62)$$

where $R(t)$ is the measured fluorescence decay, $F(t)$ is the true fluorescence response function and $I(t)$ is the instrumental or system response function. $I(t)$ is simply the directly observed excitation pulse and is determined by the time duration of the excitation pulse together with the time response of the detector, cable, and display equipment. In the majority of experiments concerned with long fluorescence lifetimes (greater than a few tens of nanoseconds) $I(t)$ can be neglected by comparison with other systematic errors associated with time linearity in the electronics. For shorter lifetimes, and particularly when measuring the time decay in the 1 ns region, $F(t)$ must be derived rather carefully from $R(t)$ and $I(t)$ and a choice of various analytic or synthetic procedures is available for doing this (Ware, 1971;

Binkert *et al.*, 1972). The most common method for extracting $F(t)$ is to synthesize a set of curves $R(t)$ from $I(t)$ and a range of chosen decay functions $F_j(t)$. A computer is used to select the best fit between measured and computed $R(t)$ and the true decay function $F(t)$ deduced (Munro and Ramsay, 1968; Binkert *et al.*, 1972). Other methods for deriving corrected results include the application of the "methods of moments" which is particularly straightforward if the fluorescence decay function is exponential (Munro and Ramsay, 1968) and taking Fourier transforms of the measured and instrumental function to derive the true decay function directly.

3.5.4. Single Photon Counting

The most popular technique using a pulsed source for lifetime determinations has arisen as an off-shoot from nuclear physics research which encouraged great advances in the production of reliable commercial fast electronic modular systems.

High energy nuclear physics research has led to the development of fast detectors, discriminators, and time-to-amplitude converters which are used to make accurate "time-of-flight" measurements. In fluorescence studies, the time of the emission of single fluorescence photons emitted from the sample is measured with respect to a zero reference time, usually associated with the light or electrical pulse from the excitation source. The method possesses several advantages. It has the highest sensitivity of any techniques since only single photons are detected and apart from the time profile of the source, it has a good time resolution (that is, a minimum time width for $I(t)$). Since the component parts are commercially available, it is easy to set up and reliable in operation.

However, the use of the method demands an understanding of the way in which systematic errors can distort the profile of the accumulated data. For high photon arrival rates (i.e. high intensities!) a distortion of the shape of the decay curve is caused by instrumental dead time effects and by fluctuations in the photon arrival rate at the photocathode. Experimental and theoretical methods have been described to discriminate against detection of multiple photon events in each cycle and to apply corrections to the distorted data (Davis and King, 1970a, b). Nevertheless, it is important to remember that distortion of the measured decay profile arises as a systematic error. Both the extent of such errors and the applicability of the single photon method demand an exact knowledge of the operating conditions of the experiments.

The theory of single photon counting (Morton, 1968) relates the number of photomultiplier output pulses per unit time to the number of photons incident on the photomultiplier detector. In fact, proportionality to better than 1% between these numbers will be achieved only if the detector registers an output signal for every 40 excitation pulses on average. Clearly, this method can be used even when the average number of photoelectrons per pulse is small unlike all other techniques where large signals and high sample efficiencies are a prerequisite for accurate measurements.

A general layout for a single photon counting experiment is shown in Fig. 3.25.

The method depends on the use of very fast pulse discriminators, usually "constant fraction" or "zero crossing" types, together with a time-to-amplitude converter (TAC) which delivers an output pulse to the multichannel analyzer, the amplitude of the pulse being accurately proportional to the time difference between the start and stop pulses at the TAC. Enhancement of the SNR can be produced by gating the discriminators. Detailed reviews of the single photon method have been given by Ware (1971), Knight and Selinger (1973), and Lewis *et al.* (1973), and the method has been applied widely using photon or electron excitation and using photon–photon or photon–electron coincidences.

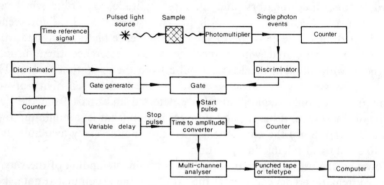

Fig. 3.25. Schematic diagram of single-photon counting apparatus for fluorescence lifetime measurements.

Recently, the pulsed light from a synchrotron storage ring has been used for lifetime studies. A source of this type with high intensity per pulse, a narrow Gaussian pulse profile and a very high repetition rate of up to 13 MHz will make the single photon technique unrivalled for sensitivity, speed, and time resolution for all wavelengths of excitation (Lopez-Delgado *et al.*, 1974, and see Section 3.5.6).

In addition to the single photon method there are several techniques which use a pulsed source for lifetime measurements. For relatively long times ($\gtrsim 100$ ns) and high intensity pulses fluorescence decay may be measured directly in real time using a fast oscilloscope. This method is commonly used when laser light is chosen for sample excitation.

A range of stroboscopic or gated methods are also used which, given a source with enough intensity and a high repetition rate, will yield fairly high sensitivity and time resolution. The simplest sampling technique is to use the photomultiplier output pulse to drive a pulse-sampling oscilloscope. When used with a high intensity, high repetition rate synchrotron radiation source, this method promises the advantages of the single photon combined with a very much faster data collection rate since every excitation pulse could be used. The reader is referred to the literature for other stroboscopic methods making use of gated detectors or image converter tubes (Bennett, 1960; Yguerabide, 1965; Sisneros, 1967).

3.5.5. Indirect Determination of Lifetime

In molecular systems the most common way to estimate the fluorescence lifetime is to assume a relationship between the lifetime, the quantum efficiency, and the absorption spectrum of the sample. As a rule, estimations of this kind are often unreliable since accurate extinction coefficient data are needed and there are many difficulties in deriving and applying a rigorous theoretical formula (Birks and Munro, 1967; Ware, 1971).

Apart from calculations based on absorption spectra, measurements of fluorescence depolarization and solvent quenching can and have been used for lifetime estimation (Stevens and Dubois, 1963; Dubois and van Hemert, 1964).

Depolarization measurements are based on an equation deduced from the theory of Brownian rotation of molecules in solution. The calculation demands a knowledge of the effective size of the molecule and is therefore only of limited application. Although depolarization measurements are of relatively low light sensitivity and are applicable only to solutions, a number of fluorescence decay time values have been reported and they compare favourably with results from phase fluorometers (Birks and Munro, 1967; Imhof and Read, 1977).

3.5.6. Time-resolved Spectroscopy

The measurement of time-resolved spectra usually requires a repetitive excitation pulse and a variable delay followed by a short sampling interval. The delay is set at the required value and the wavelength scanned slowly relative to the excitation repetition rate. Both excitation and emission spectra can be measured in this way but good time resolution requires a stable delay and a short sampling interval. For the shortest times purely electronic techniques must be used and a boxcar detector or integrator is often used for this purpose in its single-point (non-scanning) mode (Section 3.4.4).

As excitation intensities are usually large and delayed emissions often rather weak, high gain photomultipliers may require protection against overload if there is a possibility of scattered excitation light reaching the photocathode. In such circumstances the gating should take place before, or at least in, the detector rather than after it. A number of detector gating schemes have been described which allow the sensitivity to be varied over a wide range and with time characteristics from nanoseconds upwards. Bennett (1960) applied a sampling pulse to a transmission line dynode chain of a photomultiplier such that the propagation time of each section of transmission line is the same as the transit time of electrons between the dynodes (Araki et al., 1976). Another technique is to pulse the voltage on one or more dynodes to vary the gain (Bhaumik et al., 1965; De Martini and Wacks, 1967; De Marco and Penco, 1969; Elphick, 1969; Rossetto and Mauzerall, 1972; Wieme, 1973) but this requires large amplitude pulses. An alternative scheme uses bistable circuits in the dynode chain which can be triggered into either state, reversing the voltages on adjacent dynodes (Hamilton, 1971; Hamilton and Razi-Naqvi, 1973). This method requires only small trigger pulses and the system will remain in either state indefinitely as required. A fast linear gate for the output signal from a photomultiplier has been described by Albach and Meyer (1973).

Traditional pulsed light sources such as the xenon arc emit a spectral continuum which may give rise to scattered light problems. Pulsed lasers are of course better in this respect as well as providing higher excitation intensities but may be of fixed frequency or have limited tunability. It may also be necessary to use second or higher harmonic generation in order to obtain excitation of the required wavelength and the tunable wavelength range is likely to be small (Section 3.2.3). For very short times (< 1 ns) mode-locked lasers should however prove useful. Hochstrasser and Wessel (1974) have used a mode-locked frequency-doubled ruby laser as a source for the measurement of time-resolved spectra of anthracene crystals in the picosecond time domain, i.e. on the time scale of vibrational relaxation processes. Gating was done optically using a Kerr cell in which time-coincident birefringence was induced by the laser fundamental frequency.

For time delays longer than about 0·1 ms, mechanical chopping of light from a continuously operated source is commonly used. The antiphase double chopper (known as the Becquerel phosphoroscope) is a well-known and well-tried technique which has been used for both excitation and emission spectra (Thomas and Colbow, 1965; Parker, 1968; Costa et al. 1969; Langouet, 1972; Winefordner, 1973). The sample is placed between two chopper discs or cans with off-set apertures so that alternate excitation and viewing of the emission can occur (Fig. 3.26). The popularity of this method lies in its simplicity and in the fact that the background scattered light problem can be eliminated entirely. It is of course essential that the two choppers are driven so that their relative positions remain fixed. Since it is necessary to collect as much of the emitted light as possible, there are often geometrical problems relating to the actual positions of chopper blades and sample; replacement of the second chopper by a switched detector can provide more geometrical flexibility. These techniques allow the separation of overlapping emissions which have different decay times but great care must be exercised in interpreting the output signals; such interpretation has led to considerable controversy and it is clear that each case should

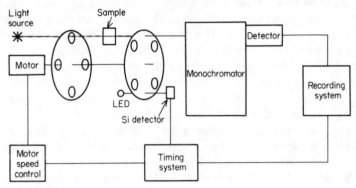

Fig. 3.26. Chopper system for recording delayed emission spectra.

be examined in detail (O'Haver and Winefordner, 1966; Parker, 1968, 1970; Razi-Naqvi, 1970; Gijzman and Razi-Naqvi, 1974). Simple integrated dc measurements can be made or a lock-in amplifier used referenced by an LED–photodiode pair across the first chopper. Chopping is usually performed at the focus of a lens or mirror where the dimensions of the light beam are very small. However, in circumstances where the excitation beam is a finite size at the point where chopping takes place, the second chopper may need to have a longer "closed" time than "open" time to avoid scattered excitation light reaching the detector. For most purposes, the pulsed-source gated-detector method is superior to mechanical chopping on account of its better time resolution and higher excitation intensity (Fisher and Winefordner, 1972).

In measuring long lifetimes the mechanical phosphoroscope has drawbacks, since a sharp cut-off of excitation requires a high chopper speed, whereas long lifetimes require long viewing periods; such problems can be overcome by using a third chopper disc rotating at a different speed (Langelaar et al., 1969) or by two counter rotating discs running at slightly different speeds. Another effective technique is to use a fast camera-type shutter together with a current-to-frequency converter and a multichannel scaler (Birks et al., 1976).

On–off "square-wave" excitation can easily be accomplished using electron beam excitation (Section 3.2.5) at modulation frequencies ranging from less than 1 Hz to several MHz. The same square-wave modulation frequency is also used as a reference for a lock-in amplifier. Alternatively, the electron beam can be repetitively pulsed and a boxcar detector used to sample at a particular delay after the pulse, as when using a pulsed light source. There is, of course, no scattered light with electron excitation and therefore protection of the detector against overload is unnecessary.

The measurement of the time decay of fluorescence using either the stroboscopic or single photon methods can be adapted easily to measure time-resolved spectra (Ware, 1971). If a conventional flash lamp is used, lamp stability is needed for the period to scan through the spectrum and the time resolution will be determined largely by the finite width of the light pulse; Q-switched or mode-locked lasers can provide very good time resolution. Recently synchrotron radiation has been used to measure time-resolved spectra and a block diagram of the apparatus is shown in Fig. 3.27 (Lopez-Delgado et al., 1974). The spectra are recorded by counting delayed coincidences between the reference signal (obtained from the storage ring accelerating radio-frequency signal) and the fluorescence signal for a given delay. Although the gate module had only a poor time resolution it was

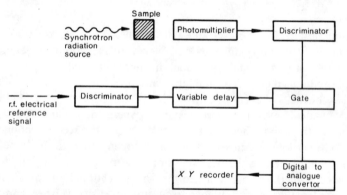

Fig. 3.27. Block diagram of apparatus for the measurement of fluorescence lifetimes and time-resolved spectra using synchrotron radiation.

easily possible to separate the decay of the fluorescent neutral species and the excited anion of 2-naphthol. Figure 3.28 illustrates the power of this method. The fast decay of the directly excited neutral species ($\tau = 5\cdot8$ ns) is seen at 350 nm while the emission from the excited anion of 2-naphthol produced by the protolytic reaction from the neutral molecule seen at 420 nm rises and subsequently decays with a rather longer lifetime (10·5 ns). Synchrotron radiation sources, which have very high repetition rates, narrow

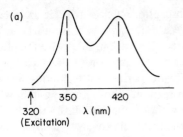

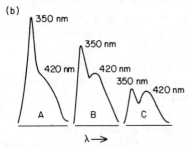

Fig. 3.28. Time-resolved spectra from 2-naphthol obtained using a synchrotron radiation source: (a) normal fluorescence of 2-naphthol; (b) time-resolved fluorescence of 2-naphthol with a delay time; $A = 3$ ns, $B = 5$ ns, $C = 9$ ns.

pulse width and are extremely stable, are ideal ultraviolet sources for this type of experiment.

3.5.7. Transient Absorption Spectroscopy

The first major advance in the field of time-resolved spectroscopy was made with the development of the technique of flash photolysis (Norrish and Porter, 1949; Porter, 1950a, b). The technique involves excitation of the specimen by a short duration, high power excitation flash of light. Transient absorption from the excited levels produced in this way is then studied using a second monitoring light flash. This second (absorption) flash should ideally be a smooth continuum, so as to reveal the absorption spectrum (when suitably dispersed) of the excited species and the reduced intensity of the spectrum of the original species. If the experiment is repeated several times with increasing known delays between the excitation absorption flashes then it is possible to measure the decay time of the transiently excited species and the rate of reappearance of the initial species. If the duration of the excitation flash used to excite an aromatic hydrocarbon is about 1 ms before it has decayed sufficiently for the scattered light not to interfere with the absorption measurements, then the specimen will contain a high proportion of molecules in the lowest triplet state and the absorption flash will thus reveal $T_1 \rightarrow T_n$ absorption. Of course, it is necessary also for the absorption flash to have a duration which is short compared with the lifetime of the transient under observation, otherwise the photographic measurements will integrate over a large portion of the decay of the transient.

A variant of this general experimental technique has been called "kinetic spectrophotometry". Instead of photographing the whole spectrum at a fixed time interval after the excitation flash, the method views a single wavelength interval of the transient absorption spectrum and then monitors the decay of the excited species at the selected wavelength after each excitation flash on an oscilloscope. The complete spectrum is scanned by repeating the experiment at different wavelengths. The output wavelength is often selected using a grating monochromator (Strong et al., 1957).

Both of these methods require repetitive flashing and gradually provide a build-up of the transient absorption as a function of time and of wavelength. In either case the results are valid only if photodecomposition does not occur during the flashing procedure.

Laser Flash Photolysis. The unique properties of laser sources have enabled the technique of flash photolysis to be extended easily into the nanosecond region (Porter and Topp, 1970) and, more recently, into the picosecond range.

For example, a high power flash lamp can be made to deliver about 5×10^{17} quanta per pulse in a pulse length of $2\ \mu s$. If a Q-switched ruby laser is used it will deliver about 7×10^{17} quanta per pulse (after frequency doubling) in a pulse length of about 15 ns. If a mode-locked laser is used, then pulse widths of about 5×10^{-12} s can be achieved. A laser source is well collimated with a small beam cross-section (less than $2\ cm^2$) and low divergence (about 1 mrad) so that all the emitted light can be used. The light is also polarized and therefore polarization studies can be carried out without any further reduction in intensity. The only difficulty can be the choice of a laser source of the required wavelength of emission to achieve excitation of the sample. However, a range of tunable dye lasers are now available in addition to a variety of argon, ion and nitrogen, and frequency-multiplied high power neodymium lasers all of which provide emission in the ultraviolet and, in a limited number of cases, in the vacuum ultraviolet.

An experimental arrangement which allows spectra to be recorded photoelectrically is shown in Fig. 3.29 (Hodgkinson and Munro, 1973). A Q-switched and frequency-doubled pulsed ruby laser is used to illuminate the sample placed at the focal spot of a quartz lens. Any excess ultraviolet light from the laser beam is removed by a filter which will, of course, set a limit to the short wavelength absorption measurements. The monitoring flash used to probe the excited state absorption spectrum of the sample is a high power xenon flash tube arranged to fire synchronously with the laser,

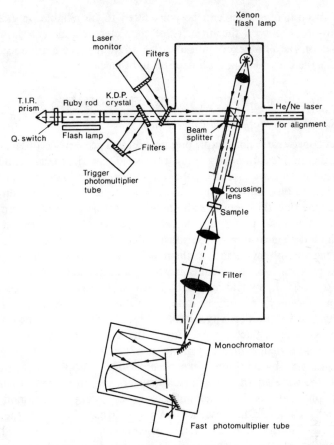

Fig. 3.29. Apparatus for photoelectric recording of transient absorption spectra. (After Hodgkinson and Munro, 1973.)

the laser flash occurring at the peak of the absorption flash to within $\pm 50 \, \mu s$. Note that in this arrangement the laser light and the absorption flash emerge from the beam splitter as parallel beams and follow identical optical paths so that, in principle at least, the absorption flash only traverses that region of the sample excited by the laser. In practice, of course, the focal spot of the laser is much smaller than that of the xenon flash tube. Spectra are recorded, point by point, using selected photomultiplier tubes whose linearity is checked carefully. The output can be photographed on a fast oscilloscope or stored and displayed on a transient recorder. All spectra have to be normalized by the laser intensity monitor to take into account fluctuations in successive laser pulses. In all techniques employing lasers, careful alignment of all optical elements is crucial to the success of the experiment. In addition, relative spectral intensities can be considered to be reliable only after the most careful checks on the linearity of the detector and recording system as well as on the photoexcitation and degradation processes induced in the sample by the excitation and absorption light flashes.

3.6. MISCELLANEOUS TECHNIQUES

3.6.1. Sample Preparation and Containment

Proper consideration of sample preparation and geometry is often neglected in experimental measurements. While published material exists in abundance to discuss the merits of spectroscopic equipment and to consider the fundamental limitations, for example, in making time-resolved measurements, it is also true that the literature abounds with errors associated with the use of impure samples, spectra not corrected for sample reabsorption effects, and with lifetime measurements seriously affected by the presence of scattered light. Some measurements such as the relative fluorescence quantum efficiency of solids as a function of excitation wavelength are notoriously difficult to undertake and often yield unreliable results. A measurement of relative quantum efficiency may necessitate corrections for the effects of light scattering, surface reflectivity, bulk transmission, and reabsorption in addition to source fluctuations during the experiment. Each correction involves a knowledge of the sample conditions which perhaps may be understood only by additional experiments—for example, to measure and eliminate the wavelength dependence of surface reflection from the sample.

Many physical and chemical effects at the sample can modify and occasionally eliminate fluorescence emission. Most important, the sample must be pure! Large molecular weight organic compounds are noted for the presence of impurities. These impurities are often isomers of the sample and provide a highly efficient mechanism for elimination of sample fluorescence by energy transfer to the impurity. All solids should be zone-refined when possible and where thermal decomposition is not a problem; liquid samples should be distilled. Any fluorescence signal will be susceptible also to trace impurities in any solvent used as well as the sample itself. Impurities in solvents which may cause quenching can arise through contamination from vacuum greases, cleansing agents, chemical reagents, filter paper, plastic wash bottles, or even as a result of direct photochemical decomposition. For measurements in solution, spectroscopic grade purity solvents at least should be used whenever obtainable.

When measurements are made in solution, the lifetimes and quantum efficiency can be considerably altered in the presence of dissolved oxygen. The simple process of bubbling oxygen-free nitrogen ("white spot") through the solution for several minutes will reduce the oxygen concentration to about 10^{-6} M. For delayed fluorescence or phosphorescence measurements, the oxygen concentration can be reduced to less than 10^{-8} M by using a freeze–pump–thaw sequence of operations (Parker, 1968).

In inorganic materials the presence of the divalent ions of Fe, Ni, and Co can seriously impair the luminescence efficiency of many phosphors in concentrations of the order of 0·1%. Other transition-metal ions such as Cr^{3+} and Mn^{2+} form very efficient luminescence centres in some crystals even when the concentration is very low (~ 10 ppm). Cr^{3+} substitutes very readily for Al^{3+} and Mn^{2+} for Ca, Mg, or Zn; there are many instances in the literature where Mn^{2+} emission is evident from a supposedly "pure" sample.

Even though the starting materials may be free from impurities, diffusion of impurity ions into the sample may occur, particularly at high temperatures, from the crucible, sample encapsulation, or furnace lining. Preparation of inorganic crystals often requires very high temperatures and sometimes high pressure. Single crystals are usually grown either from the melt, using some form of the Czochralski or Bridgman techniques, or from the vapour phase (Gilman, 1963). Considerable progress in producing crystals of III–V compounds for LED applications has been made in recent years (Bhargara, 1975).

Hydrothermal synthesis, which uses aqueous solvents under high temperature and pressure, may enable certain polymorphs to be prepared which cannot be produced by other means (e.g. α-quartz). Silicates are often prepared by hydrothermal methods at temperatures up to 900 °C and pressures up to several kilobars by enclosure in a noble-metal capsule inserted in a tube autoclave or "bomb". Single crystals of sapphire and ruby can also be produced hydrothermally (Gilman, 1963).

All fluorescence measurements, and in particular those concerned with lifetime and quantum efficiency, must be made with the sample at a fixed (known) temperature. However, hidden local heating effects may arise from the focused beam of a source—particularly a laser or electron beam— or from its proximity to a high power source lamp. Even such heating effects associated with the warm dynode chain of a photomultiplier may be important when the detector is fixed to a small enclosed sample chamber.

Low temperature measurements become increasingly important now that simple, reliable cryostats are commercially available. In general, the "simplest" cryostats to use are those which involve liquid or cold helium gas transfer by continuous flow to a heat exchanger supporting the sample. The simplicity of the design minimizes the numbers of cooled windows which are needed; the cryostat does not need to be precooled in any way and provides a sufficient rate of cooling (more than 0·5 W at 4 K) to allow deposition of layers of almost any material including solid rare gases as a sample matrix. Single crystals or polycrystalline powders should be in good thermal contact with the cooled sample mount; a thin layer of grease is often used but care is required since many greases are luminescent to some degree. An aliphatic solvent of high purity which freezes to a glass may be used as an alternative to grease provided the sample does not dissolve in it. The design of sample compartments and the geometry of samples have been discussed elsewhere (Bartholomew, 1958; Parker and Rees, 1962; Udenfriend, 1962; Hercules, 1966).

In nearly all applications front surface excitation and observation is most satisfactory. This applies especially to crystals, solid samples, and concentrated solutions. When dilute solutions ($< 10^{-5}$ M), vapour samples, or sample holders prone to scattering such as narrow cylindrical low temperature dewars are used, then right-angled viewing may be advantageous. Many commercial instruments are arranged with filters or monochromators mounted at right angles, so that the sample fluorescence may be viewed either from the front surface or at right angles. If the direction of observation of fluorescence is perpendicular to the direction of excitation it is important to remember that the relative fluorescence intensity is a very sensitive function of the absorption coefficient of the sample.

3.6.2. Quantum Efficiency

The luminescence quantum efficiency is defined as the ratio of the number of photons emitted by a luminescent material to the number initially absorbed. In most luminescent systems the emitted photon is of lower energy than the absorbed photon, so that in principle there can be efficiencies of greater than 100%. However, in most cases the conditions are such that 100% represents the maximum possible. Accurate measurements of yield are of considerable importance since, in conjunction with lifetimes, they provide a direct determination of non-radiative rate constants. It should be noted that quantum efficiencies are not always constant with variation of exciting photon energy, though they may well be over a limited range. In the latter case the substance may be used as a quantum counter which allows a detector with variable energy response to be converted to one with constant sensitivity over that range of energies. This technique is commonly used in calibrating spectrometers (Section 3.3.7) or for providing correction signals for electronic processing to obtain corrected spectra.

The measurement of absolute quantum efficiency presents considerable difficulty. The determination of relative efficiency is somewhat easier and allows absolute determination if a standard is available. Unfortunately there is still significant disagreement about the accuracy and reliability in the "standards" in use at present, though the position is improving. The techniques and problems of measurement have been well reviewed and the literature should be well studied before any measurements are attempted as there are many pitfalls and consequently much unreliable data (Melhuish, 1964, 1973; Lipsett, 1967; Demas and Crosby, 1971; Crosby *et al.*, 1973). Similar problems arise in the measurement of energy efficiencies under electron excitation (Bril and Klasens, 1952). The measurement of luminescence efficiency under particle excitation must always be made with due regard to the probable degradation of luminescence on account of radiation damage effects; such damage can reduce the efficiency of luminescence by an order of magnitude or more, particularly if a high flux density of heavy particles (e.g. protons) is used for excitation or if irradiation times are long (Hanle and Rau, 1952; Young, 1955; Van Wijngaarden and Hastings, 1967).

3.6.3. Polarized Light Techniques

The great majority of experiments in spectroscopy assume the use of unpolarized (or randomly polarized) light and frequently make no attempt to identify optical elements such as mirrors and gratings which may introduce partial polarization effects into measured spectra. Nevertheless, it is increasingly the case that either linearly or circularly polarized light is used for sample excitation in research. Knowledge of the direction of the electric vector in the excitation beam can be used to provide fundamental information about the anisotropic properties of materials on a molecular scale. This information is particularly important for the case of very large molecules, often of biological importance.

A variety of means exist for producing and detecting polarized light. At least two source types, synchrotron radiation sources and lasers using cavity reflectors at the Brewster angle, emit intrinsically polarized light and may not require any further effort to render the radiation polarized. For non-polarized light sources in the visible, "Polaroid" film and "Polacoat" polarizers can be used cheaply and easily to yield linearly polarized light. Better still are Glan prism polarizers which have high extinction ratios ($<10^{-4}$) with reasonably wide fields of view ($\sim 20°$). Prisms of this type constructed from quartz can be used down to about 200 nm. Below that wavelength various materials such as magnesium fluoride, lithium fluoride, calcite, and biotite can yield from 40% to 99% polarization either by reflection or by transmission but the associated absorption losses are usually large. Triple reflection from gold film can give about 90% polarization down to wavelengths as short as 50 nm.

Polarized light can be used to measure either linear or circular dichroism spectra as well as a magnetic circular dichroism. Polarized light from synchrotron sources has been used to determine the absorption spectrum and principal polarization of fluorescence as a function of exciting light frequency. These measurements can specify the magnitude of the transition moments in a molecule or luminescence centre along various crystal axes. The measurements may permit the separation of different but overlapping electronic transitions in organic crystals (Skobel'tsyn, 1965). Measurements of polarized absorption or excitation spectra of transition-metal ions in crystals can provide information about the site symmetry of such centres and about cation ordering among different sites (Burns, 1970).

In viscous fluid media, a measure of the rotational Brownian motion of a molecule can be obtained by measuring the depolarization of fluorescence and this is, in turn, directly related to the fluorescence lifetime of the material (Weber, 1960).

Both inter- and intra-molecular energy transfer effects can be studied in this way. In very large molecules such as proteins, it is possible to photo-select a fluorescent "tag" on the polymer chain and subsequently study its rotation about the long axis of the polymer simply by observation of the depolarization of fluorescence. Such measurements provide estimates of rotation of specific chosen segments of even the very longest chain molecules. An automatic recording polarization spectrofluorometer (Ainsworth and Winter, 1964; Cary et al., 1964; Witholt and Brandt, 1968) has been constructed and its versatility in application to a range of fluorescence studies described.

REFERENCES

General References

Thorne, A. P. (1974). "Spectrophysics." Chapman & Hall, London.

Text References

Abernethy, J. D. W. (1971). *Wireless World*, **76**, 576.
Abernethy, J. D. W. (1973). *Physics Bulletin*, **24**, 591.
Ainsworth, S. and Winter, E. (1964). *Appl. Optics*, **3**, 371.
Albach, G. G. and Meyer, J. (1973). *Rev. Sci. Instrum.* **44**, 615.
Alfano, R. R. and Ockman, N. (1968). *J. Opt. Soc. Amer.* **58**, 90.
Allison, R., Burns, J., and Tuzzolino, A. J. (1964). *J. Opt. Soc. Amer.* **54**, 747.
Ambroziak, A. (1968). "Semiconductor Photoelectric Devices." Iliffe Books, London.
Amsel, G. and Bosshard, R. (1970). *Rev. Sci. Instrum.* **41**, 503.
Anderson, L. K. and McMurty, B. J. (1966). *Appl. Optics*, **5**, 1573.
Araki, T., Uchida, T., and Minami, S. (1976). *Jap. J. Appl. Phys.* **15**, 2421.
Argauer, R. J. and White, C. E. (1964). *Anal. Chem.* **36**, 368.
Arecchi, F. T., Gatti, E., and Sona, A. (1966). *Rev. Sci. Instrum.* **37**, 942.
Aspnes, D. E. (1967). *Rev. Sci. Instrum.* **38**, 1663.
Barelli, A. E. (1973). *Electronic Design*, 19 July, p. 68.
Bartholomew, R. J. (1958). *Rev. Pure and Appl. Chem.* **8**, 265.
Baxendall, P. J. (1968). *Wireless World*, **74**, 388 and 454.

LUMINESCENCE INSTRUMENTATION 233

Beauchamp, K. G. (1973). "Signal Processing Using Analog and Digital Techniques." George Allen & Unwin, London.

Bell, D. A. (1960). "Electrical Noise." Van Nostrand, Princeton, N.J.

Bennett, R. G. (1960). *Rev. Sci. Instrum.* **31**, 1275.

Benoît à la Guillaume, C. (1976). *J. Luminescence*, **12/13**, 57.

Betts, D. B. (1965). *J. Sci. Instrum.* **42**, 243.

Betts, J. A. (1970). "Signal Processing, Modulation and Noise." English Universities Press, London.

Bhargava, R. N. (1975). *IEEE Trans. Electron Devices*, **ED-22**, 691.

Bhaumik, M. L., Clark, G. L., Snell, J., and Ferder, L. (1965). *Rev. Sci. Instrum.* **36**, 37.

Billings, B. H. (1951). *J. Opt. Soc. Amer.* **41**, 966.

Binkert, T., Tschanz, H. P., and Zinsli, P. E. (1972). *J. Luminescence*, **5**, 187.

Birks, J. B. and Dyson, D. J. (1961). *J. Sci. Instrum.* **38**, 282.

Birks, J. B., Dyson, D. J., and Munro, I. H. (1963). *Proc. Roy. Soc.* A **275**, 575.

Birks, J. B. and Munro, I. H. (1967). *Progress in Reaction Kinetics* (G. Porter, ed.), **4**, 239. Pergamon Press, New York.

Birks, J. B., Hamilton, T. D. S., and Najbar, J. (1976). *Chem. Phys. Lett.* **39**, 445.

Bloom, D. M., Bekkers, G. W., Young, J. F., and Harris, S. E. (1975). *Appl. Pshy. Lett.* **26**, 687.

Bonfiglioli, G. and Brovetto, P. (1964). *Appl. Optics*, **3**, 1417.

Bonnet, G. (1965). *Nuc. Instrum. Meth.* **37**, 217.

Bowen, E. J. (1946). "Chemical Aspects of Light." Oxford University Press, London.

Braga, C. L. and Lumb, M. D. (1966). *J. Sci. Instrum.* **43**, 341.

Breeze, R. H. and Ke, B. (1972). *Rev. Sci. Instrum.* **43**, 821.

Brewer, L., Berg, R. A., and Rosenblatt, G. M. (1963). *J. Chem. Phys.* **38**, 1381.

Bril, A. and Klasen, H. A. (1952). *Philips Res. Reports*, **7**, 401.

Budde, W. and Dodd, C. X. (1971). *Appl. Optics*, **10**, 2607.

Burns, R. G. (1970). "Mineralogical Applications of Crystal Field Theory." Cambridge University Press.

Cardona, M. (1969). "Modulation Spectroscopy" (Solid State Physics, Suppl. 11). Academic Press, New York and London.

Cary, H., Hawes, R. C., Hooper, P. B., Duffield, J. J., and George, K. P. (1964). *Appl. Optics*, **3**, 329.

Christensen, R. L. and Ames, I. J. (1961). *J. Opt. Soc. Amer.* **51**, 224.

Clarke, D. and Grainger, J. F. (1971a). "Polarised Light and Optical Measurement", p. 162. Pergamon Press, Oxford and New York.

Clarke, D. and Grainger, J. F. (1971b). "Polarised Light and Optical Measurement", p. 114. Pergamon Press, Oxford and New York.

Clark, W. G. and Kerlin, A. L. (1967). *Rev. Sci. Instrum.* **38**, 1593.

Coates, P. B. (1972). *Rev. Sci. Instrum.* **43**, 1855.

Coates, P. B. (1975). *J. Phys. E (Sci. Instrum.)*, **8**, 189.

Collins, F. G. and Katchinoski, R. (1973). *Rev. Sci. Instrum.* **44**, 1178.

Conn, G. K. T., and Avery, D. G. (1960). "Infrared Methods." Academic Press, New York and London.

Costa, L., Grum, F., and Paine, D. J. (1969). *Appl. Optics*, **8**, 1149.

Crosby, G. A., Demas, J. N., and Callis, J. B. (1973). "Accuracy in Spectrophotometry and Luminescence Measurements" (R. Mavrodineau, J. I. Schultz, and O. Menis, eds), p. 151. NBS Special Publication 378.

234   T. D. S. HAMILTON, I. H. MUNRO AND G. WALKER

Cundall, R. B. and Evans, G. B. (1968). *J. Phys. E (Sci. Instrum.)*, **1**, 305.
Danby, P. C. G. (1970). *Electronic Eng.* **42**, 36.
Dance, J. B. (1969). "Photoelectric Devices." Iliffe Books, London.
Davidson, S. M. and Rasul, A. (1977). *J. Phys. E (Sci. Instrum.)*, **10**, 43.
Davies, W. E. R. (1972). *Rev. Sci. Instrum.* **43**, 556.
Davis, C. C. and King, T. A. (1970a). *J. Phys.* **A3**, 101.
Davis, C. C. and King, T. A. (1970b). *Rev. Sci. Instrum.* **41**, 407.
Davis, C. C. and King, T. A. (1972). *J. Phys. E (Sci. Instrum.)* **5**, 1072.
De Marco, F. and Penco, E. (1969). *Rev. Sci. Instrum.* **40**, 1158.
De Maria, A. J., Glenn, W. H., Jnr, Brienza, M. J., and Mack, M. E. (1969). *Proc. IEEE*, **57**, 2.
De Martini, F. and Wacks, K. P. (1967). *Rev. Sci. Instrum.* **38**, 866.
Demas, J. N. and Crosby, G. A. (1971). *J. Phys. Chem.* **75**, 991.
De Sa, R. J. and Gibson, Q. H. (1966). *Rev. Sci. Instrum.* **37**, 900.
Drushel, H. V., Sommers, A. L., and Cox, R. C. (1963). *Anal. Chem.* **35**, 2166.
Dubois, J. T. and van Hemert, R. L. (1964). *J. Chem. Phys.* **40**, 923.
Dunning, F. B., Stokes, E. D., and Stebbings, R. F. (1972). *Opt. Commun.* **6**, 63.
Eastman, J. W. (1966). *Appl. Optics*, **5**, 1125.
Eberhardt, E. H. (1967). *Appl. Optics*, **6**, 251.
EG and G (1967). Application Note: SGD–100 and SG, Photodiodes.
EG and G (1973). Application Note: D300B–1, Silicon Photodiodes.
Ehrenberg, W. and Franks, J. (1953). *Proc. Phys. Soc.* **B66**, 1057.
Elphick, B. L. (1969). *J. Phys. E (Sci. Instrum.)*, **2**, 953.
EMI (1970). "Photomultiplier Tubes Catalogue."
EMI. Application Note: Silicon Avalanche Photodiodes.
Engstrom, R. W. (1963). RCA Technical Manual PT–60.
Ernst, R. R. (1965). *Rev. Sci. Instrum.* **36**, 1689.
Evans, B. L. and Thompson, K. T. (1969). *J. Phys. E (Sci. Instrum.)*, **2**, 327.
Evtahov, V. and Neeland, J. K. (1965). *Appl. Phys. Lett.* **6**, 75.
Faulkner, E. A. (1968). *Radio and Electronic Engnr*, **36**, 17.
Faulkner, E. A. (1975). *J. Phys. E (Sci. Instrum.)*, **8**, 533.
Fisher, R. P. and Winefordner, J. D. (1972). *Anal. Chem.* **44**, 948.
Foord, R., Jones, R., Oliver, C. J., and Pike, E. R. (1969). *Appl. Optics*, **8**, 1975.
Foord, R., Jones, R., Oliver, C. J., and Pike, E. R. (1971). *Appl. Optics*, **10**, 1683.
Franklin, M. L., Horlick, G., and Malmstadt, H. V. (1969). *Anal. Chem.* **41**, 2.
Fünfschilling, J. and Zoschokke-Gränacher, I. (1974). *Rev. Sci. Instrum.* **45**, 598.
Gabel, C. and Hercher, M. (1972). *IEEE J. Quant. Electr.* **QE-4**, 293.
Gallo, C. F. and Courtney, J. E. (1967). *Appl. Optics*, **6**, 939.
Garlick, G. F. J. (1966). In "Luminescence of Inorganic Solids" (Goldberg, ed.). Academic Press, Chapter 12. New York and London.
Garbuny, M. (1965). "Optical Physics", pp. 285–291. Academic Press, New York and London.
Geake, J. E., Walker, G., Telfer, D. J., Mills, A. A., and Garlick, G. F. J. (1973). *Proc. 4th Lunar Sci. Conf., Geochim. Cosmochim. Acta Suppl.* 4, Vol. 3, p. 3181.
Geusic, J. E., Levinstein, H. J., Singh, S., Smith, R. G., and Van Vitert, L. G. (1968). *Appl. Phys. Lett.*, **12**, 306.
Gijzman, O. L. J. and Razi-Naqvi, K. (1974). *Spectrochimica Acta*, **30A**, 59.
Gilman, J. J. (1963). "The Art and Science of Growing Crystals." Wiley, New York and London.
Hager, R. N. and Anderson, R. C. (1970). *J. Opt. Soc. Amer.* **60**, 1444.

Hager, R. N. (1973). *Anal. Chem.* **45**, 1131A.

Hamilton, T. D. S. (1966). *J. Sci. Instrum.* **43**, 49.

Hamilton, T. D. S. (1971). *J. Sci. Instrum.* **4**, 326.

Hamilton, T. D. S. (1977). "Handbook of Linear Integrated Electronics for Research." McGraw-Hill, London.

Hamilton, T. D. S. and Razi-Naqvi, K. (1973). *Anal. Chem.* **45**, 1581.

Hamstra, R. H. and Wendland, P. (1972). *Appl. Optics*, **11**, 1539.

Hanle, W. and Rau, K. H. (1952), *Z. Phys.* **133**, 297.

Haugen, G. R. and Marcus, R. J. (1964). *Appl. Optics*, **3**, 1049.

Hercules, D. M. (1966). "Fluorescence and Phosphorescence Analysis." Interscience, New York.

Hochstrasser, R. M. and Wessel, J. E. (1974). *Chem. Phys.* **6**, 19.

Hodby, J. (1970). *J. Phys. E (Sci. Instrum.)* **3**, 229.

Hodgkinson, K. A. and Munro, I. H. (1973). *J. Molec. Spectros.* **48**, 57.

Hodgson, B. W. and Keene, J. P. (1972). *Rev. Sci. Instrum.* **43**, 493.

Holzapfel, C. (1974). *Rev. Sci. Instrum.* **45**, 894.

Hughes, R. H. (1960). *Rev. Sci. Instrum.* **31**, 1156.

Hurwitz, C. E. (1966). *Appl. Phys. Lett.* **8**, 243.

Hviid, T. and Nielsen, S. O. (1972). *Rev. Sci. Instrum.* **43**, 1198.

Imhof, R. E. and Read, F. H. (1977).Measurement of lifetimes of atom molecules and ions, *Rep. Prog. Phys.* **40**, No. 1, Jan. 1977, 1–104.

Ingle, J. D. and Crouch, S. R. (1972). *Anal. Chem.* **44**, 785.

Ingle, J. D. and Crouch, S. R. (1972a). *Anal. Chem.* **44**, 777.

Institute of Physics (1973). *J. Phys. E (Sci. Instrum.)* **6**, 417.

Jacquinot, P. (1954). *J. Opt. Soc. Amer.* **44**, 761.

Jerde, R. L. I., Peterson, L. E., and Stein, W. (1967). *Rev. Sci. Instrum.* **38**, 1387.

Jones, R., Oliver, C. J., and Pike, E. R. (1971). *Appl. Optics*, **10**, 1673.

Kasha, M. J. (1948). *J. Opt. Soc. Amer.* **38**, 929.

Keene, J. P. (1963). *Rev. Sci. Instrum.* **34**, 1220.

Klobuchar, R. L., Ahumada, J. J., Michael, J. V., and Carol, P. J. (1974). *Rev. Sci. Instrum.* **45**, 1073.

Knapp, R. A. and Smith, A. M. (1964). *Appl. Optics*, **3**, 637.

Knight, A. E. W. and Selinger, B. K. (1973). *Aust. J. Chem.* **26**, 1.

Knight, W., Kohanzadeh, Y., and Lengyel, G. (1968). *Appl. Optics*, **7**, 1115.

Kortüm, G. (1969). "Reflection Spectroscopy." Springer-Verlag, New York.

Kressel, H. (1971). *In* "Lasers" (A. K. Levine and A. DeMaria, eds). Marcel Dekker, New York.

Land, P. L. (1971). *Rev. Sci. Instrum.* **42**, 420.

Land, P. L. (1972). *Rev. Sci. Instrum.* **43**, 356.

Langelaar, J., de Vries, G. A., and Bebelaar, D. (1969). *J. Phys. E (Sci. Instrum.)* **2**, 149.

Langouet, L. (1972). *Appl. Optics*, **11**, 2358.

Lawry, R., Muller, A., and Kokubun, H. (1965). *Rev. Sci. Instrum.* **36**, 1214.

Lee, Y. W. (1960). "Statistical Theory of Communication." Wiley, New York.

Lengyel, B. A. (1966). "Introduction to Laser Physics." Wiley, New York.

Lengyel, B. A. (1971). "Lasers", 2nd ed. Wiley–Interscience, New York.

Lewis, C., Ware, W. R., Doememy, L. J., and Nemzek, T. L. (1973). *Rev. Sci. Instrum.* **44**, 107.

Ley, J. M., Christmas, T. M., and Wildey, C. G. (1970). *Proc. IEE*, **117**, 1057.

Lipsett, F. R. (1967). *Progress in Dielectrics*, **7**, 217.

Lipsett, F. R., Oblinsky, G., and Johnson, S. (1973). *Appl. Optics*, **12**, 818.

Longhurst, R. S. (1973). "Geometrical and Physical Optics", 3rd ed., p. 210. Longmans, London.

Lopez-Delgado, R., Tramer, A., and Munro, I. H. (1974). *Chem. Phys.* **5**, 72.

Lush, H. J. (1965). *J. Sci. Instrum.* **42**, 597.

Marr, G. V. and Munro, I. H. (eds) (1973). International Symposium for Synchrotron Radiation Users. *Daresbury Laboratory Report, DNPL*, R36.

Marr, G. V., Munro, I. H., and Sharpe, J. C. (1975). Bibliography of Synchrotron Radiation. *Daresbury Laboratory Report, DNPL*, R24 (1972). *Supplement DL/TM127*.

Marette, G. (1976). *Appl. Optics*, **15**, 440.

Melchior, H., Fisher, M. B., and Arams, F. R. (1970). *Proc. IEEE.* **58**, 1466.

Melhuish, W. H. (1962). *J. Opt. Soc. Amer.* **52**, 1256.

Melhuish, W. H. (1964). *J. Opt. Soc. Amer.* **54**, 183.

Melhuish, W. H. (1973). "Accuracy in Spectrophotometry and Luminescence Measurements" (R. Mavrodineau, J. I. Schultz, and O. Memis, eds), p. 137. NBS Special Publication, 378.

Melngailis, I. and Mooradian, A. (1975). *In* "Laser Applications to Optics and Spectroscopy" (S. Jacobs, M. Sargent III, J. F. Scott, and M. O. Scully, eds). Addison-Wesley, Reading, Mass.

Miles, R. B. and Harris, S. E. (1973). *IEEE J. Quantum Electr.* **QE-9**, 470.

Morton, G. A. (1968). *Appl. Opt.* **7**, 1.

Morton, G. A., Smith, H. M., and Krall, H. R. (1969). *IEEE Trans.* **NS-16**, 92.

Motchenbacher, C. D. and Fitchen, F. C. (1973). "Low-noise Electronic Design." Wiley, New York.

Munro, I. H. and Ramsey, I. A. (1968). *J. Phys. E (Sci. Instrum.)* **1**, 147.

Nakamura, J. K. and Schwarz, S. E. (1968). *Appl. Optics*, **7**, 1073.

Neely, W. C., West, A. D., and Hall, T. D. (1975). *J. Phys. E (Sci. Instrum.)*, **8**, 543.

Nishida, K. and Nakajima, T. (1972). *Rev. Sci. Instrum.* **43**, 1345.

Norrish, R. G. W. and Porter, G. (1949). *Nature*, **164**, 658.

O'Haver, T. C. and Winefordner, J. D. (1966). *Anal. Chem.* **38**, 1258.

Oliver, C. J. and Pike, E. R. (1968). *J. Phys. D (Appl. Phys.)*, **1**, 1459.

Pace, P. W. and Atkinson, J. B. (1974). *J. Phys. E (Sci. Instrum.)*, **7**, 556.

Pailthorpe, M. T. (1975). *J. Phys. E (Sci. Instrum.)*, **8**, 194.

Parker, G. A. (1968). "Photoluminescence of Solutions." Elsevier, New York.

Parker, C. A. (1970). *Chem. Phys. Lett.* **6**, 516.

Parker, C. A. and Rees, W. T. (1962). *Analyst*, **87**, 83.

Pellicori, S. F., Johnson, C. A., and King, F. T. (1966). *Appl. Optics*, **5**, 1916.

Pettifer, R. E. W. and Healey, P. G. (1974). *J. Phys. E (Sci. Instrum.)*, **7**, 604.

Phelan, R. J. and Cook, A. R. (1973). *Appl. Optics*, **12**, 2494.

Phillips, H. R. and Taft, E. A. (1959). *Phys. Rev.* **113**, 1002.

Porter, G. (1950a). *Disc. Faraday Soc.* **2**, 60.

Porter, G. (1950b). *Proc. Roy. Soc.* A **200**, 284.

Porter, G. and Topp, M. R. (1970). *Proc. Roy. Soc.* A **315**, 163.

Poulson, R. E. (1964). *Appl. Optics*, **3**, 99.

Poultney, S. K. (1972). *Advances in Electronics and Electron Physics* (L. Marton, ed.) **31**, 39. Academic Press, London and New York.

Razi-Naqvi, K. (1970). *Chem. Phys. Lett.* **5**, 171.

RCA (1970). "Photomultiplier Manual", Technical Series, PT-61.

Redfield, D. (1961). *Rev. Sci. Instrum.* **32**, 557.

Rex, R. L. and Roberts, G. T. (1969). *Hewlett-Packard J.* **21**, 2.

Rieck, H. (1970). "Semiconductor Lasers." Macdonald, London.

Rolfe, J. and Moore, S. E. (1970). *Appl. Optics*, **9**, 63.

Rosen, P. and Edelman, G. M. (1965). *Rev. Sci. Instrum.* **36**, 809.

Rossetto, M. and Mauzerall, D. (1972). *Rev. Sci. Instrum.* **43**, 1244.

Rossi, J. A., Chinn, S. R., and Heckscher, H. (1973). *Appl. Phys. Lett.* **23**, 25.

Russell, A. M. and Torchia, D. A. (1962). *Rev. Sci. Instrum.* **33**, 442.

St. John, P. A., McCarthy, W. J., and Winefordner, J. D. (1966). *Anal. Chem.* **38**, 1828.

Schäfer, F. P. (ed.) (1973). "Dye Lasers: Topics in Applied Physics", Vol. 1. Springer-Verlag, Berlin, Heidelberg, and New York.

Schiff, P. (1968). *Electronics*, 5 August, p. 130.

Schurer, K. and Stoelhorst, J. (1967). *J. Sci. Instrum.* **44**, 952.

Scouler, W. J. and Mills, E. D. (1964). *Rev. Sci. Instrum.* **35**, 489.

Searles, S. K. and Hart, G. A. (1975). *Appl. Phys. Lett.* **27**, 243.

Searles, S. K. (1976). *Appl. Phys. Lett.* **28**, 602.

Seib, D. H. and Aukerman, L. W. (1973). *Advances in Electronics and Electron Physics* (L. Marton, ed.), **34**, 95. Academic Press, New York and London.

Shank, C. V. and Ippen, E. P. (1973). *In* "Dye Lasers" (F. P. Schäfer, ed.). Springer-Verlag, Berlin, Heidelberg, and New York.

Sisneros, T. E. (1967). *Appl. Optics*, **6**, 417.

Skobel'tsyn, D. V. (1965). Optical methods of investigating solid bodies. *Consultants' Review*, **25**. New York.

Smith, K. L. (1972). *Wireless World*, **78**, 367.

Smith, P. W. (1970). *Proc. IEEE.* **58**, 1342.

Smith, R. A. (1965). *Appl. Optics*, **4**, 631.

Smith, R. A., Jones, F. E., and Chasmar, R. P. (1968). "Detection and Measurement of Infrared Radiation." Oxford.

Stevens, B. and Dubois, J. T. (1963). *Trans. Faraday Soc.* **59**, 2813.

Stewart, J. E. and Galloway, W. S. (1962). *Appl. Optics*, **1**, 421.

Strong, R., Chien, J. C. W., Graf, P. E., and Wilford, J. E. (1957). *J. Chem. Phys.* **26**, 1287.

Taylor, W. B., LeBlanc, J. C., Whillans, D. W., Herbert, M. A., and Johns, H. E. (1972). *Rev. Sci. Instrum.* **43**, 1797.

Tavares, S. E. (1966). *IEEE Trans.* **IM–15**, 33.

Tekelec/Airtronic (1970). *Orbit Mag.* **21–26**, 31–34.

Thomas, D. E. and Colbow, K. (1965). *Rev. Sci. Instrum.* **36**, 1853.

Thomas, G. A. (1976). *Scientific American*, **234**, 6, 28.

Topp, J. A., Schrötter, H. W., Hacker, H., and Brandmüller, J. (1969). *Rev. Sci. Instrum.* **40**, 1164.

Topp, J. A. and Schmid, W. J. (1971). *Rev. Sci. Instrum.* **42**, 1683.

Trimble, C. R. (1968). *Hewlett-Packard J.* **19**, 2.

Tuan, V. D. and Wild, U. P. (1973). *Appl. Optics*, **12**, 1286.

Tull, R. G. (1968). *Appl. Optics*, **7**, 2023.

Udenfriend, S. (1962). "Fluorescence Essay in Biology and Medicine." Academic Press, London and New York.

Usher, M. J. (1974). *J. Phys. E* (*Sci. Instrum.*), **7**, 957.

Van Wijngaarden, A. and Hastings, L. (1967). *Can. J. Phys.* **45**, 3083 and 4039.

238   T. D. S. HAMILTON, I. H. MUNRO AND G. WALKER

Velapoldi, R. A. (1973). "Accuracy in Spectrophotometry and Luminescence Measurements" (R. Mavrodineanu, J. I. Shultz, and O. Menis, eds), p. 231. NBS Special Publication 378.

Von der Linde, D. (1973). *Appl. Phys.* **2**, 281.

Ware, W. R. (1971). "Creation and Detection of the Excited State" (A. A. Lamola, ed.), Vol. 1, p. 213. Marcel Dekker, New York.

Weber, J. (1960). *Biochem. J.* **75**, 345.

Welkowsky, M. and Braunstein, R. (1972). *Rev. Sci. Instrum.* **43**, 399.

Wendlandt, W. W. and Hecht, H. G. (1966). "Reflectance Spectroscopy." Wiley, New York.

Wieme, W. (1973). *J. Phys. E (Sci. Instrum.*), **6**, 203.

Winefordner, J. D. (1973). "Accuracy in Spectrophotometry and Luminescence Measurements" (R. Mavrodineanu, J. I. Shultz, and O. Menis, eds), p. 231. NBS Special Publication 378.

Witholt, B. and Brandt, L. (1968). *Rev. Sci. Instrum.* **39**, 1271.

Yarborough, J. M. and Massey, G. A. (1971). *Appl. Phys. Lett.* **18**, 438.

Yguerabide, J. (1965). *Rev. Sci. Instrum.* **36**, 1934.

Yguerabide, J. (1968). *Rev. Sci. Instrum.* **39**, 1048.

Young, A. T. (1966). *Rev. Sci. Instrum.* **37**, 1472.

Young, J. R. (1955). *J. Appl. Phys.* **26**, 1302.

Zatzick, M. R. (1970). *Research/Development*, **21**, 16.

Zernicke, F. and Midwinter, J. E. (1973). "Applied Non-linear Optics" (1973a), Chapter 3; (1973b) Chapter 7; (1973c) p. 45. Wiley–Interscience, New York.

Zucca, R. and Shen, Y. R. (1973). *Appl. Optics*, **12**, 1293.

4

Magnetic Field Effects in Organic Molecular Spectroscopy

NICHOLAS E. GEACINTOV and CHARLES E. SWENBERG

Chemistry Department and Radiation and Solid State Laboratory,
New York University, New York, New York 10003

4.1. INTRODUCTION

The aim of this article is to introduce the reader to recently developed concepts on magnetic field effects on the luminescence properties in organic systems. The material is developed in an elementary fashion and a minimum amount of theory is presented to explain the origin of the magnetic field effects. As background, a one-year course in elementary quantum mechanics has been assumed, while no previous knowledge of the luminescence properties of organic molecules and solids is necessary.

We begin this chapter with a brief and elementary summary on the electronic properties and luminescence of organic molecules and crystals emphasizing the most widely studied polycyclic aromatic hydrocarbons. Following a brief discussion of excitons in organic crystals, we list the basic magnetic interactions which are operative in these systems. In this article only those magnetic field effects which are presently well understood are discussed in detail. These include primarily the bimolecular effects involving excitons in aromatic hydrocarbon crystals. In the concluding section of this chapter some specialized applications of magnetic fields in the study of luminescence are discussed. The topics include brief discussions of magnetic field effects in the study of scintillation, electroluminescence, and luminescence in fluid solutions. This survey was completed in January 1976.

4.2. EXCITED STATES OF AROMATIC MOLECULES

4.2.1. Introduction

All of the magnetic field effects described in this chapter were observed with polycyclic aromatic molecules. In this section, the nature and characteristics of the electronic excited states of these molecules, which are responsible for the luminescence, are described. The structures of some typical aromatic molecules which are considered in this chapter are shown in Fig. 4.1.

The optically active electronic levels of polycyclic aromatic molecules can be described, to a good approximation, in terms of the delocalized π molecular orbitals (Birks, 1970). Electronic transitions between these levels give rise to absorption and emission of light in the near ultraviolet and visible regions of the spectrum. The overall wavefunctions Ψ_i describing each electronic

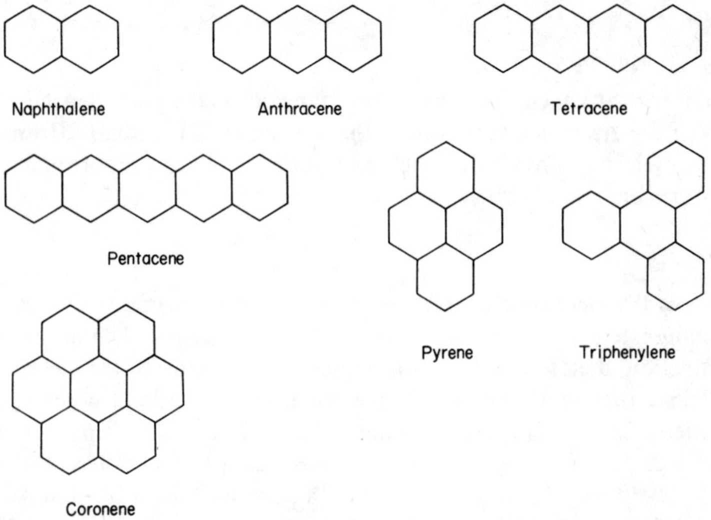

Naphthalene   Anthracene   Tetracene

Pentacene

Pyrene   Triphenylene

Coronene

Fig. 4.1. Structures of some polycyclic aromatic hydrocarbons referred to in this chapter.

state may be described in terms of product wavefunctions including the electronic part $\psi_{ie}(q, Q)$, nuclear part $\chi_{iv}(Q)$, electron spin Θ_{is} and nuclear spin Θ_{in} wavefunctions:

$$\Psi_i = \psi_{ie}(q, Q) \chi_{iv}(Q) \Theta_{is} \Theta_{in} \qquad (4.1)$$

where q and Q describe the electronic and nuclear spatial coordinates respectively. The separation of variables into electronic and nuclear coordinates and writing the wavefunction as a product of ψ_{ie} and χ_{iv} is called the Born–Oppenheimer approximation (Born and Huang, 1954; Kauzmann, 1957). The electron spin and nuclear spin wavefunctions are important in understanding the effects of magnetic fields as will be shown later. The nuclear component Θ_{in} is unimportant in determining the nature and characteristics of the optical electronic transitions

$$\Psi_i \rightleftarrows \Psi_j \qquad (4.2)$$

which we are considering in this section. The experimentally observable characteristics which are determined by the electronic, vibronic, and electron

spin components of the wavefunctions in Eq. (4.2) include the following:
(1) Wavelength of absorption and emission bands.
(2) Intensity of absorption (allowedness of the transition).
(3) Modulation of absorption and emission spectra by the vibronic overlap functions (Franck–Condon factors) (Kauzmann, 1957).
(4) Lifetimes of excited states.

4.2.2. Absorption

The energy at which the lowest-lying transitions take place depends on the details of the electronic structure of the molecules. The radiative transitions in these molecules are electric dipole in nature and are proportional to the matrix element $|M_{ij}|^2$, where

$$M_{ij} = \langle \Psi_i | \sum_q e\mathbf{r}_q | \Psi_j \rangle \tag{4.3}$$

where e is the electronic charge and $\mathbf{r}_q$ is the position vector of electron q. The sum extends over all electrons of the molecule. The ground-state wavefunction, denoted by $^1\Psi_i$, is a singlet state since all of the electrons are paired and the total spin $S = 0$; the superscript in $^1\Psi_i$ denotes the spin multiplicity $2S+1$. The excited states Ψ_j can have either singlet ($^1\Psi_j$) or triplet character ($^3\Psi_j$) since two electrons occupy two different orbitals. Since $\langle ^3\Theta_s | ^1\Theta_s \rangle = 0$, transitions $^3\Psi_j \rightleftarrows {}^1\Psi_i$ are forbidden by spin selection rules. Nevertheless, such transitions can be observed under certain conditions, even though they are extremely weak. With polycyclic aromatic molecules, a singlet → triplet absorption can be observed in the presence of the "external spin–orbit coupling effect". Spin–orbit coupling, in general, will be discussed for the *intramolecular* case in Section 4.4, in which the spin and orbital motion of electrons couple on the same molecule to produce a mixing between singlet and triplet states. In the intermolecular, or external spin–orbit coupling effect, singlet–triplet mixing within the polycyclic aromatic molecule is produced by an adjacent molecule which has paramagnetic character or contains heavy atoms. This effect is discussed in detail in the book by McGlynn *et al.* (1969). In crystals, on the other hand, it is possible to observe $^3\Psi_j \leftarrow {}^1\Psi_i$ transitions directly, without the presence of paramagnetic molecules or heavy atoms, because the molecules are densely packed and thus even extremely weak optical transitions are observable.

In contrast to the singlet–triplet transitions, the singlet–singlet transitions are allowed by spin selection rules, but may be forbidden by orbital selection rules. In centrosymmetric molecules, the parity selection rules $g \rightleftarrows u$ (allowed), $u \leftrightarrows u$ (forbidden), $g \leftrightarrows g$ (forbidden) are operative for electric dipole transitions. g and u refer to the orbital symmetries (with respect to inversion) and denote the symmetric and anti-symmetric properties of the orbital component of the overall wavefunction Ψ. As in the case of spin selection rules, there is some relaxation of the orbital selection rules in that non-totally-symmetric vibrations are capable of distorting the symmetry of the electronic states (vibronic mixing); such transitions are weak compared to the orbitally allowed transitions.

Most aromatic hydrocarbons have three rather well-defined band systems in the visible and ultraviolet region of the spectrum. It is convenient, for our purposes, to classify these lower-lying electronic transitions in terms of Platt's perimeter free-electron orbital model (Platt, 1964), see also Birks (1970) for a pertinent summary. We will not stress here the symmetry properties of the different electronic states, but will briefly discuss the origin of the three lower-lying electronic transitions in aromatic hydro-carbons in terms of the molecular orbitals, and relate these to experimental absorption spectra.

A diagram showing the energy levels of the different molecular orbitals in a typical polycyclic aromatic hydrocarbon such as naphthalene, anthracene, etc., is shown in Fig. 4.2 (*cata*-condensed aromatic hydrocarbons). In a molecule with n π electrons, there will be $n/2$ filled π bonding molecular orbitals and $n/2$ unfilled π^* anti-bonding molecular orbitals in the ground state. In the first $\pi-\pi^*$ transition, an electron is promoted from the $n/2$ orbital to the first empty π^* orbital denoted $n/2+1$ (Fig. 4.2b). This is the

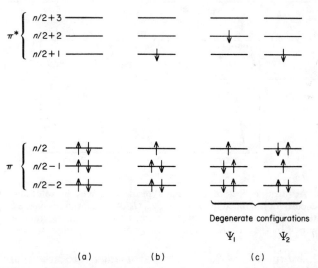

Degenerate configurations

Ψ_1 Ψ_2

(a)   (b)   (c)

Fig. 4.2. Energy levels of aromatic hydrocarbons in the Hückel molecular orbital approximation; n is the number of π-electrons; π and π^* are bonding and anti-bonding molecular orbitals respectively: (a) ground state, (b) first excited state ^1L_a (p state), (c) these two degenerate configurations give rise to the ^1B_b (β) and ^1L_b (α) states (see the text).

^1L_a state in Platt's notation. The energetically next higher excitation corresponds to the promotion of an electron from the $n/2$ π orbital to the $n/2+2$ π^* orbital. However, this transition has the same, or nearly the same, energy as the transition in which an electron is removed from the penultimate $n/2-1$ level and is promoted to the $n/2+1$ level. These two configurations ψ_1 and ψ_2, which are shown in Fig. 4.2c, are degenerate or nearly degenerate and are mixed by configuration interaction (Ham and Ruedenberg, 1956; Streitwieser, 1961). The result is that two excited states are possible:

$$\psi_+ = (1/\sqrt{2})\{a\psi_1 + b\psi_2\} \quad (^1B_b) \tag{4.4}$$

$$\psi_- = (1/\sqrt{2})\{a\psi_1 - b\psi_2\} \quad (^2L_b) \tag{4.5}$$

a and b are mixing coefficients ($<1\cdot0$). The energies of the transitions corresponding to excitation to either one of these two states are different.

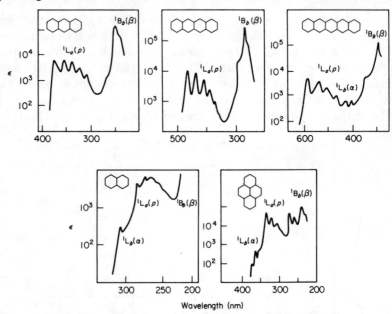

Fig. 4.3. Absorption spectra of some typical molecules in solutions. The molar extinction coefficient is defined in terms of $\log_{10}(I_0/I) = \varepsilon cl$, where c is the concentration in moles/liter, l is the optical path length in cm, and I/I_0 is the fraction of the incident light transmitted by the cuvette of length l containing the solution of concentration c.

Typical absorption spectra of several aromatic hydrocarbons are shown in Fig. 4.3. The assignment of the absorption bands in terms of Platt's nomenclature (^1B_b, ^1L_b, and ^1L_a) is also indicated. Frequently, Clar's p, α, and β nomenclature (Clar, 1964) will also be encountered in the literature, where ^2L_a is the p-band, ^1B_b the β-band and ^1L_b the α-band. The ^1B_b excitation is a strongly allowed transition and occurs at higher energies than the ^1L_a transition. The ^1L_b transition is orbitally forbidden and is therefore a weakly allowed transition. In some molecules it occurs at lower energies than the ^1L_a transition, in other molecules it appears to be buried under the ^1L_a band, while in pentacene it occurs at an energy higher than that of the ^1L_a transition.

The vibrational structure is quite pronounced in the ^1L_a band of anthracene, tetracene, and pentacene. The different peaks within the ^1L_a bands correspond to excitations to different vibronic levels of the $n/2+1$ electronic state. The relative intensities of these different vibronic bands are determined by Franck–Condon factors, i.e. the vibrational overlap integrals.

4.2.3. Fluorescence and Phosphorescence

The excited states decay back to the ground state of the molecules by a combination of processes. We neglect here photochemical decay channels in which the structure of the molecules is altered.

Regardless of which electronic state is excited, fluorescence usually occurs from the lowest excited singlet state denoted by S_1 (the upper singlet excited states will be denoted by S_2, S_3, ..., S_q, in anthracene, for example, S_1 corresponds to the ^1L_a or p state). This is known as *Kasha's rule* (see Birks, 1970). However, important exceptions to this rule have been noted, particularly in recent years (Birks, 1975). Azulene, in which the fluorescence occurs from S_2, is a well-known example.

If S_2, S_3, or S_q is excited directly, a radiationless transition $S_q \leadsto S_1$ will occur rapidly on time scales of 10^{-12} s. This process is called *internal conversion* and the excess energy $\Delta E(S_q - S_1)$ is dissipated by the vibrational lattice modes of the medium in which the excited molecule is dissolved. The S_1 state can decay either by *fluorescence*

$$S_1 \longrightarrow S_0 + photon$$

or by *intersystem* crossing to the triplet manifold

$$S_1 \xrightarrow{k_{TM}} T_q \leadsto T_1 \qquad (4.6)$$

which is also a radiationless process. An upper triplet state T_q may be formed initially, but it rapidly ($\sim 10^{-12}$ s) decays to the lowest triplet T_1 by internal conversion. The T_1 state may decay to the ground state by a radiationless process:

$$T_1 \xrightarrow{k_{GT}} S_0 + heat \qquad (4.7)$$

or by a radiative decay called *phosphorescence*:

$$T_1 \xrightarrow{k_{PT}} S_0 + photon \qquad (4.8)$$

The lifetimes of the lowest singlet excited states S_1 are determined by the intersystem crossover rates k_{TM} and by the radiative fluorescence decay rate k_{FM}. The radiationless transitions $S_1 \leadsto S_0$ with rate constant k_{GM} can also occur, but there is considerable evidence (Birks, 1970) that $k_{FM} + k_{TM} > k_{GM}$. We can therefore write an expression for the quantum efficiency of fluorescence q_{FM} (quanta of fluorescence photons emitted by a molecule per quanta absorbed):

$$q_{FM} \cong \frac{k_{FM}}{k_{FM} + k_{TM}} \tag{4.9}$$

The radiative decay rates $S_1 \rightarrow S_0$ are related to the probabilities of absorption $S_1 \leftarrow S_0$, which in turn are related to the transition moment matrix elements in Eq. (4.3). Thus, in naphthalene and pyrene the lowest excited states S_1 correspond to the forbidden transitions ^1L_b and the decay times are ~ 100 ns and 400–500 ns respectively for these two molecules. For anthracene and tetracene the S_1 states are allowed (^1L_a type) and the fluorescence decay times are of the order of 5 ns.

Although intersystem crossing $S_1 \leadsto T_q$ is forbidden by spin selection rules, the spin–orbit coupling is sufficiently large so that k_{TM} is of the same order of magnitude as k_{FM}, even if the fluorescence is fully allowed. The reader is referred to accounts by Birks (1970) and by McGlynn et al. (1969) for a discussion of this interesting effect, which is beyond the scope of this chapter. Correspondingly, the fluorescence quantum yields of polycyclic aromatic hydrocarbons can be as low as several per cent, the formation of triplets being very efficient. Thus, while triplets cannot be formed efficiently (in molecules in dilute solutions) by direct absorption ($T_1 \leftarrow S_0$), they can be generated indirectly quite easily by first exciting S_1, which is followed by intersystem crossing $S_1 \leadsto T_q$.

The quantum yield of phosphorescence is given by

$$\Phi_{PT} = \left(\frac{k_{PT}}{k_{PT} + k_{GT}}\right)\left(\frac{k_{TM}}{k_{TM} + k_{FM}}\right) \tag{4.10}$$

which is equal to the quantum efficiency per triplet excited state generated, multiplied by the quantum yield of triplets generated per photon absorbed. A unit efficiency has been assumed for the internal conversion process $T_q \leadsto T_1$.

With polycyclic aromatic molecules dissolved in rigid glasses at low temperatures, or dissolved in rigid plastic matrices at room temperature, the triplet lifetimes are within the range of 0·1–10 s, due to the spin forbiddenness of processes (Eqs 4.7 and 4.8). In fluid solutions or in the gas phase, the lifetimes are considerably shorter. Presumably, the reason is that small traces of impurities which are present in all solvents are capable of quenching

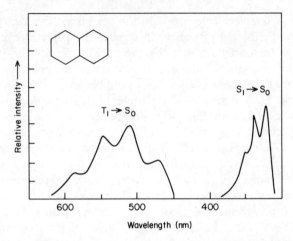

Fig. 4.4. Typical fluorescence ($S_1 \to S_0$) and phosphorescence spectrum ($T_1 \to S_0$) showing the differences in emission wavelengths. Naphthalene at room temperature (the $T_1 \to S_0$ spectrum was obtained using a rigid polystyrene solution of naphthalene). The fluorescence and phosphorescence spectra are plotted on different intensity scales.

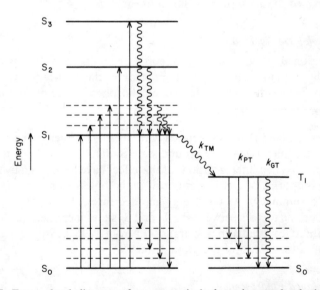

Fig. 4.5. Energy level diagram of an aromatic hydrocarbon molecule (or crystal). The electronic levels S_0, S_1, S_2, S_3, and T_1 are denoted by solid lines, the vibronic levels by dashed lines. Absorption is indicated by solid arrows pointing up, fluorescence ($S_1 \to S_0$) and phosphorescence by solid arrows pointing down. Radiationless transitions are indicated by wavy arrows. k_{PT} and k_{GT} are the radiative and non-radiative triplet decay rates respectively.

the triplet by a bimolecular process. In the gas phase, the triplets are also deactivated by collisional processes. In crystals the triplet decay times are also considerably shorter because the triplets can migrate throughout the crystal (see the next section) and can be annihilated at impurities.

The radiative decay rates of triplets, k_{PT} are believed to be of the order of 0.03 s^{-1} or less. Thus in most cases the triplet lifetimes are dominated by the non-radiative rate k_{GT} in Eq. (4.7). For molecules dissolved in plastics, Φ_{PT} is of the order of several per cent or less. Nevertheless, at room temperature in plastics (in the absence of oxygen, see below), or at low temperatures, phosphorescence can be quite easily observed visually. The phosphorescence of aromatic molecules such as naphthalene, chrysene, coronene, a,h-dibenzanthracene is particularly bright. For molecules such as anthracene, tetracene, and pentacene, the phosphorescence decay times are short and the emission lies in the red or near infrared region of the spectrum in which the sensitivity of the eye is low.

The triplet T_1 in aromatic hydrocarbons is always lower in energy than S_1. Consequently the phosphorescence emission occurs at longer wavelengths than the fluorescence. As an example, the emission spectra of naphthalene are shown in Fig. 4.4. An energy level diagram depicting the various transitions discussed in this section is shown in Fig. 4.5.

4.2.4. Triplet Decay Times

As mentioned above, the triplet lifetime of aromatic molecules dissolved in dilute and rigid solutions is dominated by the non-radiative constant k_{GT}. Thus, if τ_T is the triplet lifetime, or phosphorescence decay time,

$$\tau_T \sim k_{GT}^{-1}$$

It has been noted that τ_T decreases with decreasing energy gap $\Delta E(T_1-S_0)$ between the triplet T_1 and the ground state S_0. This is known as the energy gap law and detailed interpretations of this effect have been given by Siebrand and others in terms of recent theories of radiationless transitions (see Birks, 1970, for a summary). As the energy gap decreases, the vibrational overlap Franck–Condon factors between the T_1 and S_0 states increase, thus increasing the probability of a radiationless transition from T_1 to an isoenergetic highly excited vibrational level of the ground state S_0.

4.2.5. Delayed Fluorescence

The $S_1 \rightarrow S_0$ fluorescence decays on time scales of 10^{-9}–10^{-7} s and is called the *prompt* fluorescence. However, in concentrated solutions, in the gas phase, and particularly in crystals, an emission is observed which decays on time scales of milliseconds and has the same spectrum as the prompt fluorescence. This emission is called *delayed* fluorescence. There are several types and origins of delayed fluorescence, but we shall be concerned only with one type here, namely the process

$$T_1 + T_1 \longrightarrow S_1 + S_0 \longrightarrow 2S_0 + photon \qquad (4.11)$$

which is magnetic field dependent. In this mechanism two excited molecules in the T_1 triplet states collide and pool their energy to produce one excited singlet S_1. This singlet can subsequently decay via its normal decay channels such as fluorescence. The delayed fluorescence therefore displays the characteristic wavelength dependence of the prompt fluorescence, but its decay time, following a brief flash of exciting light, displays the characteristics of the triplet. Since two excited triplets must diffuse and collide with each other to form a singlet, the concentration of T_1 states must be sufficiently high. The exciting light intensities therefore must be sufficiently elevated in order to observe the delayed fluorescence. The required intensity of the exciting light source to observe the delayed fluorescence depends on numerous parameters which include the triplet lifetime, the rate of diffusion of the triplets, the quantum efficiency of generation of triplets, and the sensitivity of the apparatus.

In crystals it is customary to excite the triplets by direct $T_1 \leftarrow S_0$ absorption with relatively long wavelength light (red light in anthracene) and then to observe the delayed light (blue in the case of anthracene) at shorter wavelengths. The excitation can thus be carried out in steady state, taking care to isolate properly the emission optical paths from contamination with light corresponding to the excitation wavelengths. Alternately, a brief (nanosecond or microsecond duration) flash of light may be used in which the triplets are excited indirectly via $S_1 \leftarrow S_0$ absorption and subsequent intersystem crossing, and the decay is monitored following the flash. Furthermore, delayed light may be measured with a phosphoroscope arrangement in which the sample is illuminated and the delayed fluorescence is viewed alternately—the fluorescence is viewed when the excitation beam is blocked and the excitation impinges on the sample when the viewing path is blocked. For additional details on how to measure delayed light, the original references should be consulted. For organic crystals, references may be found in the reviews by Avakian (1974), Sokolik and Frankevich (1973, 1974), and by Swenberg and Geacintov (1973). For solutions the book by Parker (1968) may be consulted.

4.2.6. Quenching of Excited States

So far we have only considered the intramolecular decay mechanisms in an aromatic molecule. Intermolecular radiationless decay processes are also possible which prevent either the S_1 or T_1 states from decaying radiatively. These processes are termed *quenching* and the molecules which catalyze these processes are called *quenchers*.

Quenchers can be roughly subdivided into two classes, paramagnetic and non-paramagnetic quenchers. We first consider non-paramagnetic quenchers.

1. *Quenching by Non-paramagnetic Molecules* (X). Molecules X which have lower energy levels than the excited molecule M* can quench the latter by an energy transfer mechanism described by

$$M^* + X_0 \longrightarrow M_0 + X^* \tag{4.12}$$

where M* is the excited molecule, M_0 is the same molecule in the ground state, and X_0 and X* are the quencher molecule in the ground and electronically excited states respectively. This quenching process may proceed by a collision of X_0 and M* in which case M* may be either a singlet or a triplet. Energy transfer to X_0 can occur if the requirements of conservation of energy are met. If M* is a singlet state S_1, then energy transfer and quenching can also proceed by a long range resonance energy transfer (Förster) mechanism (Birks, 1970).

Quenching of fluorescence can also occur by a charge transfer mechanism in which an electron is transferred between M* and X_0 (Beens *et al.*, 1967, Aloisi *et al.*, 1974); the direction of this electron transfer depends on the ionization potentials and electron affinities of these species.

Quenching of fluorescence by intersystem crossing catalyzed by external heavy atoms has also been observed. The external quenchers include xenon, iodide ion, bromobenzene, etc. The heavy-atom quenching is accompanied by a corresponding increase in the triplet state production (Wilkinson, 1975). This is another example of an external spin–orbit coupling effect.

2. *Quenching by Paramagnetic Molecules*. The best known and ubiquitous quencher of singlet and triplet excited states of aromatic hydrocarbons is molecular oxygen. In the ground state $^3O_2$ is a triplet, but upon quenching triplets of aromatic hydrocarbons, it appears to be excited to the singlet $(^1O_2)$ state:

$$^3M^* + {}^3O_2 \longrightarrow {}^1M_0 + {}^1O_2 \tag{4.13}$$

As will be discussed later, this type of reaction is fully spin-allowed and is analogous to the delayed fluorescence ("fusion") reaction of Eq. (4.11). However, in contrast to this process, we have shown that quenching of triplets by molecular oxygen is not magnetic field sensitive (Geacintov and Swenberg, 1972). Claims to the opposite, however, have also been made (Frankevich and Sokolik, 1971a, b).

The quenching of the triplets by molecular oxygen occurs with an efficiency between 0·10 and 0·01 per collision (Benson and Geacintov, 1973, 1974;

Gijzeman *et al.*, 1973), while the quenching of singlets proceeds with nearly unit efficiency (Parmenter and Rau, 1969; Berlman, 1971). The reasons for these effects are discussed in detail in the references quoted.

Nitric oxide has a spin of $\frac{1}{2}$ and is also an effective quencher of excited states. In general, the quenching of triplets by spin $S = \frac{1}{2}$ particles (^2R), e.g. radicals or charge carriers which are doublets, is described by

$$^3T_1 + {}^2R \longrightarrow {}^1S_0 + {}^2R + \text{heat} \tag{4.14}$$

and is magnetic field sensitive. It is one of the important magnetic field-dependent processes which can occur in solutions and in crystals, and which is described in this chapter.

4.3. EXCITED STATES IN MOLECULAR CRYSTALS

4.3.1. Introduction

Molecular crystals are characterized by high vapor pressures, weak intermolecular forces, and small cohesive energies. The optical and electronic properties of an organic solid can be divided into two distinct groups. The first group consists of those properties directly traceable to the properties displayed by molecules in the gas phase, whereas the latter group constitutes those properties which are a direct manifestation of the aggregated phase. The optical properties, such as the fluorescence, phosphorescences, and absorption, or organic solids constitute examples of the first group whereas electronic conductivity is an aggregate property. It is not our intention to present here a detailed discussion of the nature and properties of the excited states of organic solids since our primary concern involves the effects of external magnetic fields on their luminescence. Furthermore, the subject has been adequately discussed in texts by Davydov (1962) and Craig and Walmsley (1968) to which the reader is referred. We restrict our brief discussion only to those properties of excited states in the condensed phase which are necessary for clarity, coherence, and understanding of organic magneto-optical effects.

Consider an organic solid, composed of, say, N molecules. According to quantum mechanics this crystal, when in its ground state, is characterized by a total energy $E_0 = N\varepsilon_0$ and a wavefunction Ψ_0 (which is a function of all electronic and nuclear position variables and their spins, as can be seen from Eq. (4.1)). The difference between ε_0 and the free molecule ground-state energy, ω_0, is defined as the cohesive energy per molecules, $\delta\varepsilon$, which for typical organic solids, such as anthracene, tetracene, and pyrene, is considerably smaller than that exhibited by ionic solids or metals (Gutmann and Lyons, 1967). The actual physical state described by Ψ_0 is one in which all molecules are in their lowest (ground) state. If we let ϕ_0^i denote the free molecular ground state for the ith molecule, then an excellent approximation to Ψ_0 is

$$\Psi_0 \approx \overline{\Psi}_0 \equiv \prod_{i=1}^{N} \phi_0^i \tag{4.15}$$

(where $\overline{\Psi}_0$ is defined as the product of wavefunctions) since $\delta\varepsilon$ is small. This approximation in conventional solid-state terminology is called the tight binding approximation. $\overline{\Psi}_0$ is exact only in the case of non-interacting molecules.

In order to specify the excited states of our system of N-oriented molecules, which for reasons of simplicity we consider as forming a rigid lattice, we let ϕ_m^j represent the mth excited state of the jth molecule. Here m denotes a collection of quantum numbers specifying the electronic state, spin, and vibrational modes of the individual molecule. The crystal Hamiltonian operator, $\mathcal{H}$, assuming only electrostatic interactions, is a sum of Hamiltonians $\mathcal{H}_j$ for the free molecules and potential energy terms V_{ij} specifying the intermolecular interaction between molecules j and i. Thus

$$\mathcal{H} = \sum_j \mathcal{H}_j + \sum_{i<j} V_{ji} \equiv \mathcal{H}_0 + V \tag{4.16}$$

It is important to realize that from a fundamental point of view Eq. (4.16) is only an approximation to the true state of affairs in that magnetic interactions have been neglected. However, as discussed later in this chapter, magnetic terms are generally small and can be adequately handled within the context of perturbation theory.

In addition to the state $\overline{\Psi}_0$, which consisted of a product of ground state molecular wavefunctions we have an infinite number of states of the form

$$\overline{\Psi}(n1, n'2, ..., n''N) = \phi_1^n \phi_2^{n'} ... \phi_N^{n''} \tag{4.17}$$

However, due to the existence of the intermolecular interaction V it is obvious that the function given by Eq. (4.17) is *not* an eigenstate of the crystal Hamiltonian (Eq. 4.16). Nevertheless, it has been found that, at least for low-lying crystal states, the manifold of states with one, two,..., etc. excited molecules constitutes a "reasonably" sufficient basis (although not complete in a mathematical sense) in which to interpret the optical properties of the crystal. Equation (4.17) neglects the anti-symmetrization principle; however, this can be easily handled by using an anti-symmetrizing operator for intermolecular exchange (Kauzmann, 1957). In order to complete the basis set for perturbation calculations it is essential to include states in which the excited electrons are transferred to neighboring or more distant molecules.

In the case where an electron is transferred to a neighboring molecule the residual hole and excess electron are still under the Coulomb forces of each other and thus constitute a bound, localized electron–hole pair. In this configuration such a state is called a localized charge-transfer state. Obviously there are "many" such states; however, as the separation of the electron and hole increase their mutual intermolecular interaction decreases and eventually becomes screened out with the subsequent result that both the electron and hole move uncorrelated with respect to each other. In this way charge-transfer states of the crystal lie in the low energy region of the conduction band.

In the preceding discussion we have not yet constructed stationary states of the crystal but have only indicated the form of non-stationary basis states which are appropriate for the calculation. For clarification we shall discuss a one-dimensional lattice model with one molecule per unit cell. From this analysis the concept of the exciton as an excited mobile, neutral non-conducting state of an insulating solid will be introduced which in turn will allow us to outline briefly the differences in spectra manifested in the crystal phase as compared to the vapor phase (isolated molecule case).

4.3.2. One-dimensional Exciton Model

We consider a linear array of N equally spaced molecules which we assume have non-degenerate excited states. We assume periodic boundary condition such that the molecule labelled 1 is fully equivalent to the $N+1$ molecule. The generalization to degenerate excited states constitutes no new physical concepts other than the use of degenerate perturbation theory; in addition, attention must be paid to a proper book-keeping of the various subscripts. It should be obvious that all states of the form

$$\overline{\Psi}(01, 1n, ..., 0N) = \phi_1^n \prod_{i \neq n} \phi_0^i = \overline{\Psi}_n' \tag{4.18}$$

where $n = 1, 2, ..., N$ are degenerate since an electron excited to its first excited state on the nth molecule, where the remaining molecules are in their ground state, are entirely equivalent energetically to having the mth molecule excited with the other molecules in the ground state. States described by $\overline{\Psi}_n'$ are localized and are non-stationary due to the presence of V. In Fig. 4.6 two such states are schematically illustrated.

As is well known from quantum mechanics, degeneracy can be lifted by forming appropriate linear combinations of the above non-stationary states by minimizing the energy for the system (Kauzmann, 1957). Thus we form a set of states of the form

$$\Psi = \sum_n a_n \overline{\Psi}_n' \tag{4.19}$$

and then demand within the context of the variational principle that

$$\delta(\langle \Psi | \mathscr{H} | \Psi \rangle) = 0 \tag{4.20}$$

subject to the condition that the normalization of the wavefunction given by Eq. (4.19) is preserved under a variation of the coefficients a_n. It is not difficult to see that the aggregate of N energy levels associated with each free-molecule state of the aggregate of N identical molecules spreads into a band of states.

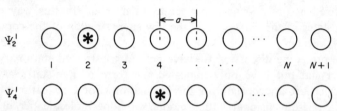

Fig. 4.6. Numbers 1, 2, ... label the molecules in this one-dimensional lattice with lattice spacing a. Cyclic boundary conditions are assumed such that molecule 1 is entirely equivalent to molecule $N+1$. Two non-stationary electronically neutral bound pair states are represented. ✳ denotes the excited molecule, whereas all other molecules are taken to be in their ground state.

If we assume periodic boundary conditions and retain only nearest-neighbor interactions, then the solution to Eq. (4.20) is

$$\Psi_k = \frac{1}{\sqrt{N}} \sum_{l=1}^{N} \exp\left(2\pi i k l/N\right) \overline{\Psi}'_l \tag{4.21}$$

where $k = 0, \pm 1, ..., +N/2$ serves to classify the states and is the one-dimensional wavevector. The eigenvalues corresponding to these N states are

$$\varepsilon(k) = \delta\omega + 2\beta \cos\left(2\pi k/N\right), \tag{4.22}$$

where β is the nearest neighbor interaction matrix element and is negative in magnitude. When exchange effects are neglected

$$\beta = \langle\, \phi_0^i(r_1)\, \phi_1^{i+1}(r_2) \,|\, V_{12} \,|\, \phi_1^i(r_1)\, \phi_0^{i+1}(r_2) \,\rangle \tag{4.23}$$

The term $\delta\omega$ constitutes an energy term independent of the k-value. In Fig. 4.7 we summarize the basic effects on the energy spectrum. One of the basic properties, characteristic of all organic crystals, is that the crystal fluorescence (represented by line 2 in Fig. 4.7) and absorbance occurs at longer wavelength ("red shifted") than in the free molecule case. This red shift can be of the order of several hundred cm^{-1} for singlet spin states.

States described by Eq. (4.21) are called Frenkel excitons (tight-binding approximation or neutral excitons) after Frenkel (1931) who initially introduced the concept. However, its application to organic crystals was initially given by Davydov (1948). The label k, introduced as only an index for solutions to Eq. (4.20), can be shown (Knox, 1963) to correspond to the exciton's momentum, i.e. those excitons with large k are moving faster than those with small k values, whereas for $k = 0$ the exciton is stationary. Since the wavelength of visible radiation is long compared to the intermolecular separation, it follows that only transitions to states near $k = 0$ are directly accessible from the ground state. Hence transitions, such as that represented by line D in Fig. 4.8 are not allowed in first order theory unless phonon–exciton interactions are considered explicitly.

States of a given k value travel in a fixed direction; however, on collision with lattice impurities or phonons, a transition to another k state as represented by the dotted line C in Fig. 4.8 is possible. It is convenient to

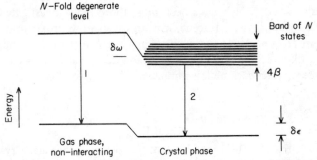

Fig. 4.7. General level scheme illustrating shifts in ground and first excited eigenstates for the crystal states in relationship to the vapor phase states. The N-fold degenerate level for non-interacting molecules ($v = 0$) forms a band of states with an energy spread of $4|\beta|$, each state labelled by a distinct k value representing the crystal momentum. In general, a photon emitted from the crystal states, represented by line 2, is longer in wavelength than the corresponding transition (line 1) in the gas phase.

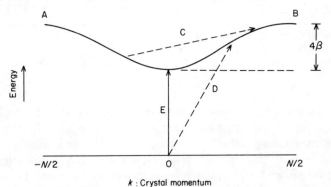

Fig. 4.8. Energy band structure assuming only nearest-neighbor interactions whose strength is represented by β in this figure (Eq. 4.23, in text). State A ($k = -N/2$) is fully equivalent to state B ($k = N/2$). The dotted line, labelled C, represents intraband scattering between two Frenkel exciton states (see text). Line E denotes the transition from the ground state to the $k = 0$ state. The transition represented by line D is forbidden in first order and only occurs with the inclusion of phonon interactions.

define the length of time an exciton remains in a given k-state as the "coherence time" for that state, $\tau(k)$, where the distance it travels as its coherence distance, $l(k)$. If $\tau(k)$ is large or $l(k)$ is much larger than several lattice constants we generally say that the excitons for that system move coherently; whereas incoherent motion implies a short scattering time or coherence distance. In the incoherent limit, where $l(k) \approx$ lattice constant, then it is convenient to consider the excited state or "localized" exciton as hopping in a stochastic manner from one molecular site to another at a rate proportional to $|\beta|^2$, where β is given by Eq. (4.23). Typical hopping times between neighboring molecules for singlet excitons are $\sim 10^{-13}$ s where the transfer rates for triplet excitons are of the order of 10^{11}–10^{12} s^{-1}.

In organic solids which have been most thoroughly investigated, such as anthracene, pyrene, and tetracene, it is an excellent approximation at room temperature to consider the low-lying singlet or triplet states as migrating in random walk-like manner. However at low temperatures, where phonon–exciton interactions become less important, the degree of coherence increases

TABLE I

Triplet Exciton Diffusion Constants for Several Molecular Crystals
(units are cm²/s and refer to room temperature values)

Naphthalene [a]	$D_{aa} = 3\cdot3 \times 10^{-5}$	$D_{bb} = 2\cdot7 \times 10^{-5}$	
Anthracene [b]	$D_{aa} = 1\cdot5 \times 10^{-4}$	$D_{bb} = 1\cdot8 \times 10^{-4}$	$D''_{cc} = 1\cdot2 \times 10^{-5}$
Pyrene [c]	$D_{aa} \approx 0\cdot3 \times 10^{-4}$	$D_{bb} = 1\cdot25 \times 10^{-4}$	$D''_{cc} \approx 0\cdot3 \times 10^{-4}$
Stilbene [a]	$D_{aa} = 0\cdot9 \times 10^{-4}$	$D_{bb} = 0\cdot7 \times 10^{-4}$	
1,4 DBN [a]	$D_{aa} \approx 10^{-5}$	$D_{bb} \approx 10^{-5}$	$D_{cc} \approx 3\cdot5 \times 10^{-4}$

[a] Ern, V. (1972). *J. Chem. Phys.* **56**, 6259.
[b] Ern, V. (1969). *Phys. Rev. Lett.* **22**, 8.
[c] Arnold, S., Fave, J. L., and Scott, M. (1974). *Chem. Phys. Lett.* **28**, 412.

allowing the exciton wave-packet to travel distances of hundreds of lattice spacings before being scattered. Our discussions in this chapter of bi-molecular magneto-optical effects is presented within the incoherent limit.

Because of their random walk behavior it is possible to characterize the motion of singlet and triplet exciton states by a diffusion tensor. The tensorial nature is a direct consequence of the anisotropy of the transfer integrals, β. Singlet exciton diffusion coefficients vary from 10^{-2}–10^{-4} cm²/s whereas triplet excitons, which migrate by exchange interaction due to the vanishing of the direct Coulomb interaction, have diffusion coefficients of $\sim 10^{-3}$–10^{-6} cm²/s. Table I lists the diffusion coefficients for several crystals.

4.3.3. Exciton Properties for Crystals with more than one Molecule per Unit Cell

In addition to the red shift in the fluorescence, other spectral differences manifested by the crystalline phase as compared to the vapor state are: (1) the multiplication of the number of lines in the absorption spectra; (2) changes in the light absorption intensities caused by the "borrowing" of intensity by one transition from another; and (3) the relaxation of spectral selection rules with the appearance of totally new bands.

In his application of exciton theory to molecular crystals, Davydov showed that for crystals with more than one inequivalent molecule per unit cell, a given molecular energy level is split into as many components as there are inequivalent molecules per unit cell. This splitting is in addition to the band splitting as discussed above. For a crystal with N unit cells containing two inequivalent molecules per unit cell, the crystal spectra exhibit two distinct

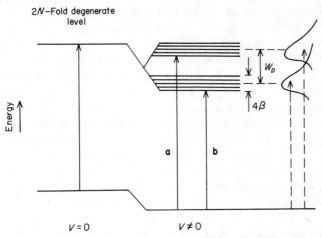

Fig. 4.9. Schematic illustration of the splitting of the $2N$-fold degenerate level for a crystal with two molecules per unit cell into two distinct Davydov bands; **a** and **b** denote the corresponding polarizations associated with each of the Davydov bands. W_D is defined as the Davydov splitting and its magnitude arises from inequivalent interactions only.

bands each containing N possible k-states as shown in Fig. 4.9. The splitting, W_D, between the crystal components is known as the Davydov splitting and depends only upon the interactions between molecules not translationally equivalent, whereas the mean shift downward depends on equivalent interactions. The Davydov splitting is typically of the order of 10 cm^{-1} for triplet states (because of the short range of exchange interactions) whereas for singlet excitons it can vary from a few hundred cm^{-1} to several thousand cm^{-1} for highly excited singlet states (Gutmann and Lyons, 1967; Birks, 1970).

In addition to this spectral splitting, the two Davydov bands of states have different polarization properties. Either from symmetry or direct diagonalization of the crystal Hamiltonian within the manifold of states $\overline{\Psi}_n^1$ it can be shown (Davydov, 1962) that for monoclinic crystals (such as anthracene and naphthalene) with two crystallographic equivalent molecules in the unit cell, the crystal states for the Davydov bands are

$$\Psi_b(k) = \frac{1}{\sqrt{2}}[\Psi_k^{(1)} - \Psi_k^{(2)}] \tag{4.24}$$

$$\Psi_a(k) = \frac{1}{\sqrt{2}}[\Psi_k^{(1)} + \Psi_k^{(2)}] \tag{4.25}$$

Here $\Psi_k^{(1)}$ and $\Psi_k^{(2)}$ are given by Eq. (4.21) where the subscript refers to the sum in Eq. (4.21) over the corresponding sublattices. These functions are sometimes called site excitons. For free molecular transitions which are dipole allowed, it can be shown that the symmetric combination of site functions transition moment lies in the ac-plane whereas the antisymmetric combination is b-polarized. In no case, however, is the strict forbiddenness of gerade to gerade transition broken in the condensed phase. Calculation of the intensities of these a- and b-polarized transitions constitutes a formidable task since good agreement between theory and experiment is not obtainable in the *oriented gas* model and necessitates inclusion of higher order effects of intermolecular coupling (Craig and Walmsley, 1968). In Fig. 4.10 we show a plot of the absorption spectrum of crystalline anthracene which illustrates both the energy and intensity dependence of the two Davydov components. The dotted curve in Fig. 4.10 is the absorption spectra in hexane solution which on comparison with the solid phase illustrates the general lowering of the energy levels, i.e. red shift in the crystalline phase.

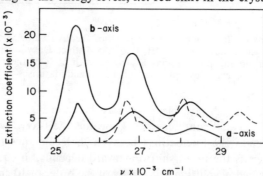

Fig. 4.10. The pure crystal spectrum of anthracene at room temperature with the electric vector along the **a** and **b** directions of the monoclinic crystal. The dotted curve corresponds to anthracene spectrum in hexane. (Data from T. A. Claxton, Ph.D. Thesis, University of London, 1961.)

4.4. BASIC MAGNETIC INTERACTIONS

The Hamiltonian of a molecule may be written

$$\mathcal{H} = \mathcal{H}_{el} + \mathcal{H}_{spin} + \mathcal{H}_{so} \tag{4.26}$$

where $\mathcal{H}_{el}$ is independent of spin and includes electronic and nuclear co-ordinates, thus determining the overall electronic energy of the molecule. $\mathcal{H}_{spin}$ includes all of the spin degrees of freedom (electronic and nuclear) while $\mathcal{H}_{so}$ is the spin–orbit coupling term. It is important to understand the nature of the spin Hamiltonian, which also contains the Zeeman terms, for an appreciation of the magnetic field effects described here. The spin–orbit coupling term $\mathcal{H}_{so}$ plays no role in the magnetic field effects, but is nevertheless important in mixing singlet and triplet states intramolecularly. Before proceeding to the spin Hamiltonian, we shall make some brief comments on spin–orbit coupling.

4.4.1. Spin–Orbit Coupling Hamiltonian

In the absence of spin–orbit interaction, one may speak of pure singlet and pure triplet states of a molecule; the spin selection rules prohibiting transitions between states of different multiplicity are rigorously obeyed in this case. Spin–orbit coupling is a relativistic phenomenon and may be visualized as the coupling of the spin angular momentum of an electron with its orbital motion. In the frame of reference of the electron it is actually the nucleus which is in motion and which gives rise to a magnetic field with which the electron spin interacts. It can be shown that the spin–orbit coupling operator has the form (McGlynn et al., 1969)

$$\mathcal{H}_{so} = \zeta \mathbf{L} . \mathbf{S} \tag{4.27}$$

where ζ is the spin–orbit coupling constant which is proportional to Z^4, where Z is the atomic number. L and S are the orbital and spin operators respectively. This is the origin of the heavy-atom effect which can operate intramolecularly or intermolecularly as pointed out in connection with the quenching effects discussed in the previous section.

The spin–orbit coupling operator provides a small admixture of triplet states to the singlets and gives rise to singlet character within the triplet states. According to perturbation theory the mixing coefficients C_{ij} are given by

$$C_{ij} = \frac{\langle {}^3\Psi_i^0 | \mathcal{H}_{so} | {}^1\Psi_j^0 \rangle}{|{}^3E_i - {}^1E_j|} \tag{4.28}$$

where ${}^3E_i$ and ${}^1E_j$ are the energies of the pure triplet and singlet zero-order states ${}^3\Psi_i^0$ and ${}^1\Psi_j^0$ respectively. The perturbed wavefunctions can then be written

$$^3\Psi'_i = {}^3\Psi'^0_i + \sum_j C_{ij}{}^1\Psi'^0_j \tag{4.29}$$

and

$$^1\Psi'_j = {}^1\Psi'^0_j + \sum_i C_{ji}{}^3\Psi'^0_i \tag{4.30}$$

The summations in Eqs (4.29) and (4.30) run over all possible singlets and triplets respectively. In polynuclear aromatic hydrocarbons the three $S_{1,2,3} \leftarrow S_0$ transitions (corresponding to an excitation of the ^1L_a, ^1L_b, and ^1B_b states) are polarized in the planes of the molecules. The phosphorescence $T_1 \rightarrow S_0$ in these molecules is polarized primarily normal to the plane of the molecules. It thus appears that at least in the case of phosphorescence, the dominant perturbing singlet states in Eq. (4.30) are $\sigma\pi^*$ or $\pi^*\sigma$ states.

The non-radiative phenomenon of intersystem crossing $S_1 \rightsquigarrow T_q$ is, of course, also mediated by $\mathcal{H}_{so}$, but the coupling routes can be quite complicated. Thus, the population of each of three triplet sublevels of T_q from S_1 are highly selective, i.e. the three sublevels are not populated equally (El-Sayed, 1971).

We would like to stress at this point that not all processes in which triplets are produced from singlets are mediated by the spin–orbit coupling operator. One of the important processes, which is discussed in detail in this chapter, is singlet exciton fission. In this phenomenon one singlet fissions into two triplets: $S_1 \rightarrow T_1 + T_1$. It is completely allowed by spin selection rules and so the intervention of the spin–orbit coupling mechanism is not necessary in this highly special case of intersystem crossing.

4.4.2. The Spin Hamiltonian

The various relevant terms in the spin Hamiltonian for molecules in triplet states are:

$$\mathcal{H}_{\text{spin}} = \underbrace{\mathcal{H}_{\text{SS}} + \mathcal{H}_{\text{SZ}}}_{\text{electron spin terms}} + \underbrace{\mathcal{H}_{\text{HF}}}_{\substack{\text{electron–nuclear} \\ \text{interaction}}} + \underbrace{\mathcal{H}_{\text{NZ}}}_{\substack{\text{nuclear Zeeman} \\ \text{term}}} \tag{4.31}$$

We discuss the significance of each of these terms separately.

(a) *Zero-field Splitting.* $\mathcal{H}_{\text{SS}}$ is the fine structure operator whose origin is the spin–spin interaction of the two unpaired electrons in the triplet state which can be shown (Bersohn and Baird, 1966) to have the form

$$\mathcal{H}_{\text{SS}} = g_e{}^2 \beta_e{}^2 \left\{ \frac{S_1 \cdot S_2}{r^3} - \frac{3(S_1 \cdot r)(S_2 \cdot r)}{r^5} \right\} = S \cdot D \cdot S \tag{4.32}$$

where $g_e = 2\cdot002322$ for electrons, $\beta_e = 0\cdot927 \times 10^{-20}$ erg/gauss (or $4\cdot64 \times 10^{-5}$ cm^{-1}/gauss) is the Bohr magneton, r is the vector joining the two electrons and S_1 and S_2 are the spin operators. D is known as the zero-field splitting tensor which can be brought into diagonal form by a proper choice of principal axes (Carrington and McLachan, 1967; McGlynn *et al.*, 1969). In the case of the polycyclic aromatic hydrocarbons, these principal axes correspond to the molecular symmetry axes. After diagonalizing Eq. (4.32), one obtains

$$\mathscr{H}_{SS} = D(S_z^2 - \tfrac{1}{3}S^2) + E(S_x^2 - S_y^2) \qquad (4.33)$$

The molecular symmetry axes x, y, z are indicated for naphthalene in Fig. 4.11. It is evident that z is the normal to the plane and corresponds to the twofold axis of rotation.

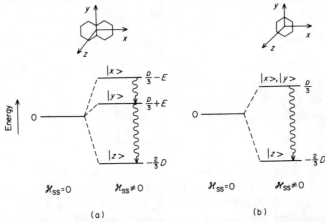

Fig. 4.11. Zero-field splitting of triplet magnetic substates in aromatic hydrocarbons. (A) Relatively low symmetry: naphthalene, anthracene, tetracene, pentacene. (B) Relatively high symmetry (equal to or greater than C_3): benzene, triphenylene, coronene; the degeneracy is only partially lifted in this case. $|x\rangle$, $|y\rangle$, and $|z\rangle$ are the spin functions. The wiggly lines denote the spin relaxation processes between the magnetic sublevels.

D and E are the zero-field splitting parameters. If ψ_j describes the orbital part of the triplet wavefunctions, then

$$D = \tfrac{3}{4}g_e^2\beta_e^2 \left\langle \psi_j \Big| \frac{r_{12}^2 - 3z_{12}^2}{r_{12}^5} \Big| \psi_j \right\rangle \qquad (4.34)$$

and

$$E = \tfrac{3}{4}g_e^2\beta_e^2 \left\langle \psi_j \Big| \frac{y_{12}^2 - x_{12}^2}{r_{12}^5} \Big| \psi_j \right\rangle \qquad (4.35)$$

with $r_{12}^2 = x_{12}^2 + y_{12}^2 + z_{12}^2$ being the distance between the two electrons.

Due to the spin–spin interaction described by Eqs (4.33) to (4.35), the degeneracy of the three triplet sublevels is lifted even in the absence of an external magnetic field. This zero-field splitting arises from an "internal" magnetic field which is due to the spin of each electron and its effect on the other. In planar molecules, when the symmetry axis is chosen as the perpendicular z-axis, and when the symmetry is equal to or higher than C_3 (threefold axis of rotation), $E = 0$ since x- and y-axes cannot be differentiated from each other and $x_{12}^2 = y_{12}^2$. Thus for molecules such as benzene, coronene, and triphenylene, the degeneracy among the three triplet sublevels is not completely lifted and there are only two triplet sublevels rather than three. The splitting into magnetic sublevels is illustrated in Fig. 4.11. In aromatic hydrocarbons D and E are generally of opposite sign, with $D/E \sim -10$ while D is of the order of ~ 0.1 cm^{-1} and is positive.

(b) *Zeeman Terms*. The electron spin Zeeman term is

$$\mathscr{H}_{SZ} = \beta_e \mathbf{H} . \mathbf{g}_e . \mathbf{S} \tag{4.36}$$

Due to the small anisotropy of the g-factor, Eq. (4.36) can be rewritten as

$$\mathscr{H}_{SZ} = g_e \beta_e \mathbf{H} . \mathbf{S} \tag{4.37}$$

where $\mathbf{S} = \mathbf{S}_1 + \mathbf{S}_2$ represents the total spin operator for the two electrons and $\mathbf{H}$ is the external magnetic field.

The nuclear Zeeman term is

$$\mathscr{H}_{NZ} = \sum_i g_N \beta_N \mathbf{H} . \mathbf{I}_i \tag{4.38}$$

where $\mathbf{I}_i$ is the spin operator for nucleus i and $\beta_N = 0.505 \times 10^{-23}$ erg/gauss $= 2.53 \times 10^{-8}$ cm^{-1}/gauss is the nuclear magneton. For protons the g-value is $g_N = 5.59$. Thus $g_N \beta_N / g_e \beta_e \approx 1.5 \times 10^{-3}$ and we can usually ignore the nuclear Zeeman term in comparison to the electronic Zeeman term. The term $\mathscr{H}_{NZ}$ will thus not enter further into our consideration of magnetic field effects.

(c) *Hyperfine Interactions*. The hyperfine term $\mathscr{H}_{HF}$ in Eq. (4.31) describes the interaction between the nuclei and the electrons. It consists of two parts: an isotropic part $a_{ij} \mathbf{I}_i . \mathbf{S}_j$ which is due to the Fermi contact interaction and an anisotropic part $\mathbf{S}_j . T . \mathbf{I}_i$ which is due to the electron–nuclear dipolar interaction. We thus have

$$\mathscr{H}_{HF} = \sum_{i,j} a_{ij} \mathbf{I}_i . \mathbf{S}_j + \sum_{i,j} \mathbf{S}_j . T . \mathbf{I}_i \tag{4.39}$$

The magnitudes of a_{ij}, the hyperfine contact terms, are of the order of 10^{-4}–10^{-3} cm^{-1}. The electron–nuclear dipolar interactions are similar in magnitude.

4.4.3. Magnetic Field Effects and Magnitudes of Terms in the Spin Hamiltonian

The electronic Zeeman term is $g_e \beta_e \approx 10^{-4}$ cm^{-1}/gauss. Thus at fields of 1000 gauss, the electronic Zeeman term $\mathscr{H}_{SZ}$ in the spin Hamiltonian is of the same size as the zero-field splitting term $\mathscr{H}_{SS}$. This term is characterized by the parameters D and E $(D \sim 0 \cdot 1 \text{ cm}^{-1})$; magnetic field effects which are due to an interplay of the electron spins with each other and with the external magnetic field will thus manifest themselves at magnetic fields around 1000 gauss. The fusion and fission effects described below belong to this category of effects.

Magnetic field effects on luminescence arising from the hyperfine interaction term $\mathscr{H}_{HF}$ in Eq. (4.31) are also possible. These effects have been interpreted in terms of the hyperfine contact interaction a_{ij} in Eq. (4.39). For anthracene the values of a_{ij} are of the order of 10^{-3} cm^{-1} which corresponds to a magnetic field of ~ 10 gauss $(g_i \beta_e H = 10^{-3} \text{ cm}^{-1})$. Thus, magnetic field effects in which hyperfine effects manifest themselves should be observable at field strengths of the order of 10 gauss or less.

4.4.4. Molecular Triplet Spin Functions and Energies

For an understanding of magnetic field effects it is necessary to understand first the nature of the molecular triplet eigenfunctions, and the changes in the energies of the triplet sublevels in the presence of an external magnetic field.

We consider the zero-field case and neglect, to a first approximation, the small perturbation due to the hyperfine term $\mathscr{H}_{HF}$. By choosing the usual triplet basis states

$$\Theta_j = |\alpha_1 \alpha_2 \rangle, \quad |\beta_1 \beta_2 \rangle, \quad \text{and} \quad (2)^{-\frac{1}{2}}\{|\alpha_1 \beta_2 \rangle + |\beta_1 \alpha_2 \rangle\} \qquad (4.40)$$

the eigenstates and corresponding eigenvalues of $\mathscr{H}_{SS}$ are found by considering the 3×3 matrix

$$\langle \Theta_i | \mathscr{H}_{SS} | \Theta_j \rangle \qquad (4.41)$$

This is done explicitly in the books by Carrington and McLachlan (1967) and McGlynn *et al.* (1969). It is shown that the matrix (4.41) can be brought into diagonal form by choosing an orthogonal coordinate system with the principal axes of the tensor $\mathbf{D}$ along the x, y, and z molecular axes of the aromatic hydrocarbon molecules. The eigenfunctions which bring this matrix into diagonal form are

$$|x\rangle = (2)^{-\frac{1}{2}}\{|\beta_1 \beta_2 \rangle - |\alpha_1 \alpha_2 \rangle\} \qquad (4.42)$$

$$|y\rangle = i(2)^{-\frac{1}{2}}\{|\beta_1 \beta_2 \rangle + |\alpha_1 \alpha_2 \rangle\} \qquad (4.43)$$

$$|z\rangle = (2)^{-\frac{1}{2}}\{|\alpha_1 \beta_2 \rangle + |\beta_1 \alpha_2 \rangle\} \qquad (4.44)$$

Using these basis functions and the Hamiltonian in Eq. (4.33), the eigenvalues may be calculated. The eigenvalues corresponding to the eigenstates

$|x\rangle, |y\rangle$, and $|z\rangle$ are

$$W_x^0 = D/3 - E$$
$$W_y^0 = D/3 + E \qquad (4.45)$$
$$W_z^0 = -2/3D$$

These are the eigenvalues indicated in Fig. 4.11. Each of these states and corresponding eigenvalues may be associated with one of the principal axes. The physical meaning of these three triplet substates may be viewed as follows. It is easily seen that

$$S_z^2|z\rangle = S_y^2|y\rangle = S_x^2|x\rangle = 0 \qquad (4.46)$$

The z state corresponds to a precession of electron spins in the xy-plane with a zero component of spin angular momentum along the z-axis. Similarly, in the $|y\rangle$ state the spins precess about the y-axis in the xz-plane. In the $|x\rangle$ state the spins precess in the yz-plane.

We now consider the energies of the triplets in the presence of an external magnetic field H. If H is directed along the z-axis of the molecule, the energy of the $|z\rangle$ state remains unchanged, while the energies of the other two states $|x\rangle$ and $|y\rangle$ increase and decrease respectively with increasing field strengths. The energies of the triplets can be calculated at any field strength by finding the roots of the secular determinant

$$|H_{ij} - W\delta_{ij}| = 0 \qquad (4.47)$$

with

$$H_{ij} = \langle \Theta_i | \mathscr{H}_{SS} + \mathscr{H}_{SZ} | \Theta_j \rangle \qquad (4.48)$$

However, for our purposes it is sufficient to calculate the eigenvalues in the limiting case when $\mathscr{H}_{SZ} \gg \mathscr{H}_{SS}$ and the $\mathscr{H}_{SS}$ term can be treated in the context of first-order perturbation theory. In this case, it is convenient to use the functions in Eq. (4.40) as the zero-order, unperturbed states. If we denote by α, β, and γ the angles between the magnetic field $\mathbf{H}$ and the molecular axes x, y, and z respectively, the energy levels are according to first-order perturbation theory

$$\left. \begin{aligned} W_+ &= g\beta H + \tfrac{1}{2}W' \\ W_0 &= -W' \\ W_- &= -g\beta H + \tfrac{1}{2}W' \end{aligned} \right\} \qquad (4.49)$$

where

$$W' = D(\cos^2\gamma - \tfrac{1}{3}) + E(\cos^2\alpha - \cos^2\beta) \qquad (4.50)$$

4.4.5. Triplet Exciton Spin Functions and Energies

In most of the crystals which concern us here, there are two translationally inequivalent molecules per unit cell. The spin Hamiltonian of the triplet excitons is the average over the Hamiltonians $\mathscr{H}_1$ and $\mathscr{H}_2$ of the two inequivalent molecules (Sternlicht and McConnell, 1961)

$$\mathscr{H}^* = \tfrac{1}{2}\{\mathscr{H}_1 + \mathscr{H}_2\} = \tfrac{1}{2}\{\mathscr{H}_{SZ}(1) + \mathscr{H}_{SS}(1) + \mathscr{H}_{SZ}(2) + \mathscr{H}_{SS}(2)\} \quad (4.51)$$

It can be shown that $\mathscr{H}^*$ can be expressed in the form similar to that of the molecular Hamiltonian:

$$\mathscr{H}^* = \mathscr{H}^*_{SS} + \mathscr{H}^*_{SZ}$$

$$= g_e \beta_e \mathbf{H} . \mathbf{S} + D^*(S_{z*}^{\ 2} - \tfrac{1}{3}S^2) + E^*(S_{x*}^{\ 2} - S_{y*}^{\ 2}) \quad (4.52)$$

x^*, y^*, and z^* are no longer the molecular axes, but the principal axes of the crystal fine-structure tensor which bring $\mathscr{H}^*$ into a diagonal form. D^* and E^* are the fine-structure zero-field splitting parameters for the triplet exciton state. For crystals with monoclinic symmetry (e.g. naphthalene and anthracene), the z^*-axis coincides with the crystallographic b-axis. The x^* and y^* lie in the ac-plane; their orientation in this plane depends on the values of D^* and E^*. For tetracene, which is triclinic, the z^*-axis is close to the crystallographic b axis but does not coincide with it.

Values of D, E, D^*, and E^* have been summarized for several molecules by Swenberg and Geacintov (1973). It should be noted that for both anthracene and tetracene the values of D^* are smaller than the molecular terms D by a factor of about 10, while E^* is about 4–5 times larger than the corresponding molecular parameter E. It follows therefore that the equivalence of $\mathscr{H}^*_{SS}$ and $g_e \beta_e H$ is reached at a lower magnetic field for crystal excitons than for the individual molecules (~ 350 gauss in anthracene crystals).

The triplet exciton magnetic sublevel energies as a function of the orientation of $\mathbf{H}$ with respect to x^*, y^*, and z^* principal axes are given by expressions analogous to those in Eqs (4.49) and (4.50) with D and E replaced by D^* and E^* respectively. The triplet exciton wavefunctions are also denoted by $|x^*\rangle$, $|y^*\rangle$, and $|z^*\rangle$ and have the same form as the wavefunctions given in Eqs. (4.42)–(4.44).

4.5. MAGNETIC FIELD EFFECTS ON LUMINESCENCE

4.5.1. Overview

The initial discovery of a magnetic field effect on the luminescence of an organic crystal was made in 1967 by Johnson et al. (1967). They found that the delayed fluorescence in anthracene, which is due to triplet–triplet annihilation (fusion), can be modulated by external magnetic fields of the order of several hundred to several thousand gauss. The magnitude of the field effect is dependent on both the field strength and the orientation of the magnetic field strength with respect to the crystallographic axes. These effects were successfully interpreted by Merrifield (1968) in terms of the triplet exciton spin Hamiltonian. More elaborate explanations were later provided by Johnson and Merrifield (1970) and by Suna (1970). Lucid summaries of the research carried out by this group have been published (Merrifield, 1971; Avakian, 1974).

Swenberg and Stacy (1968) predicted that the opposite of the fusion of two triplet excitons, namely the fission of one singlet exciton into two triplets, could occur in crystalline tetracene. Experimentally, fission was first observed in tetracene by Geacintov et al. (1969) and by Merrifield et al. (1969). This effect is readily observable in tetracene crystals where the energy of two triplet excitons is energetically about ~ 1700 cm^{-1} above that of the singlet. Thermally induced fission is thus possible and the prompt fluorescence of tetracene is magnetic field sensitive at room temperature, but not at low temperature where fission is suppressed (Geacintov et al., 1969).

In later papers (Pope et al., 1969; Groff et al., 1970a) the existence of fusion in tetracene crystals was also demonstrated. In tetracene, the fission can also occur from the short-lived upper vibrationally excited levels (Klein et al., 1972, 1973; Moller and Pope, 1973; Arnold, 1974; Swenberg et al., 1974). This case is termed "hot fission" and the effect of magnetic field is usually small ($\sim$ several per cent at most) and dependent on the wavelength of excitation. This is in contrast to thermally activated fission in tetracene, which increases by 15-35% in the presence of the magnetic field and where the effect does not depend on the wavelength of excitation.

The phenomena of heterofission (Geacintov et al., 1971) and heterofusion (Groff et al., 1970c; Chabr et al., 1974; Kalinowski and Godlewski, 1974) in which the interacting triplets are on different types of molecules in a crystal has been observed also. In these experiments doped crystals are used in which the concentration of guest molecules is less than 10^{-4} M. In heterofusion, which has been observed in tetracene-doped anthracene, a triplet is first trapped at a tetracene guest site. Another triplet exciton, belonging to the host anthracene crystal, collides with a tetracene-trapped triplet and fusion takes place to produce one tetracene singlet which is then also trapped. The tetracene molecule of course constitutes a trapping site for both singlet and triplet excitons because the energy levels of the tetracene are lower than those of the host anthracene. Heterofission has been observed so far only in tetracene crystals doped with pentacene. A tetracene host singlet exciton diffuses throughout the crystal and is trapped by a pentacene impurity molecule. This pentacene is energetically capable of fissioning into one tetracene and one pentacene triplet.

Another magnetic field effect involving triplet states includes the quenching of triplets by paramagnetic doublet states (Ern and Merrifield, 1968). The quenching efficiency is reduced in the presence of an external magnetic field.

The magnetic effects described here constitute the class of effects which take place in organic crystals and are those best understood. These effects are due to an interplay between the zero-field part $\mathcal{H}^*_{SS}$ and the Zeeman part $\mathcal{H}^*_{SZ}$ of the spin Hamiltonian, (Eq. 4.52), and exhibit characteristic features which may be used to recognize the origin of these particular magnetic field effects. These features include the effect of magnetic field strengths on either the prompt fluorescence (fission), delayed fluorescence (fusion) and triplet–doublet quenching, which are summarized in Fig. 4.12. Fission and fusion display a characteristic inversion at low fields, while no such inversion is

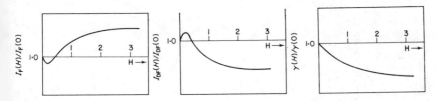

Fig. 4.12. Schematic magnetic field (H in kilogauss) dependence of the prompt fluorescence I_F (thermally stimulated and hot fission, maximum effect 1·35 and minimum $\sim 1·01$–1·02 respectively), delayed fluorescence I_{DF} (maximum $\sim 1·05$, minimum $\sim 0·80$) and triplet–doublet interaction characterized by the rate constant γ.

exhibited in triplet–doublet quenching. Another readily observed diagnostic test of this type of magnetic field effects is the effect of the orientation of the magnetic field with respect to the crystallographic axes on the magnitude of the effect; this is illustrated in Fig. 4.13. Two maxima or minima are observed for every 180° degree of rotation of the magnetic field. The triplet–doublet quenching effect displays a similar orientation dependence.

A second class of magnetic field effects are the hyperfine effects which are due to the hyperfine term $\mathcal{H}_{HF}$ in the spin Hamiltonian in Eq. (4.39). These types of effects were observed by Groff et al. (1972, 1974) and by Frankevich and Rumyantsev (1972, 1973) and manifest themselves at low magnetic fields (< 100 gauss). Experimentally, it is observed that dye-sensitized delayed fluorescence is modulated by magnetic fields as low as 10 gauss. Magnetic field orientation effects are also observed, but they appear to be less clear cut than in the effects involving triplet–triplet or triplet–doublet interactions.

We now proceed to a detailed discussion of each of these effects.

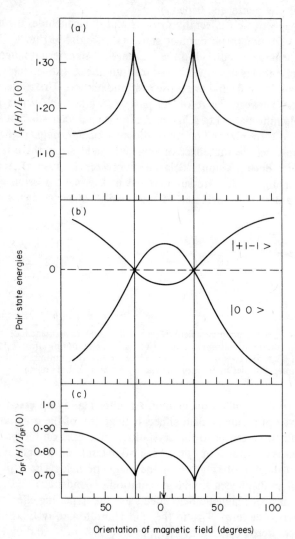

Fig. 4.13. Magnetic field effects in crystalline tetracene as a function of the orientation of the magnetic field **H** in the *ab*-plane. The extinction direction *c* closest to the *b* axis is located at 0° and is easily found using a polarizing microscope; the location of the crystallographic *b*-axis is shown by an arrow. (a) Prompt fluorescence I_F at 300 K. (c) Delayed fluorescence I_{DF} at 192 K, adapted from Groff *et al.* (1970c). (b) Calculation of pair state energies according to Eq. (4.76) using a ratio of $D/E = -13$ for the molecular zero-field splitting parameters of tetracene.

4.5.2. Delayed Fluorescence (Fusion)

The fusion of two triplet excitons can produce either a single triplet exciton or a singlet exciton according to the scheme:

$$^3T_1 + {}^3T_1 \underset{k_{-1}}{\overset{k_1}{\rightleftharpoons}} {}^{1,3,5}(T_1T_1) \quad \begin{array}{l} \overset{k_S}{\longrightarrow} \quad {}^1S_1 + {}^1S_0 \longrightarrow 2S_0 + h\nu_{DF} \\[2ex] \overset{k_T}{\longrightarrow} \quad {}^1S_0 + {}^3T_2 \longrightarrow {}^1S_0 + {}^3T_1 \end{array} \tag{4.53}$$

The free triplet excitons $^3T_1$ diffuse freely in the crystal and collide with each other with rate constant k_1 forming the pair state (T_1T_1) in which the triplet excitons are on adjacent molecules. These pair states can dissociate with rate constant k_{-1}, or undergo fusion to produce either a singlet (rate constant k_S) or a single triplet excited to the T_2 level (rate constant k_T). The singlet S_1 subsequently decays by emitting a "delayed fluorescence" photon $h\nu_{DF}$.

When the pair state (T_1T_1) is formed from two spin $S = 1$ particles, the overall spin character of the pair state can be singlet, triplet, or quintet in character which is designated by the left-hand superscript in Eq. (4.53). Fusion to form quintets is usually not possible, because single-particle quintets would involve the simultaneous excitation of more than one electron. These single-molecule quintet states are expected to lie at energies which are higher than those of the (T_1T_1) pair state. Energy, of course, is conserved in Eq. (4.53).

It has been shown that k_{-1}^{-1} is of the order of $\sim 10^{-9}$ (Johnson and Merrifield, 1970) which is much shorter than the residence time of $\sim 10^{-11}$ (Maier et al., 1967) of a triplet exciton on a given molecule. Suna (1970) has shown that the pair state should be viewed as a state in which the two excitons diffuse with respect to each other and collide about five times during the 10^{-9} s interval and thus have that many chances to undergo fusion.

It is important to realize that the triplet excitons preserve their spin character during this time interval. While the triplet exciton population in the crystal as a whole is in thermal Boltzmann equilibrium, the two triplets in the pair state are not. Thus, an $|x^*\rangle$ triplet entering a pair state will remain in this state; spin–lattice relaxation phenomena which are responsible for transitions $|x^*\rangle \rightleftharpoons |y^*\rangle \rightleftharpoons |z^*\rangle$ are slower (rate constants $< 10^8$ s^{-1}) than the lifetime of the triplet pair state. If the opposite were true, no magnetic field effects of the type described here would be observable.

After about $\sim 10^{-9}$ s the excitons diffuse away from each other which is a consequence of random walk behaviour of the triplet excitons in the incoherent limit. If two particles encounter each other the probability that these same two particles will re-encounter each other is very high and can be calculated (Suna, 1970). This is the reason for the long apparent lifetime of the pair state. Its lifetime of $\sim 10^{-9}$ s is determined by a hop of one of the triplet excitons to an adjacent crystallographic plane. In anthracene and tetracene the diffusion of excitons is essentially two-dimensional—the diffusion coefficient being about ten times smaller perpendicular to the ab-planes than within these planes (see Table I). When one of the excitons jumps to another plane, the pair state is no longer correlated—the interactions between two triplets in different planes is much smaller than when both triplets are in the same plane (Swenberg, 1969).

The rate of fusion events, and thus the intensity of the delayed fluorescence, depends on the density of triplet excitons, which is governed by Eq. (4.54)

$$\frac{\partial T_1(x)}{\partial t} = G(x) - \frac{T_1(x)}{\tau_T} - \gamma_{\text{TOT}} T_1^2(x) + D\nabla^2 T_1(x) \tag{4.54}$$

This equation gives the concentration of triplet excitons at position x and at time t. $G(x)$ is the rate of generation of triplets, τ_T is the triplet life-time, γ_{TOT} is the overall bimolecular annihilation (fusion) rate including both the singlet and triplet channels in Eq. (4.53). The last term describes the diffusion of triplet excitons (diffusion coefficient D) and is non-zero only if the concentration of triplet excitons is inhomogeneous with respect to x (i.e. unequal light absorption at different points within the crystal). In the experiments described here we shall be concerned only with homogeneous excitation of triplets so that the last term in Eq. (4.54) can be set equal to zero, i.e. $D\nabla^2 T_1(x) = 0$.

Normally, the experiments on delayed fluorescence are done by exciting the triplets directly by the $T_1 \leftarrow S_0$ process. As pointed out earlier, the absorption coefficient α (cm^{-1}) for this process is extremely small although finite in organic crystals ($\alpha \approx 10^{-3}$–10^{-4} cm^{-1}). Thus using visible light ($\lambda > 500$ nm), triplets can be excited homogeneously within the crystal. The intensity of the exciting light $I(x)$ at a position x within the crystals is given by $I(x) = I_0 \exp(-\alpha x)$. For typical crystal thicknesses $x = 0 \cdot 1$–1 cm, it is evident that $I(x) = I_0$ anywhere within the crystal. Furthermore, the light sources in these experiments are steady state (CW helium–neon lasers, $\lambda = 6328 \cdot 1$ Å, are commonly used), thus Eq. (4.54) can be written

$$\frac{\partial T_1}{\partial t} = \alpha I_0 - \frac{T_1}{\tau_T} - \gamma_{\text{TOT}} T_1^2 = 0 \tag{4.55}$$

for the steady-state conditions. We distinguish two limiting cases in writing down the intensity of the delayed fluorescence I_{DF}:

(a) The triplet lifetime is determined by the monomolecular term in Eq. (4.55), i.e.

$$T_1/\tau_T \gg \gamma_{\text{TOT}} T_1^2 \tag{4.56}$$

and (b) the triplet exciton density is sufficiently high so that the bimolecular term dominates:

$$\gamma_{\text{TOT}} T_1^2 \gg T_1/\tau_T \tag{4.57}$$

Only a certain fraction of the bimolecular annihilations lead to singlet states S_1 and thus to delayed fluorescence. If we denote the bimolecular rate constant for generating singlets by γ_S, the delayed fluorescence intensity is

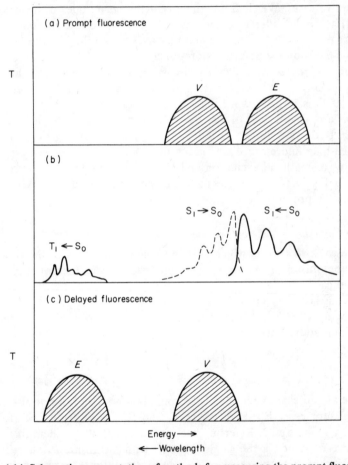

Fig. 4.14. Schematic representation of methods for measuring the prompt fluorescence or delayed fluorescence in organic crystals. Spectra for absorption into the singlet band ($S_1 \leftarrow S_0$), into the triplet band ($T_1 \leftarrow S_0$) and the fluorescence ($S_1 \rightarrow S_0$) are shown for a hypothetical organic crystal in (B). (A) Mode of excitation E and wavelength range for viewing V the prompt fluorescence; C excitation E and viewing V in a delayed fluorescence experiment. T is the relative transmittance of hypothetical optical filters which can be used for isolating the different wavelength regions.

given by the following two expressions in the two limiting cases (Avakian and Merrifield, 1968)

(a) $I_{DF} = \frac{1}{2}\gamma_S \alpha I_0^2 \tau_T^2$ $(T_1/\tau_T > \gamma_{TOT} T_1^2$, low light intensities)   (4.58)

and

(b) $I'_{DF} = \frac{1}{2}(\gamma_S/\gamma_{TOT}) \alpha I_0$ $(\gamma_{TOT} T_1^2 > T_1/\tau_T$, high light intensities)   (4.59)

It is important to note the differences in the light intensity dependences in these two cases. We shall note later that the magnitudes of the magnetic field effects will be different in these two cases. The two rate constants γ_S and γ_{TOT} are magnetic field sensitive and this gives rise to the observed effects of magnetic fields on the delayed fluorescence.

Before we proceed to an explicit discussion of these effects, we make some general remarks on the experiments involved. In anthracene, the energy of the triplet T_1 is 1.83 eV, while that of the singlet S_1 is 3.15 eV. A triplet pair state has a pooled energy of 3.66 eV which is more than enough to produce a singlet. In delayed fluorescence one thus excites with red light and observes the violet-blue fluorescence emanating from S_1. A schematic representation for the wavelengths of excitation and viewing of the delayed fluorescence is shown in Fig. 4.14 and is compared to these parameters for exciting and detecting the prompt fluorescence.

4.5.3. Magnetic Field Effects on Homofusion

When the two triplets which undergo fusion are identical, the process is called homofusion, in contrast to heterofusion where the two triplets are not identical.

We consider the zero-field case. There are three triplet excitons with wavefunctions $|x^*\rangle$, $|y^*\rangle$, and $|z^*\rangle$ and $3 \times 3 = 9$ possible combinations for the pair state. These are

$$|x^* x^*\rangle, |y^* y^*\rangle, |z^* z^*\rangle, |x^* y^*\rangle, |y^* x^*\rangle, |x^* z^*\rangle,$$
$$|z^* x^*\rangle, |y^* z^*\rangle, |z^* y^*\rangle \quad (4.60)$$

These are all eigenfunctions of the spin Hamiltonian, but not of the total spin operator $S^2 = (S_1 + S_2)^2$, where S_1 and S_2 are the spin operators for triplet excitons 1 and 2. Referring back to Eq. (4.53), it is evident that fusion can proceed either via the singlet channel with a transition rate k_S, or via the triplet channel with a rate k_T. The rate constants for these two annihilation channels expressed in terms of the kinetic parameters in Eq. (4.53) are (Merrifield, 1968; Swenberg and Geacintov, 1973),

$$\gamma_S = \frac{1}{9} k_1 \sum_{l=1}^{9} \frac{k_S/k_{-1} |C_S^l|^2}{1 + (k_S/k_{-1}) |C_S^l|^2} \quad (4.61)$$

and

$$\gamma_T = \frac{1}{9} k_1 \sum_{l=1}^{9} \frac{k_T/k_{-1} |C_T^l|^2}{1 + (k_T/k_{-1})|C_T^l|^2} \tag{4.62}$$

The symbols are defined in Eq. (4.53). C_S^l denotes the singlet amplitude of the lth pair state, while C_T^l is the triplet amplitude of the lth pair state.

We immediately note two important facts. No magnetic field effects are observed either when $k_S \leqslant k_{-1}$ or when the opposite is true, i.e. $k_{-1} \ll k_S$ (Geacintov and Swenberg, 1972). It is therefore remarkable that magnetic field effects can be observed only if the rather delicate condition $k_S/k_{-1} \sim 1$ is maintained. We shall show below that γ_T is magnetic field insensitive.

One of the essential features of the magnetic field effects is that C_S^l changes as a function of field strength. The variation of C_S^l with H arises because the annihilation process leading to the singlet manifold is spin-conserving and the total spin operator does not commute with the pair state Hamiltonian. C_S^l also evolves as a function of time within the $\sim 10^{-9}$ s lifetime of a triplet pair state. This feature is also an essential and subtle condition for the observation of magnetic field effects, but we will not describe this aspect of the theory further and refer the interested reader to the original references (Avakian and Suna, 1971; Avakian, 1974).

The singlet amplitudes C_S^l are found by first finding the wavefunction of the pure singlet. Since $\mathscr{H}_{SS}$ and S^2 do not commute, i.e. $[\mathscr{H}_{SS}^*, S^2] \neq 0$ the pair state functions in Eq. (4.60) are not eigenfunctions of the total spin. The eigenfunctions of the spin may however be found by diagonalizing the 9×9 matrix

$$\langle \Theta_l | (S_1 + S_2)^2 | \Theta_m \rangle \tag{4.63}$$

where Θ_l and Θ_m are any of the pair state spin functions in Eq. (4.60). Nine eigenfunctions are obtained, one of which is a singlet $|S\rangle$, three are triplets $|T\rangle$ and five are quintets:

$$|S\rangle = 3^{-\frac{1}{2}}\{|x^*x^*\rangle + |y^*y^*\rangle + |z^*z^*\rangle\} \tag{4.64}$$

$$\left.\begin{array}{l} |T_1\rangle = 2^{-\frac{1}{2}}\{|y^*z^*\rangle - |z^*y^*\rangle\}, \quad |T_2\rangle = 2^{-\frac{1}{2}}\{|x^*y^*\rangle - |y^*x^*\rangle\}, \\ |T_3\rangle = 2^{-\frac{1}{2}}\{|x^*z^*\rangle - |z^*x^*\rangle\} \end{array}\right\} \tag{4.65}$$

$$\left.\begin{array}{l} |Q_1\rangle = 2^{-\frac{1}{2}}\{|x^*z^*\rangle + |z^*x^*\rangle\}, \quad |Q_2\rangle = 2^{-\frac{1}{2}}\{|x^*y^*\rangle + |y^*x^*\rangle\}, \\ |Q_3\rangle = 2^{-\frac{1}{2}}\{|y^*z^*\rangle + |z^*y^*\rangle\}, \quad |Q_4\rangle = 2^{-\frac{1}{2}}\{|x^*x^*\rangle - |y^*y^*\rangle\}, \\ |Q_5\rangle = 6^{-\frac{1}{2}}\{|x^*x^*\rangle + |y^*y^*\rangle - 2|z^*z^*\rangle\} \end{array}\right\} \tag{4.66}$$

The singlet amplitudes are given by

$$|C_S^l| = |\langle \Theta_l | S \rangle| \tag{4.67}$$

It is evident that when $H = 0$ only three out of the nine possible pair states have singlet character by comparing the wavefunctions in Eq. (4.60) with Eq. (4.64). However, these same wave functions have quintet character since $\langle \Theta_l | Q_4 \rangle$ and $\langle \Theta_l | Q_5 \rangle \neq 0$ for $\Theta_l = |x^*x^*\rangle$, $|y^*y^*\rangle$, and $|z^*z^*\rangle$. Thus, these three pair functions are singlet–quintet mixtures, but contain no triplet character. The other six spin pair functions in Eq. (4.60) can be shown to be either pure quintets or pure triplets. For example, $|x^*y^*\rangle$ is degenerate with $|y^*z^*\rangle$. Small residual interactions, not (necessarily) included in the spin Hamiltonian given here, will lift this degeneracy yielding the two functions

$$\Theta_\pm = 2^{-\frac{1}{2}}\{|x^*y^*\rangle \pm |y^*x^*\rangle\} \tag{4.68}$$

The symmetric combination Θ_+ is seen to be identical with the quintet $|Q_2\rangle$ whereas Θ_- is identical with the pure triplet $|T_2\rangle$. Similar results are obtained by considering the doubly degenerate pair states $|x^*z^*\rangle$ and $|y^*z^*\rangle$.

We next consider the case when a magnetic field is present and is of sufficient strength so that $\mathcal{H}^*_{SZ} \gg \mathcal{H}^*_{SS}$. This limit is commonly called the high field limit and corresponds to field strengths $\gtrsim 1$ kG. In this case the triplet excitons are described, to a good approximation, by the wavefunctions $|0\rangle$, $|+1\rangle$, and $|-1\rangle$ where 0, $+1$, and -1 correspond to the standard magnetic sublevel quantum numbers. The nine possible pair states in this limit are:

$$|00\rangle, |+1-1\rangle, |-1+1\rangle, |01\rangle, |10\rangle, |-10\rangle,$$
$$|0-1\rangle, |+1+1\rangle, |-1-1\rangle \tag{4.69}$$

The pure singlet, triplet, and quintet intermediate states are found by diagonalizing the operator $(S_1 + S_2)^2$ within the space spanned by the basis function in Eq. (4.69). These are

$$|S(H)\rangle = 3^{-\frac{1}{2}}\{|00\rangle - |+1-1\rangle - |-1+1\rangle\} \tag{4.70}$$

$$\left.\begin{aligned}
|T_1(H)\rangle &= 2^{-\frac{1}{2}}\{|+1-1\rangle - |-1+1\rangle\}, \\
|T_2(H)\rangle &= 2^{-\frac{1}{2}}\{|+10\rangle - |0+1\rangle\}, \\
|T_3(H)\rangle &= 2^{-\frac{1}{2}}\{|-10\rangle - |0-1\rangle\}
\end{aligned}\right\} \tag{4.71}$$

$$\left.\begin{aligned}
|Q_1(H)\rangle &= 2^{-\frac{1}{2}}\{|+10\rangle + |0+1\rangle\}, \\
|Q_2(H)\rangle &= 2^{-\frac{1}{2}}\{|-10\rangle + |0-1\rangle\}, \\
|Q_3(H)\rangle &= |+1+1\rangle, \quad Q_4(H) = |-1-1\rangle, \\
|Q_5(H)\rangle &= 6^{-\frac{1}{2}}\{2|00\rangle + |+1-1\rangle + |-1+1\rangle\}
\end{aligned}\right\} \tag{4.72}$$

Out of the nine pair states in Eq. (4.69) only three have a non-zero singlet amplitude. However, the $|+1-1\rangle$ and $|-1+1\rangle$ states are degenerate in first order and the degeneracy is lifted by the weak inter-triplet interaction and necessitates forming new linear combinations given by

$$\Theta_\pm(H) = 2^{-\frac{1}{2}}\{|+1-1\rangle \pm |-1+1\rangle\} \tag{4.73}$$

On comparison of these states with Eqs (4.70) and (4.71) it is easily seen that the antisymmetric state $\Theta_-(H)$ is a pure triplet, while the symmetric state $\Theta_+(H)$ has a non-zero singlet amplitude. We also note, in general, that the singlet and quintet states are symmetric upon interchange of particles whereas the triplet states are antisymmetric.

What happens at intermediate field strengths, when the triplet excitons cannot be described by either the eigenfunctions of $\mathcal{H}_{SS}$ as in Eq. (4.60), or by the eigenfunctions of $\mathcal{H}_{SZ}$ as in Eq. (4.69)? This occurs when the magnetic field strength is such that $g\beta H \sim D^*, E^*$. Under these conditions the eigenfunctions of $\mathcal{H}_{SS}^* + \mathcal{H}_{SZ}^*$ must be found by direct diagonalization using any complete basis set and, in general, the resulting eigenfunctions have the form:

$$|\Theta_i'(H)\rangle = C_1|x^*\rangle + C_2|y^*\rangle + C_3|z^*\rangle \tag{4.74}$$

where C_1, C_2, and C_3 are coefficients which are functions of both the field strength and its orientation. Thus, when $\mathcal{H}_{SS}^* \sim \mathcal{H}_{SZ}^*$ the magnetic field tends to mix the three triplet exciton functions. Therefore in the most general case all nine possible pair states acquire some singlet character. In the case of homofusion and homofission only six possible states can acquire singlet character due to the identical nature of the two particles involved.

4.5.4. High Field Resonances

Special orientations of the magnetic field with respect to the crystallo-graphic axes exist in which the $|00\rangle$ state is degenerate with the $|+1-1\rangle$ and $|-1+1\rangle$ states in the high field limit. At these orientations the small intertriplet interaction lifts this degeneracy to give only *one* intermediate state with singlet character, namely the pure singlet

$$|S(H) = 3^{-\frac{1}{2}}\{|00\rangle - |+1-1\rangle - |-1+1\rangle\} \tag{4.75}$$

Referring to the expression Eqs (4.49) and (4.50) for triplet excitons, these orientations occur when the following conditions are satisfied:

$$W_+^* + W_-^* = W_0^* + W_0^* \tag{4.76}$$

Straightforward algebraic manipulation shows this condition is satisfied when

$$D^*(\cos^2\gamma^* - \tfrac{1}{3}) + E^*(\cos^2\alpha^* - \cos^2\beta^*) = 0 \tag{4.77}$$

where the angles α^*, β^*, and γ^* refer to the angles between the field **H** and the fine-structure tensor axes x^*, y^*, and z^*. If the orientation of these axes is not known exactly, the orientation of the resonances (called level-crossing resonances) can also be calculated from the molecular parameters D and E. The orientations of **H** with respect to the *molecular* symmetry axes x_1, y_1, z_1 and x_2, y_2, z_2 are given by the angles $\alpha_1, \beta_1, \gamma_1$ and $\alpha_2, \beta_2, \gamma_2$, where the subscripts 1 and 2 refer to the two inequivalent molecules in naphthalene, anthracene, or tetracene. Using these parameters, the orientations of the level-crossing resonances are given by

$$D(\cos^2\gamma_1 + \cos^2\gamma_2 - \tfrac{2}{3}) + E(\cos^2\alpha_1 + \cos^2\alpha_2 - \cos^2\beta_1 - \cos^2\beta_2) = 0 \quad (4.78)$$

4.5.5. Variation of γ_S with Field Strength and Orientation

We are now ready to summarize the effects of magnetic field on the delayed fluorescence. Since $\sum |C_S^l|^2 = 1$ and constant, it can be shown under most conditions, that γ_S is larger, the larger the number of pair states which have singlet character, i.e. the larger the number of terms in Eq. (4.61) for which $C_S^l \neq 0$. Our previous discussions can be summarized as follows:

Field strength	Number of pair states with $C_S^l \neq 0$	
0	3	
Intermediate ($\mathscr{H}_{SS}^* \approx \mathscr{H}_{SZ}^*$)	9, generally 6	(4.79)
High ($H > 1000$ gauss)	2 (in general)	
High field, level crossings	1	

Thus, as the field strength is increased, the delayed fluorescence intensity should first rise and then should decrease as the field is further increased. This behaviour is precisely what is observed as is shown in Fig. 4.12. Furthermore, the orientations at which the level-crossing resonances occur can be calculated using Eqs (4.77) and (4.78). Such a calculation is shown in Fig. 4.13 in which the energies of the pair states $|00\rangle$ and $|+1 -1\rangle$ are plotted as a function of the orientation of **H** using tetracene as an example. At the orientation at which the levels cross there is only one pair state with singlet character and the delayed fluorescence should be at a minimum as is indeed observed.

The simplified explanations presented here were originally advanced by Merrifield (1968). The success of this theory in explaining a set of rather complicated results constitutes a beautiful application of quantum mechanics. It shows, furthermore, that the spin Hamiltonian $\mathscr{H}^* = \mathscr{H}_{SS}^* + \mathscr{H}_{SZ}^*$ is an excellent description of the spin properties of the triplet excitons.

While exact agreement between the calculated and observed level crossings is not obtained for tetracene (Swenberg and Geacintov, 1973), the agreement is much better for anthracene which constitutes the only example where extensive calculations have been performed (Johnson and Merrifield, 1970; Suna, 1970). Analogous results have been also obtained with pyrene (Yarmus et al., 1973) and naphthalene (Port and Wolf, 1968; Chabr, 1975).

4.5.6. Effects on High Triplet Exciton Densities

We noted previously (see Eqs 4.58 and 4.59) that in the limiting case when the triplet lifetime is limited by bimolecular collisions, the intensity dependence of the delayed fluorescence is linear. In the limit where the triplet lifetime is limited by the monomolecular decay term, T_1/τ_T, the intensity dependence is quadratic. At intermediate triplet exciton densities the intensity dependence varies between linear and quadratic (exponent in I_0 between 1·0 and 2·0). Experimentally it is easy to verify whether the exciting light intensity corresponds to one or the other limiting regime.

We consider the high intensity case. A number of complications can occur, such as bimolecular electron spin relaxation effects, etc. which cannot be considered here. A detailed discussion has been presented by Chabr (1975). We shall confine ourselves only to any possible variation with respect to the magnetic field of γ_{TOT}. In the low intensity limit, I_{DF} is directly proportional to the magnetic field sensitive rate constant γ_S, while in the high intensity regime it is proportional to γ_S/γ_{TOT}.

Groff et al. (1970b) have shown that γ_{TOT} is given by

$$\gamma_{TOT} = \tfrac{1}{2}[(1+q_{FM})\gamma_S + \gamma_T] \qquad (4.80)$$

where η is the quantum efficiency of fluorescence (the only non-radiative decay term of singlet excitons considered here is intersystem crossing $S_1 \leadsto T_1$ with a quantum efficiency of $1 - q_{FM}$). We have also neglected in Eq. (4.80) the fusion process $T_1 + T_1 \to S_0 + S_0 + \text{heat}$, in which only ground-state molecules are formed; this mechanism is considered much less likely than fusion to produce $S_1 + S_0$, because of the unfavourable Franck–Condon factors which arise when large amounts of electronic energy have to be converted to vibrational energy.

The magnetic field dependence of γ_{TOT} arises only from the magnetic field dependence of γ_S. Referring back to Eq. (4.62), γ_T should remain constant as long as the number of triplet pair states remains unchanged as a function of magnetic field. This is easy to show for the high field and zero field limits (Pope et al., 1969). In the zero field limit, the $|y^*z^*\rangle$ and $|z^*y^*\rangle$, the $|x^*y^*\rangle$ and $|y^*x^*\rangle$, and the $|x^*z^*\rangle$ and $|z^*x^*\rangle$ are degenerate. Thus linear combinations between the degenerate states are formed as usual. The antisymmetric (odd) states are pure triplets, while the symmetric (even) combinations are pure quintets as may be seen by examining the pure spin states in Eqs (4.65) and (4.66). There are thus three pair states in zero field which have triplet character.

In the high field case, considering the degeneracy of the states $|0+1\rangle$, $|-10\rangle$, and $|-1+1\rangle$ it can also be shown that there are only three pair states with triplet character in the high field limit. Furthermore, it can be shown that because of the differences in symmetry between triplet and singlet states, there is no mixing between them in general. Thus, the overall distribution of triplet amplitude remains independent of magnetic field strength and thus there is no magnetic field effect on γ_T.

Since γ_S is the only magnetic field sensitive term in γ_{TOT}, it follows that

$$\frac{\gamma_{TOT}(H)}{\gamma_{TOT}(0)} > \frac{\gamma_S(H)}{\gamma_S(0)} \tag{4.81}$$

This inequality implies that the change in γ_{TOT} with magnetic field is less than that in γ_S. In anthracene (Groff *et al.*, 1970b)

$$\frac{\gamma_S}{\gamma_{TOT}} = 0 \cdot 36 \pm 0 \cdot 02 \tag{4.82}$$

Since the quantum efficiency of fluorescence of crystalline anthracene is nearly unity ($q_{FM} \approx 1$), we find from Eqs (4.80) and (4.81) that

$$\gamma_T/\gamma_S \approx 3 \cdot 6 \tag{4.83}$$

Thus, fusion to produce one triplet exciton is more probable than fusion to produce one singlet exciton. If we take a typical example for anthracene corresponding to a field orientation (in the high field limit) where

$$\left(\frac{I_{DF}(H)}{I_{DF}(0)}\right)_{\text{low } I_0} = \frac{\gamma_S(H)}{\gamma_S(0)} = 0 \cdot 80 \tag{4.84}$$

then in the high intensity limit this ratio will be larger. Using Eqs (4.80) and (4.83) with $q_{FM} = 1$, we find

$$\left(\frac{I_{DF}(H)}{I_{DF}(0)}\right)_{\text{high } I_0} = \frac{\gamma_S(H)}{\gamma_S(0)} \times \frac{\gamma_{TOT}(0)}{\gamma_{TOT}(H)} = 0 \cdot 86 \tag{4.85}$$

In summary, magnetic-field-induced changes on the delayed fluorescence (in the high field limit) tend to be smaller the higher the light intensity. We will refer back to this phenomenon later, when we discuss magnetic field effects on the scintillation produced by ionizing particles in organic crystals.

4.5.7. Homofission

Fission of one singlet exciton into two triplet excitons is represented by

$$
\begin{array}{c}
S_1 \xrightarrow{\ k_{tis}\ } T_1 + T_1 \\[4pt]
\!\!\xrightarrow{\ k_{FM}\ } S_0 + h\nu_{FM}
\end{array}
\tag{4.86}
$$

Fission is a radiationless decay process which competes with the emission of a photon ($h\nu_{FM}$) by fluorescence. Fission can be viewed as the opposite of exciton fusion and therefore an external magnetic field is expected to modulate the fission rate constant in the same manner as the rate constant leading to delayed fluorescence. In this type of experiment, the experimental parameter which is monitored as a function of magnetic field strength is the prompt fluorescence emanating from the singlet exciton S_1. In the high field limit, the magnetic field causes the distribution of singlet character to condense into few pair states (as can be seen from Eq. 4.79) ($k_{fis}(H) < k_{fis}(0)$) and the prompt fluorescence *increases*. The delayed fluorescence originating from $T_1 + T_1$ fusion, on the other hand, *decreases* under these same conditions. Thus, the magnetic field has opposite effects on the prompt and delayed fluorescence which is demonstrated for the field orientation dependence in Fig. 4.13 and for the field strength dependence in Fig. 4.12.

As mentioned previously, two types of fission may be distinguished: (1) S_1 is in a thermally relaxed lowest vibrational energy level, and (2) S_1^* is

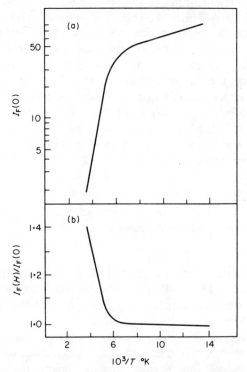

Fig. 4.15. (A) Temperature dependence of the prompt fluorescence I_F in crystalline tetracene. (B) Magnetic field dependence of the prompt fluorescence as a function of temperature.

not relaxed vibrationally and is therefore short-lived ($\sim 10^{-12}$ s). The asterisk indicates that S_1 is vibrationally excited. In principle, "hot fission" involving the process $S_1^* \rightarrow T_1 + T_1$ can occur in any crystal since it is possible to choose the excitation energy E so that $E_{S_1^*} > E_{2T_1}$. Fission from thermally relaxed levels S_1, on the other hand, requires that $E_{S_1} \approx E_{2T_1}$. There are few crystals which meet this condition. Tetracene is one of these, and fission has been documented most thoroughly in this material. Frankevich has reported that fission can also occur in solid rubrene (Frankevich et al., 1973).

Since the energy difference between S_1 and two triplet levels in tetracene is only ~ 1700 cm^{-1}, fission at room temperature is more than 98% efficient and the fluorescence quantum yield is low. As the temperature is lowered, the fluorescence quantum yield increases dramatically and the magnetic field effect decreases correspondingly and finally goes to zero below 160 K (see Fig. 4.15).

Thus, at low temperatures thermally activated fission is completely suppressed (Geacintov et al., 1969) in tetracene. However, hot fission is observable at low temperatures (Klein et al., 1972, 1973; Moller and Pope, 1973); the effect of magnetic field on the prompt fluorescence at low temperatures in tetracene is observed only when the excitation energy exceeds 2·5 eV, which is the energy of two triplets. Furthermore, the magnetic effect is less than $\sim 1\%$, whereas at room temperature with thermally activated fission in tetracene, the magnetic effects are as high as 35%.

In anthracene the difference in energy between S_1 and $2T_1$ is $\sim 0·5$ eV (~ 4000 cm^{-1}) while kT (k is Boltzmann's constant) at room temperature is only $\sim 0·025$ eV or 200 cm^{-1}. Thus, thermally induced fission is not observed in anthracene, while hot fission with magnetic field effects of the order of $\sim 1\%$ has been reported (Klein et al., 1972, 1973; Arnold, 1974).

4.5.8. Coexistence of Fission and Fusion in Tetracene

Exciton fusion can also occur in tetracene. The fission is thus partially reversible according to the kinetic scheme

$$S_1 \xrightleftharpoons{\hspace{1.5cm}} T_1 + T_1 \tag{4.87}$$

At room temperature fusion is difficult to observe using an experimental scheme as outlined in Fig. 4.14(C). The reason is that any singlet S_1 formed by fusion of two triplets is capable of fissioning back to the two triplets with high efficiency. Thus, at room temperature, while delayed fluorescence is observed, the magnetic field effect is negligible (Groff et al., 1970a). As the temperature is lowered, fission is suppressed, and the normal magnetic field effect on the delayed fluorescence appears. Below 160 K, the delayed fluorescence in tetracene appears and has the same qualitative features as delayed fluorescence in anthracene.

The excitation intensity dependence of the prompt fluorescence in tetracene at room temperature also reveals the coexistence of fusion and fission. Below $I_0 \sim 10^{15}$ quanta $cm^{-3} s^{-1}$ the fluorescence quantum yield is constant (fluorescence photons emitted/photons absorbed ≈ 0.002 in tetracene at room temperature). As I_0 exceeds this value the quantum yield rises (Pope et al., 1969) and eventually reaches a plateau (Ern et al., 1971). This increase in quantum yield at high light intensities is due to the reverse fusion reaction in Eq. (4.87). Since fission of S_1 is not 100% efficient at room temperature, the fusion reaction can contribute to the total fluorescence yield. As expected, the magnitude of the magnetic field effect decreases with intensity above $I_0 \gtrsim 10^{15}$ quanta $cm^{-2} s^{-1}$ due to the presence of the delayed component in the total observed fluorescence, and since the delayed component is not magnetic field sensitive (Pope et al., 1969).

4.5.9. Triplet–Doublet Quenching

A direct manifestation of triplet–exciton migration in pure organic solids is the observed enhancement of the exciton's lifetime in the presence of an external magnetic field. Initially this was observed in high energy X-irradiated anthracene crystals by Ern and Merrifield (1968) and subsequently by Ern and McGhie (1971) in anthracene crystals subject to internal beta-irradiation. The triplets which are not quenched can be detected by using delayed fluorescence as a tool. Recently Arnold and co-workers (1976a) have shown how this magneto-optical effect can be used to elucidate the distribution of both paramagnetic and diamagnetic quenchers in organic crystals subject to alpha-particle irradiation.

The radiations (α-, X-, or γ-rays) introduce paramagnetic spin $\frac{1}{2}$ quenchers, ^2R, which quench the mobile triplets according to the spin-conserving reaction

$$^3T + {}^2R \xrightleftharpoons[k_{-1}]{k_1} \quad {}^{2,4}(TR) \xrightarrow{k_2} \quad {}^2R + {}^1S_0 \qquad (4.88)$$

where ^3T and $^1S_0$ denote the triplet and crystal ground states respectively. The bracket pair, $^{2,4}(TR)$, represents diagrammatically the intermediate "correlative" state which has double–quartet character. As discussed for fusion, in order to observe a magnetic field effect, it is essential that the pair state lifetime be short compared with both the triplet lifetime and the spin-relaxation rate between the magnetic sublevels (see Fig. 4.11). Furthermore k_2/k_{-1} should be comparable to unity. The k_i are rates for formation and dissociation of the pair state. The observed increase in the lifetime, approximately 8% for on-resonance orientations as compared to 3% in the off-resonance position at high magnetic fields ($H \gtrsim$ few kilogauss) for anthracene, arises from a decoupling of the triplet state with the paramagnetic quenching centers.

The picture which emerges is that triplet excitons because of the large distances which they cover encounter a free radical, i.e. a trapped hole or electron or a paramagnetic structurally damaged site, leading to the formation of a pair state at a rate k_1. This rate of formation is assumed to be spin independent in the sense that it is independent of the state of R and the magnetic sublevels of T involved. Similarly the back scattering rate, k_{-1} is assumed spin-independent whereas the conversion of the triplet energy to heat, i.e. the transition rate to final state, occurs at a rate which is spin-dependent.

The six spin states of the triplet–doublet pair can be analyzed in a manner similar to that employed for a pair of triplets. The intermediate pair Hamiltonian consists of the Zeeman interaction for both the triplet and radical in addition to the zero-field splitting terms for the triplet excitons, i.e.:

$$\mathscr{H}_{\text{pair}} = g_e \beta \mathbf{H}.\boldsymbol{\sigma} + g\beta \mathbf{H}.\mathbf{S} + D^*(S_z^{2*} - \tfrac{1}{3}S^2) + E^*(S_x^{2*} - S_y^{2*}) \qquad (4.89)$$

where g_e is the radical g-factor, assumed isotropic, and $\boldsymbol{\sigma}$ and $\mathbf{S}$ are the spin operators for the radical and triplet exciton, respectively. Similar to fission and fusion, the total pair spin operator $(\boldsymbol{\sigma}+\mathbf{S})^2$ does not commute with $\mathscr{H}_{\text{pair}}$ and thereby causes the spin states, for a general orientation of H and field strength to be doublet–quartet mixtures. Since the transition rate from a given pair state to the final state, k_2, is proportional to the doublet character (the overall reaction is taken to be spin conserving), it follows that the quenching process, Eq. (4.88), exhibits a magnetic field dependence from the change in the distribution of doublet character among the six possible states. In general, the more states there are with doublet spin character, the larger is the quenching rate constant, i.e. the shorter the triplet lifetime.

We leave it for the reader to construct the appropriate spin states for the special cases of $H = 0$, H large for both on- and off-resonance positions, and to show that the number of pair states with doublet character changes as follows:

Field strength	Number of pair states with non-zero doublet character	
0	6	
Intermediate	<6	
High ($H > 1$ kG)	4	(4.90)
High-on-resonance	2	

In addition to the general monotonic decrease in the quenching rate with increasing field (see Fig. 4.12), there are high field anisotropies due to level-crossing resonances of the pair states similar to those observed in the delayed fluorescence and in the tetracene prompt fluorescence (see Fig. 4.13). The resonances occur when $|0 + \tfrac{1}{2}\rangle$ and $|-1 + \tfrac{1}{2}\rangle$ are degenerate, respectively, with the $|1-\frac{1}{2}\rangle$ and $|0-\frac{1}{2}\rangle$ states. This occurs when

$$W_0 + W_{\frac{1}{2}} = W_{+1} + W_{-\frac{1}{2}} \tag{4.91}$$

and

$$W_1 + W_{-\frac{1}{2}} = W_0 + W_{\frac{1}{2}} \tag{4.92}$$

where $W_{\pm\frac{1}{2}}$ are the energies of the doublets, which are equal to $\pm g_e \beta H/2$. The special orientations occur at exactly the same positions as the resonances observed in triplet–triplet fusion which are formed by solving Eq. (4.77). By monitoring the time dependence of the delayed fluorescence (in the low intensity limit of Eq. (4.58)) after the introduction of paramagnetic centers by either irradiation or other means, the monotonic increase in τ_T can be observed with increasing magnetic field strength. In addition, cusp-like increases in τ_T occur at those orientations where Eqs (4.91) and (4.92) are satisfied.

It is important to realize that the size of the magnetic field effect on $\tau_T = \beta^{-1}$ is a function not only of k_2/k_{-1} but also depends on the density of paramagnetic quenchers, N_p, and the triplet exciton lifetime, β_0^{-1}, in the unirradiated crystal according to the equation

$$\frac{\Delta\beta}{\beta(0)} = \frac{\beta(H) - \beta_0}{\beta(0)} = \frac{N_p \gamma_p(H)}{\beta_0 + \gamma_p(0) N_p} \tag{4.93}$$

Here $\gamma_p(0)$ is the quenching rate in zero applied field, $\sim 10^{-11}$ cm^3/s. For high quality anthracene crystals, β_0 is approximately 40 s^{-1}, thus Eq. (4.93) implies that for $N_p > 4 \times 10^{12}$ cm^{-3}, an easily obtainable defect concentration, $\Delta\tau_T/\tau_T = -\Delta\gamma_p/\gamma_p$. When β_0 is comparable or greater than $\gamma_p(0) N_p$, the magnetic field effect enhancement of the triplet exciton lifetime will be diluted. This will be the case for tetracene for $N_p \lesssim 10^{15}$ cm^{-1} since $\beta_0 \sim 2 \times 10^4$ s^{-1}.

In addition to the quenching of triplet excitons by paramagnetic centers, excitons play a major role in the photoconductivity of organic solids. The magnetic field dependence of photoconductivity in anthracene and tetracene was initially described by Frankevich et al. (1960, 1966a, b, c), Frankevich and Sokolik (1967a, b) and subsequently investigated by Geacintov et al. (1970) and Delanney and Schott (1969) within the context of Merrifield's theory. When extensive trapping of either holes or electrons occurs in the bulk, the quenching of excitons, Eq. (4.88), may yield free carriers and thereby give rise to an "enhanced" current. If triplet excitons play the predominant detrapping role, the enhanced photocurrent will display the anisotropy characteristics and field strength behavior as discussed above for the triplet–doublet interaction. For increasing field strength the enhanced photoconductivity should decrease whereas in the high field limit ($H > 1$ kG) at those special orientations, given by Eqs (4.91) and (4.92), where the distribution of doublet spin character is minimal, there should be an additional decrease similar to that displayed by delayed fluorescence. Figure 4.16 illustrates this characteristic anisotropy in the photoconductivity for the case of solid anthracene. It should be stressed that this type of conductivity experiment is not necessarily quite this simple, e.g. in tetracene. In this crystal the enhanced photoconductivity is opposite (for some wavelengths) to that given in Fig. 4.16. and is indicative of the role played by singlet excitons (Geacintov *et al.*, 1970). The reader should consult the original literature for further details.

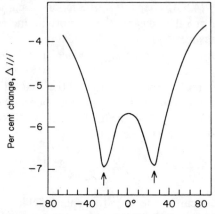

Fig. 4.16. Magnetic field effect on photoconductivity. Hole photocurrent i in an anthracene crystal. Effect of magnetic field orientation in the *ab*-plane; the *b*-axis is at 0°. The calculated level-crossing resonances are indicated by the two arrows. Magnetic field strength 3400 gauss. (After Geacintov *et al.*, 1970.)

4.5.10. Hetero-exciton Processes

In addition to pure crystal effects discussed above, it is well known that small concentrations of guest molecules in an otherwise perfect crystal can profoundly effect the luminescence properties of the host crystal. The most illustrative and most extensively studied mixed crystal is tetracene-doped anthracene where for small concentrations of tetracene (a few ppm) most absorbed photons are emitted by the guest (see Birks, 1970, for an excellent review). This is due to the fast transfer of the host singlet exciton to the deep trap site, i.e. tetracene ($\sim 2\cdot 2$ eV singlet compared to $3\cdot 15$ eV for anthracene singlet). In addition to singlet excitons, triplet excitons can also be trapped at guest sites. If a subsequent host triplet migrates to a site already occupied, i.e. the trap is blocked, then, if energetically possible, the following reactions can occur

$$T_1(\text{host}) + T_1(\text{guest}) \longrightarrow \begin{cases} S_1(\text{host}) + S_0(\text{guest}) \\ S_1(\text{guest}) + S_0(\text{host}) \end{cases} \tag{4.94}$$

Such processes are called heterofusion since dissimilar parent triplets are involved.

In the special case where the energy of S_1(guest) is comparable to the sum of the host and guest triplet energies we have the additional reactions

$$S_1(\text{guest}) \longrightarrow T_1(\text{host}) + T_1(\text{guest}) \tag{4.95}$$

This process is termed *heterofission* and has been observed only in pentacene-doped tetracene due to the necessity of near degeneracy in energy levels demanded by Eq. (4.95). Both hetero-processes are magnetic field sensitive and may be understood in terms of the pair state and overall spin conservation as discussed for exciton fusion.

Experimentally the magnetic field decreases tetracene-delayed green fluorescence in the anthracene lattice due to a decrease in the distribution of spin singlet character among the intermediate pair states. As expected, in the case of fission, the red pentacene fluorescence is *visibly* increased for pentacene-doped tetracene crystals saturating at fields $H > 3$ kG with $F(H)/F(0)|\text{red} \approx 1{\cdot}38$, where $F(H)$ is the fluorescence efficiency in the presence of a magnetic field H and $F(0)$ is the efficiency in the absence of the field. At low fields, ~ 220 gauss, the value of $F(H)/F(0)$ obtains its minimum value of $\sim 0{\cdot}93$. For both heterofusion and heterofission, four high field resonances should be observed every $180°$ in contrast to two for homo-processes, since the guest molecule in the host lattice can occupy either of the two inequivalent sites. (We are assuming only two inequivalent molecules per unit cell. Obviously additional structure occurs where there are more inequivalent molecules per unit cell.) If the guest molecule enters the host lattice without distortion, then the zero-field resonances should occur when the pair states $|+ -\rangle$ and $|0\,0\rangle$ are degenerate. Their orientations are given by solutions of the equations:

$$W_0(\text{host}) + W_0(\text{guest}) = W_-(\text{guest}) + W_+(\text{host}) \tag{4.96}$$

and

$$W_0(\text{host}) + W_0(\text{guest}) = W_+(\text{guest}) + W_-(\text{host}) \tag{4.97}$$

All of the W's in Eqs (4.96) and (4.97) refer to the *molecular* triplet energies. Similar equations are valid in the case of heterofission (Swenberg and Geacintov, 1973).

In the special case of pentacene-doped tetracene crystal (~ 10 ppm), the green (tetracene) fluorescence displays two maxima, one at $+20°$ and another at $-30°$ with respect to the crystallographic b-axis in agreement with the results of the undoped crystals as illustrated by Fig. 4.13A. However the red (pentacene) fluorescence exhibits two prominent maxima at $+14°$ and $-48°$ in addition to two very weak peaks at approximately $+45°$ and $-15°$ which is directly traceable (Geacintov et al., 1971) to the two results, one for each of the two inequivalent molecular orientations in the solutions of Eqs (4.96) and

(4.97). Both the green and red fluorescence efficiencies increase as the temperature is lowered. In the case of tetracene this is known to be due to the suppression of the thermally activated heterofission, but also due to the temperature-dependent populating rates of S_1(pentacene).

4.5.11. Hyperfine Field Modulation

It is well known that triplet exciton production can be sensitized by using an adsorbed dye layer on the crystal surface (Nickel et al., 1969). There are two processes by which triplet excitons can be produced in the bulk crystal by absorbed photons of the dye. The indirect process (line a in Fig. 4.17) whereby the excited dye singlet state internally converts to ^3D and subsequently

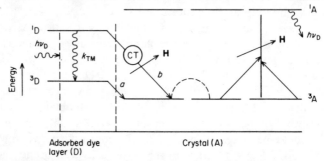

Fig. 4.17. Energy levels of absorbed dye, D, on an organic crystal whose lowest triplet ^3A is below that of the dye ^3D whereas the reverse holds for the lowest singlet states. $h\nu_0$ and $h\nu_D$ denote absorbed photon and emitted delayed photon, respectively. k_{TM} is the standard intersystem crossing process. The two processes denoted by arrows and H refer to the two processes which are magnetic field sensitive. CT is an intermediate charge-transfer state where the hole is located in the crystal, near the surface, while the electron is located on the dye molecule.

transfers its energy to the triplet state of the crystal. The efficiency of the transfer is a function of the energy separation, being larger for small energy differences. Since this is a $T \to T$ transfer and k_{TM}, the internal inversion rate, is known to be field independent, it follows that the indirect process is magnetic field independent. An alternative process is the so-called "direct process" where the state ^3A is generated "directly" from ^1D by passing through an intermediate charge-transfer state (CT). This process is represented by line b in Fig. 4.17. State ^1D dissociates into a CT-state bound to the interface and arises by injection of a hole in the crystal.

The CT-state is created in a pure singlet spin state; however, due to the interaction of both the electron and hole with the protons of the lattice, in addition to possible different g-factor values, the spin character evolves in time. At any instant of time, the spin character of a given CT-state is

$$\phi_{CT}(t) = C_S(t) |S\rangle + C_T(t) |T\rangle \qquad (4.98)$$

where $C_S(t)$ and $C_T(t)$ are the singlet and triplet amplitudes at time t and $C_S(0) = 1$, $C_T(0) = 0$. The spin Hamiltonian describing the state consists of the Zeeman interaction of both the electron and hole, the hyperfine interaction between the hole with the protons of the anthracene molecule and the hyperfine interaction of the unpaired electron with the proton in the dye radical. Other terms, such as the Zeeman interaction of the protons and the electrostatic interaction between the hole and electron, are neglected.

$|C_T(t)| \neq 0$ for $t > 0$ arises because the spin Hamiltonian in the presence of an external field does not commute with the total spin operator and thus since the lifetime of the CT-state is sufficiently long, significant spin motion can occur before either dissociation or recombination takes place. The average of $|C_T(t)|^2$ over the pair-state lifetime is *non*-zero, and is a function of the external field giving rise to a field-sensitive production of ${}^3A$ triplet excitons in the bulk of the crystal. These triplet excitons can then diffuse into the interior of the crystal. If a given triplet encounters another triplet, one possible outcome is the production of delayed fluorescence. This is illustrated in Fig. 4.17.

The magnitudes of the hyperfine terms are of the order of 10^{-4}–10^{-3} cm^{-1}, thus for fields greater than 10 gauss the Zeeman interaction is comparable to the unsymmetric term, i.e. hyperfine interaction given by Eq. (4.39), which gives rise to single–triplet mixing. Experimentally, Groff et al. (1972) observed for rhodamine-B sensitized delayed fluorescence in anthracene an initial increase of approximately 1% in the delayed fluorescence for $H \sim 5$ gauss and a subsequent decrease of over 60% at a field strength of ~ 50 gauss with an eventual saturation for $H > 300$ gauss. The shape of the magnetic field effect is quite similar, except for a change in scale, to that given in Fig. 4.12. In contrast to other magneto-optical effects, the magnetic hyperfine modulation anisotropies are much more difficult to explain since there are crystallographic planes where the field effect is independent of the orientation of the field (Avakian, 1974; Groff et al., 1974).

4.6. APPLICATIONS

In this section some specialized applications of magnetic field effect studies are described. Emphasis is placed on those topics which are most familiar to the authors, and a number of other interesting applications have been regretfully omitted for the sake of brevity.

4.6.1. Scintillation

The mechanisms of scintillation in organic crystals have been extensively discussed by Birks (1964), Voltz (1971), and more recently by Schott (1972).

Basically, the scintillation is produced by the decay of singlet excitons. When a fast particle, β, α, or heavy ion, passes through an organic solid the particle deposits its kinetic energy through the creation of plasmons, which are highly excited "collective" states of the system, and secondary electrons, called δ-rays. The average deposition is ~ 10–20 eV/Å. These plasmon states subsequently decay and within a time period $\approx 10^{-12}$ s, the α-particle track consists of a sea of triplet and singlet excitons in addition to electrons and holes. These "quasi-particles" can undergo several reactions: singlet–singlet, triplet–triplet annihilation, and singlet, triplet, and carrier quenching, and in addition we have exciton fission in the case of tetracene crystals.

The total scintillation originating from the α-particle track consists of a prompt component, which arises from singlet excitons generated directly and a delayed component which results from the bimolecular reaction between two triplet excitons. In the case of crystalline tetracene, where exciton fission is operative both components are magnetic field dependent, whereas only the latter is field dependent for anthracene solids, if we disregard the small positive field effect on internal conversion, i.e. the "hot" fission channel. The effects of external magnetic fields on the scintillation of anthracene crystals bombarded by α- and β-particles were first observed by Klein and Voltz (1975). Recent extensive studies by Geacintov et al. (1975) have illustrated how magnetic field modulation of the scintillation allows for a clarification of the processes involved.

In tetracene, at room temperature, the scintillation displays a characteristic positive magnetic effect at high fields with the characteristic inversion at approximately 430 gauss (same behavior as illustrated in Fig. 4.12). This functional form provides evidence that directly generated singlet excitons, formed either from internal conversion of the plasmon states or by electron–hole recombination, are contributing to the net scintillation. Although the scintillation has the same field strength behavior as the prompt fluorescence arising from ultraviolet excitation, the magnitude of the high field enhancement is substantially reduced from $\sim 35\%$ to $\sim 3\%$.

In contrast to tetracene, crystalline anthracene scintillation due to α-particle irradiation displays the characteristic fusion behavior with increasing field strength. This is expected since thermal fission is inoperative in anthracene, leaving only "hot fission" whose field effect is small and positive, and triplet fusion as the only magnetic channels. Similar to the reduction in the high field enhancement for tetracene crystals, the magnetic field effect for anthracene close to saturation at 4000 gauss is only $\sim -7 \cdot 5\%$ while it is about -22% when the external field is oriented in the ab-plane on a resonance direction and where singlets are generated by triplet–triplet annihilation using triplets generated directly by $S_0 \to T$ absorption.

When the temperature is lowered to 148 K, the magnetic field characteristic of scintillation for solid tetracene exhibits a fusion-like behavior which is consistent with the "freezing out" of the thermally activated fission channel. From a careful analysis, Geacintov and co-workers (1975) have inferred that for tetracene the delayed component contributes $\sim 10\%$ at room temperature to the total scintillation compared to $\sim 50\%$ at 150 K; while for anthracene the triplet–triplet fusion channel contributes ~ 50–70% to the total scintillation and is insensitive to temperature over the range from 150 K to 300 K. These authors interpret the lowering of the percentage of the magnetic field enhancement (or its decrease in the case of anthracene) in the scintillation, L, as compared to singlet emission following ultraviolet or red light excitation as arising from (a) the presence of the high triplet exciton densities within the α-particle track which causes $L \propto \gamma_S/\gamma_{TOT}$ (see previous section of this chapter); (b) dilution of the triplet-fusion channel by hot fission; (c) the high density of transient singlet exciton quenchers in the alpha particle track. Using values for the fission rate in the absence of an external field and the known quenching rate of singlet excitons by either carriers or triplet excitons, Geacintov *et al.* (1975) were able to estimate the effective density of transient quenchers in the α-particle track as $\sim (3 \pm 5) \times 10^{17}$ cm^{-3} which they identified as triplet excitons.

As a further indication of the usefulness of external magnetic field modulation of organic luminescence we note here the recent investigation of Arnold *et al.* (1976a) on the paramagnetic defect profile in solid anthracene subjected to α-irradiation. Anthracene crystals were irradiated by mono-energetic beams of 30 MeV α-particles and subsequently interrogated by a focussed He–Ne laser beam so as to create triplet excitons within a narrow rectangular region of the crystals with a resolution of approximately 50 μm. Both the steady-state, delayed fluorescence and its lifetime in the presence and absence of an external field was measured as a function of position from the irradiated surface. Figure 4.18 shows a plot of the magnetic field dependence of the delayed fluorescence as a function of position for the case of 30 MeV α-particles. The field effect on the delayed fluorescence arises from both the increased lifetime of triplet excitons due to the decoupling with the paramagnetic quenching centers and the decrease in the triplet–triplet annihilative process. As discussed in the preceding section, in the low triplet exciton density regime, the delayed fluorescence, I_{DF}, is proportional to γ/β^2, thus

$$\frac{I_{DF}(H)}{I_{DF}(0)} = \frac{\gamma_{rad}(H)}{\gamma_{rad}(0)} \left\{ \frac{1 + \gamma_{TD}(H) N_p/\beta}{1 + \gamma_{TD}(0) N_p/\beta} \right\}^2 \tag{4.99}$$

The first term in Eq. (4.99) only adjusts the base line for the magnetic field effect, $\gamma_{rad}(H)/\gamma_{rad}(0) = 0.865$ for this field orientation, whereas the observed structure in Fig. 4.18 is due to the functional dependence of N_p, the density

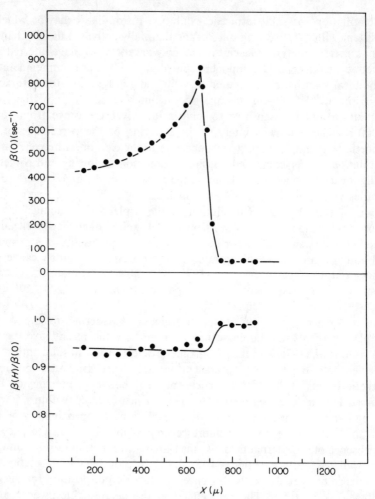

Fig. 4.18. Upper curve: Plot of reciprocal of triplet exciton lifetime as a function of position from the irradiated surface. Lower curve: The ratio of the triplet exciton lifetimes in the presence and absence of an external magnetic field as a function of penetration depth. Anthracene crystals irradiated with 30 MeV α-particles directed along the c'-direction. Experiments performed at room temperature with H oriented in the ac-plane, on resonance under high field conditions. Experimental resolution is no worse than 71 μ. (Results courtesy of Dr S. Arnold.)

of paramagnetic quenchers, on the penetration depth. $\gamma_{TD}(H)$ and $\gamma_{TD}(0)$ are the triplet–doublet quenching rates in the presence and absence of an external field; $\beta = \beta_0 + \gamma_n N_n$ where γ_n and N_n are the quenching rate and density of induced diamagnetic defect sites and β_0^{-1} is the lifetime of the triplet excitons in the unirradiated sample. An analysis of $I_{DF}(H)/I_{DF}(0)$

versus x, the penetration depth, shows that $\gamma_n N_n$ contributes negligibly to Eq. (4.99). In addition, Arnold *et al.* (1976a) were able to calculate the penetration depth profile. The functional dependence of $N_p(x)$ exhibited the expected monotonic increase from the irradiated surface (Friedlaner *et al.*, 1964) and a rather abrupt fall to a negligible density within ~ 50 µm of the Bragg peak. There is little doubt that further studies of this nature are needed, particularly with heavy ions, since measurements of the spatial distribution of damage, as well as the efficiency for producing damage, is of great importance in the therapeutic effects of high energy radiation.

4.6.2. Exciton Caging

In their analysis of delayed fluorescence in anthracene crystals, Kepler *et al.* (1963) added a quadratic term, $-\gamma T_1^2$, to the equation governing the density of free triplet excitons, T_1, in order to account for the disappearance of triplets due to triplet fusion (see Eq. (4.55)). Subsequently, in his extensive theory of the bimolecular rate constant, Suna (1970) discussed the validity of adding the term γT_1^2 to the kinetic equation describing the evolution of the "sea" of triplets. He showed that in the *very* high density regime, γ itself should be a function of T_1, i.e.

$$\gamma = \gamma_0 + \gamma_1 T_1 + O(T_1^2) \tag{4.100}$$

where $\gamma_1 < 0$. Thus excitonic processes, such as fusion and fission, are not only functions of the dimensionality of the system, as predicted by Suna (1970), but also depend on the domain size in which the reactions occur. This follows directly from Eq. (4.100) since $T_1 \propto V^{-1}$, where V is the volume (or area) of the domain in which the sea of triplets exist. However, due to the smallness of γ_1 extremely high densities are needed and are generally not obtainable. Nevertheless, Arnold and co-workers (1976b, c) by using highly doped crystals, with the triplet and singlet energy levels of the guest higher than those of the host were able to probe the consequences of Eq. (4.100). In this mixed crystal system energy cannot be transferred from host to guest and thus the guest molecules act as "obstacles" to the diffusion of the triplet excitons; that is "anti-traps" to host excitons. The introduction of anti-traps allows for the formation of small domains, or exciton cages, of varying sizes depending on the mole fraction X of the guest molecules. These "cages" are leaky to be sure, but the presence of a high concentration of guest molecules may be viewed as constraining the motion of excitons in small domains. The particular system studied was tetracene doped with relatively high concentrations of 2,3-benzocarbazole (BC). Using laser excitation of tetracene singlet excitons, the introduction of BC was found to increase the host prompt fluorescence lifetime at room temperature monotonically from 100 psec for the neat crystal to 400 psec for dopant levels of 49% ($X = 0.49$). It is well known (Merrifield *et al.*, 1969; Groff *et al.*, 1970a) that fusion and thermally induced fission are thermodynamically related by the equation,

$$\gamma_{\text{fission}}/\gamma_{\text{fusion}} = \tfrac{9}{2}\exp(-\Delta E/kT) \tag{4.101}$$

where ΔE is the thermal activation energy for the fission channel. Any decrease in γ_{fusion} (with increasing T_1 as given by Eq. 4.100) will be reflected in a decrease in γ_{fission}. The observation of the enhanced tetracene singlet lifetime, in the heavily doped crystals, partially reflects a decrease in the non-radiative fission channel, and is a direct manifestation of the high density regime or alternatively (and fully equivalent) can be viewed as enhanced germinate recombination of the triplet pair produced by singlet fission. As a further indication of exciton caging, Arnold *et al.* (1976b) observed the expected *decrease* in the enhancement of the fluorescence quantum yield with *increasing* excitation intensity with increasing BC concentration. This is in accord with the expected decrease in γ_{fusion} according to Eq. (4.101) since the exciton density, T_1, within a cage increases as the cage size decreases with increasing mole fraction of BC. For further details the reader should consult the original papers (Arnold *et al.*, 1976b, c).

Magnetic field modulation of the room temperature prompt fluorescence of tetracene-2,3-benzocarbazole mixed crystals has greatly clarified the excitonic processes which are operative. These magneto-optical studies (Arnold *et al.*, 1976c) have shown that $\varepsilon = k_S/k_{-1}$, where the rate parameters k_S and k_{-1} are those given in Eq. (4.61), is an increasing function of guest concentration for mole fractions greater than 9%. This increase in ε is consistent with the reduced domain size and enhanced geminate recombination rate of the correlative $(T_1 T_1)$ intermediate pair state. Of greater utility is the possibility of inferring crystallographic phase transitions in organic crystals from the marked orientational magnetic field dependence of the prompt fluorescence (or delayed fluorescence). From Eq. (4.78) it is easily seen that the angular separation of the position of the peaks (in the case of prompt fluorescence where fission is operative) or dips (in the case of delayed fluorescence) depends only on the molecular spin Hamiltonian parameters and the relative orientation of the two inequivalent molecules with respect to each other; thus any change in the on-resonance angular separation is a direct indication of molecular reorientation. In fact, in BC/T system, Arnold *et al.* (1976c) have shown that for BC concentrations less than $\sim 9\%$ the angular separation between the high field resonance was approximately 52° whereas for $\sim 9\%$ this value decreased to 41° and remained constant for concentrations of BC up to 22.7%. From Eq. (4.78) one can infer that the BC/T mixed crystal undergoes a phase transition at a $\sim 9\%$ dopant concentration.

4.6.3. Magnetic Field Effects on Luminescence in Solutions

Triplet–triplet annihilation, or exciton fusion, followed by the usual delayed fluorescence phenomenon is observed in fluid media (Parker, 1968) as well as in crystals. Faulkner and Bard (1969) first reported the effect of magnetic fields on anthracene triplet–triplet annihilation in fluid solutions. Numerous papers on this subject have been published since then and we refer the reader to some of the more recent articles by Tachikawa and Bard (1974), by Faulkner et al. (1972), and by Van Willigen (1974). Quenching of triplets by doublets also occurs in solution and in the presence of a magnetic field the delayed fluorescence intensity rises because the triplet lifetime increases; this effect is similar to the one observed in organic crystals which is described above. However, the magnetic field effect is larger in solutions and approaches $+30\%$ (Faulkner et al., 1972). In the absence of triplet–doublet quenching, the usual negative magnetic field effect on the delayed fluorescence is observed.

In many ways, these magnetic field effects in fluid solutions are more varied and perhaps more interesting than the ones observed in crystals. Most of the basic phenomena appear to be explainable in terms of the general theoretical framework of Merrifield and his co-workers (Johnson et al., 1967; Merrifield, 1968; Johnson and Merrifield, 1970). However, in solutions the molecules are tumbling rapidly and an important tool for verifying the origin of the magnetic effects, namely the orientation dependence as observed in crystals, is lost. The random orientation of the molecules in the intermediate triplet pair state has been taken into account by Avakian et al. (1971). A detailed treatment based on the Merrifield picture of triplet–triplet and triplet–doublet interactions and taking into account the dynamical processes which take place in liquids has been provided by Atkins and Evans (1975).

While the basic features of the magnetic effects in solutions are similar to those in crystals, a number of unusual and interesting observations have been made in fluid media. Thus Wyrsh and Labhart (1971) have observed different field dependences and noted that the magnetic effects on the monomer- and dimer-type delayed fluorescence of benz(a)anthracene were different. Furthermore, the luminescence is polarized with respect to the magnetic field direction which indicates that the fusion annihilation rate constant may depend on the relative orientations of the two molecules and their orientation with respect to the magnetic field direction. Van Willigen has found that the magnetic field effect on the delayed fluorescence of pyrene is the same for the monomer and the dimer-type (excimer) emission; unusual solvent effects are observed which are attributed to photochemical doublet-radical formation (Van Willigen, 1974).

Interesting effects on the fluorescence yield in pulse radiolysis of solutions of organic fluorophors has been observed (Brocklehurst *et al.*, 1974). Highly energetic electrons (4 MeV) produce excited cations and ions in solutions. Recombination of these ions can yield either singlet or triplet excited states of the fluorophors depending on the spins of the two unpaired electrons in the recombining ions. If a triplet is formed, no fluorescence can take place, while the opposite is true if a singlet is formed. In the presence of a magnetic field the ratio of singlets/triplets formed is changed from its zero-field value and an explanation has been provided by Brocklehurst (1974). These effects are apparently different from the usual Merrifield-type phenomena involving an interplay between the Zeeman and zero-field splitting parts of the spin Hamiltonian. Instead the hyperfine electron–nuclear interaction and its effect on the electron spin precession in the radical-ion pair gives rise to a time-dependent oscillation of the triplet and singlet character. The similarity to the hyperfine solid-state effects described above should be noted.

Magnetic field effects have also been observed in the case of gamma-irradiated solutions of fluorene (Dixon *et al.*, 1975). The results were interpreted in terms of Brocklehurst's theory and the magnitude of the effect was shown to be dependent on the nature of the solvent. This solvent dependence is due to changes in the recombination rate of the ions.

From this short summary of magnetic effects on the luminescence in fluid media, it should be evident that magnetic field studies will continue to provide important information on dynamic processes involving excited states in solutions.

4.6.4. Electroluminescence

Recombination of charge carriers takes place in solids as well as in solutions. The ionized state may be created by high energy particles or by injection of holes and electrons from opposite sides of the crystal. Metallic contacts are usually employed. Upon the recombination of holes and electrons, singlet and triplet excited states are formed via intermediate short-lived charge-transfer (CT) states. The CT states are believed to be neutral and mobile exciton states in which the hole and electron are on neighboring molecules in the crystal, or several molecules apart. The point is that the electron and hole are correlated to each other via the Coulomb potential. Recombination of charge carriers in organic crystals gives rise to the well-known electroluminescence (Helfrich, 1967). The electroluminescence can be divided into a prompt component, which is due to the decay of singlets formed directly from CT excitons and to a delayed component which is due to triplet–triplet fusion. The electroluminescence is magnetic field sensitive (Schwob and Williams, 1972, 1973; Kalinowski and Godlewski, 1975). Generally, the prompt component exhibits a fission-type magnetic field effect both in anthracene (Schwob and Williams, 1972, 1973), in tetracene, and in pentacene-doped tetracene crystals (Kalinowski and Godlewski, 1975); in anthracene thermally induced fission is not observable, and the small fission-type magnetic effect which is observed is probably due to the formation of vibrationally excited singlet states from CT states with subsequent "hot" fission into two triplets. In tetracene, the magnetic effects on the prompt component of the electroluminescence are comparatively large (up to $+24\%$), and are due to thermally induced fission of singlet excitons produced from CT states.

The delayed electroluminescence components exhibit fusion-type magnetic field effects (Schwob and Williams, 1972, 1973) and are interpretable in terms of triplet–triplet annihilation (Schwob and Williams, 1972, 1973; Kalinowski and Godlewski, 1975). Quenching of triplets by holes or electrons is also present. The magnetic field sensitivity of the delayed electroluminescence thus displays the effect of two magnetic field-sensitive processes: triplet–doublet quenching and fusion. A careful analysis of the magnetic field effects on the prompt and delayed components of the electroluminescence can thus provide valuable information about the origins of the excitonic processes involved (Kalinoswki and Godlewski, 1975).

ACKNOWLEDGEMENT

The authors wish to thank Dr S. Arnold for friendly discussions and the data reported in Fig. 4.18. We also acknowledge partial support from a DOE grant to the Radiation and Solid State Laboratory.

REFERENCES

Aloisi, G. G., Masetti, F., and Mazzucato, U. (1974). *Chem. Phys. Lett.* **29**, 502.
Arnold, S. (1974). *J. Chem. Phys.* **61**, 431.
Arnold, S., Hu, W., and Pope, M. (1976a). *Mol. Cryst. Liq. Cryst.* **36**, 179.
Arnold, S., Swenberg, C. E., and Pope, M. (1976b). *J. Chem. Phys.* **64**, 5115.
Arnold, S., Alfano, R. R., Pope, M., Yu, W., Ho, P., Selsby, R., Tharrats, J., and Swenberg, C. E. (1976c). *J. Chem. Phys.* **64**, 5104.
Atkins, P. W. and Evans, G. T. (1975). *Mol. Phys.* **29**, 921.
Avakian, P. (1974). *Pure Appl. Chem.* **37**, 1.
Avakian, P. and Merrifield, R. E. (1968). *Mol. Cryst.* **5**, 37.
Avakian, P. and Suna, A. (1971). *Mat. Res. Bull.* **6**, 891.

Avakian, P., Groff, R. P., Kellogg, R. E., Merrifield, R. E., and Suna, A. (1971). "Organic Scintillators and Liquid Scintillation Counting", p. 499. Academic Press, New York and London.

Beens, H., Knibbe, H., and Weller, A. (1967). *J. Chem. Phys.* **47**, 1183.

Benson, R. and Geacintov, N. E. (1973). *J. Chem. Phys.* **59**, 4428.

Benson, R. and Geacintov, N. E. (1974). *J. Chem. Phys.* **60**, 3251.

Berlman, I. B. (1971). "Handbook of Fluorescence Spectra of Aromatic Molecules." Academic Press, New York and London.

Bersohn, M. and Baird, J. C. (1966). "An Introduction to Electron Paramagnetic Resonance", Appendix B, p. 208. Benjamin, New York.

Birks, J. B. (1964). "The Theory and Practice of Scintillation Counting." Pergamon Press, Oxford.

Birks, J. B. (1970). "Photophysics of Aromatic Molecules." Wiley, London.

Birks, J. B. (1975). "Organic Molecular Photophysics" (J. B. Birks, ed.), Vol. 2. Wiley, London.

Born, M. and Huang, K. (1954). "Dynamical Theory of Crystal Lattices." Clarendon Press, Oxford.

Brocklehurst, B. (1974). *Chem. Phys. Lett.* **28**, 357.

Brocklehurst, B., Dixon, R. S., Gardy, E. M., Lopata, V. J., Quinn, M. J., Singh, A., and Sargent, F. P. (1974). *Chem. Phys. Lett.* **28**, 361.

Carrington, A. and McLachlan, A. D. (1967). "Introduction to Magnetic Resonance." Harper & Row, New York.

Chabr, M., Fünfschilling, J., and Zschokke-Gränacher, I. (1974). *Chem. Phys. Lett.* **25**, 387.

Chabr, M. (1975). Ph.D. Thesis, University of Basel.

Clar, E. (1964). "Polycyclic Hydrocarbons." Academic Press, London and New York; Springer-Verlag, Berlin.

Craig, D. P. and Walmsley, S. H. (1968). "Excitons in Molecular Crystals." Benjamin, New York.

Davydov, A. S. (1948). *Zhur. Eksptl. i Theoret. Fiz.* **18**, 210.

Davydov, A. S. (1962). "Theory of Molecular Excitons" (translated by M. Kasha and M. Oppenheimer, Jnr). McGraw-Hill, New York.

Delannoy, P. and Schott, M. (1969). *Phys. Lett.* **30A**, 357.

Dixon, R. S., Gardy, E. M., Lopage, V. J., and Sargent, F. P. (1975). *Chem. Phys. Lett.* **30**, 463.

El-Sayed, M. A. (1971). *Accts Chem. Res.* **4**, 23.

Ern, V. and Merrifield, R. E. (1968). *Phys. Rev. Lett.* **21**, 609.

Ern, V. and McGhie, A. R. (1971). *Mol. Cryst. Liq. Cryst.* **15**, 277.

Ern, V., Saint-Clair, J. L., Scholt, M., and Delacote, G. (1971). *Chem. Phys. Lett.* **10**, 287.

Faulkner, L. R. and Bard, A. J. (1969). *J. Am. Chem. Soc.* **21**, 6495.

Faulkner, L. R., Tachikawa, H., and Bard, A. J. (1972). *J. Am. Chem. Soc.* **94**, 691.

Frankevich, E. L. and Rumyantsev, B. (1972). *Zh. Eksp. Teor. Fiz.* **63**, 2015.

Frankevich, E. L. and Rumyantesev, M. (1973). *Sov. Phys. JETP*, **36**, 1064.

Frankevich, E. L. and Sokolik, I. A. (1967a). *Zh. Eksp. Teor. Fiz.* **52**, 1189.

Frankevich, E. L. and Sokolik, I. A. (1967b). *Sov. Phys. JETP*, **25**, 790.

Frankevich, E. L. and Sokolik, J. A. (1971a). *Zh. Eksp. Teor. Fiz., Pisma*, **14**, 577.

Frankevich, E. L. and Sokolik, J. A. (1971b). *JETP Lett.* **14**, 401.

Frankevich, E. L., Balabanov, E. I., and Vselyubskaya, G. V. (1960). *Fiz. Tverd. Tela*, **8**, 1970.

Frankevich, E. L., Balabanov, E. I., and Vselyubskaya, G. V. (1966a). *Soviet Phys. Solid St.* **8**, 1567.

Frankevich, E. L., Balabanov, E. I., and Vselyubskaya, G. V. (1966b). *Fiz. Tverd. Tela*, **8**, 855.

Frankevich, E. L., Balabanov, E. I., and Vselyubskaya, G. V. (1966c). *Soviet Phys. Solid St.* **8**, 682.

Frankevich, E. L., Rumyantsev, B. M., and Lesin, V. I. (1973). "6th Molecular Crystal Symposium", Schloss Elmau (unpublished).

Frenkel, J. (1931). *Phys. Rev.* **37**, 17, 1276.

Friedlaner, G., Kennedy, J. W., and Miller, J. M. (1964). "Nuclear and Radio-chemistry." Wiley, New York.

Geacintov, N. E. and Swenberg, C. E. (1972). *J. Chem. Phys.* **57**, 378.

Geacintov, N. E., Pope, M., and Vogel, F. (1969). *Phys. Rev. Lett.* **22**, 593.

Geacintov, N. E., Pope, M., and Fox, S. (1970). *J. Phys. Chem. Solids*, **31**, 1375.

Geacintov, N. E., Burgos, J., Pope, M., and Strom, C. (1971). *Chem. Phys. Lett.* **11**, 504.

Geacintov, N. E., Binder, M., Swenberg, C. E., and Pope, M. (1975). *Phys. Rev.* **B12**, 4113.

Gijzeman, O. L. J., Kaufman, F., and Porter, G. (1973). *J. Chem. Soc. Farad. Trans. II*, **69**, 708.

Groff, R. P., Avakian, P., and Merrifield, R. E. (1970a). *Phys. Rev.* **B1**, 815.

Groff, R. P., Merrifield, R. E., and Avakian, P. (1970b). *Chem. Phys. Lett.* **5**, 168.

Groff, R. P., Merrifield, R. E., Avakian, P., and Tomkiewicz, Y. (1970c). *Phys. Rev. Lett.* **25**, 105.

Groff, R. P., Merrifield, R. E., Suna, A., and Avakian, P. (1972). *Phys. Rev. Lett.* **29**, 429.

Groff, R. P., Suna, A., Avakian, P., and Merrifield, R. E. (1974). *Phys. Rev.* **B9**, 2655.

Gutmann, F., and Lyons, L. E. (1967). "Organic Semiconductors." Wiley, New York.

Ham, N. S. and Ruedenberg, K. (1956). *J. Chem. Phys.* **25**, 13.

Helfrich, W. (1967). *In* "Physics and Chemistry of the Organic Solid State", Vol. 3 (D. Fox, M. M. Labes, and A. Weissberger, eds). Wiley, New York.

Johnson, R. C. and Merrifield, R. E. (1970). *Phys. Rev.* **B1**, 896.

Johnson, R. C., Merrifield, R. E., and Avakian, P. (1967). *Phys. Rev. Lett.* **19**, 285.

Kalinowski, J. and Godlewski, J. (1974). *Chem. Phys. Lett.* **25**, 499.

Kalinowski, J. and Godlewski, J. (1975). *Chem. Phys. Lett.* 345.

Kauzmann, W. (1957). "Quantum Chemistry." Academic Press, New York and London.

Kepler, R. G., Caris, J. C., Avakian, P., and Abramson, E. (1963). *Phys. Rev. Lett.* **10**, 400.

Klein, G. and Voltz, R. (1975). *Int. J. Radiat. Phys. Chem.* **7**, 155.

Klein, G., Voltz, R., and Schott, M. (1972). *Chem. Phys. Lett.* **16**, 340.

Klein, G., Voltz, R., and Schott, M. (1973). *Chem. Phys. Lett.* **19**, 391.

Knox, R. S. (1963). "Theory of Excitons." Academic Press, New York and London.

McGlynn, S. P., Azumi, T., and Kinoshita, M. (1969). "Molecular Spectroscopy of the Triplet State." Prentice-Hall, Englewood Cliffs, N.J.

Maier, G., Haeberlein, V., and Wolf, H. C. (1967). *Phys. Lett.* **25A**, 323.

Merrifield, R. E. (1968). *J. Chem. Phys.* **48**, 4318.

Merrifield, R. E. (1971). *Pure and Appl. Chem.* **27**, 481.

Merrifield, R. E., Avakian, P., and Groff, R. P. (1969). *Chem. Phys. Lett.* **3**, 155.
Moller, W. M. and Pope, M. (1973). *J. Chem. Phys.* **59**, 2760.
Nickel, B., Staerk, H., and Weller, A. (1969). *Chem. Phys. Lett.* **1**, 27.
Parker, C. A. (1968). "Photoluminescence of Solutions." Elsevier, Amsterdam.
Parmenter, C. S. and Rau, D. J. (1969). *J. Chem. Phys.* **51**, 2242.
Platt, J. R. (1964). "Free Electron Theory of Conjugated Molecules: A Source Book." Wiley, New York.
Pope, M., Geacintov, N. E., and Vogel, F. (1969). *Mol. Cryst. Liq. Cryst.* **6**, 83.
Port, H. and Wolf, H. C. (1968). *Z. Nature,* **239**, 315.
Schott, M. (1972). Thesis, Université de Paris. (Unpublished).
Schwob, H. P. and Williams, D. F. (1972). *Chem. Phys. Lett.* **13**, 581.
Schwob, H. P. and Williams, D. F. (1973). *J. Chem. Phys.* **58**, 1542.
Sokolik, I. A. and Frankevich, E. L. (1973). *Uspekh. Fiz. Nauk,* **111**, 261.
Sokolik, I. A. and Frankevich, E. L. (1974). *Engl. Transl. Sov. Physics, Uspekh.* **16**, 687.
Sternlicht, H. and McConnell, H. (1961). *J. Chem. Phys.* **35**, 1793.
Streitwieser, A., Jr (1961). "Molecular Orbital Theory for Organic Chemists", Chapter 8. Wiley, New York.
Suna, A. (1970). *Phys. Rev.* **B1**, 1716.
Swenberg, C. E. (1969). *J. Chem. Phys.* **51**, 1753.
Swenberg, C. E. and Stacy, W. T. (1968). *Chem. Phys. Lett.* **2**, 327.
Swenberg, C. E. and Geacintov, N. E. (1973). *In* "Organic Molecular Photophysics", Vol. I (J. B. Birks, ed.). Wiley, London.
Swenberg, C. E., Ratner, M., and Geacintov, N. E. (1974). *J. Chem. Phys.* **60**, 2152.
Tachikawa, H. and Bard, A. J. (1974). *Chem. Phys. Lett.* **26**, 568.
Van Willigen, H. (1974). *Chem. Phys. Lett.* **26**, 568.
Voltz, R. (1971). "International Discussion on Progress and Problems in Contemporary Radiation Chemistry", Vol. 1, p. 139. Academia, Prague.
Wilkinson, F. (1975). *In* "Organic Molecular Photophysics", Vol. 2 (J. B. Birks, ed.). Wiley, London.
Wyrsh, D. and Labhart, H. (1971). *Chem. Phys. Lett.* **8**, 217.
Yarmus, L., Swenberg, C. E., Rosenthal, J., and Arnold, S. (1973). *Phys. Lett.* **43A**, 103.

5

Magneto-optical Investigations of Recombination Radiation in Inorganic Crystals

B. C. CAVENETT

Department of Physics, The University, Hull

5.1. INTRODUCTION TO RECOMBINATION RADIATION

5.1.1. General Introduction

The luminescence resulting from the annihilation of electron–hole pairs in crystalline solids is called recombination radiation. This is a general term which covers both short and long wavelength luminescence whereas the description "edge emission" refers, in particular, to luminescence with energy just less than the band gap. When crystals are excited by laser or other radiation, electrons are removed from the valence band and placed in the conduction band leaving holes in the valence band. Because of the mutual attraction between the electrons and the holes, these particles can exist as pairs or excitons which can move throughout the crystal. The excitons can be trapped at donors or acceptors in either neutral or ionized states or can be localized such that the electron is on a donor and the hole is on an acceptor. These various electron–hole recombination processes produce an emission spectrum with many different components and the purpose of this section is to briefly introduce the reader to the interpretation of the spectra.

In general, recombination luminescence and, particularly, edge emission is produced by excitation of the crystal with radiation having an energy greater than the band gap. Since the absorption coefficient is very large for this radiation, the luminescence will come from a volume of the crystal which is only a few microns thick. Thus, the state of the surface is usually important for the observation of edge emission and, if a crystal cannot be cleaved, a suitable surface must be prepared by mechanical polishing and etching.

Many excitation sources have been used to produce recombination luminescence. The most common in the past were the high pressure mercury and xenon lamps, but now these discharge sources are being replaced by argon ion, krypton, helium–neon, GaAs, and other lasers. Cathode rays and X-rays are also used, especially in the case of wide band gap materials. However, one of the most important processes is the electrical excitation of radiation from a p–n junction in forward bias. The recombination radiation comes from the plane of the junction and the devices are called light emitting diodes or LEDs. These visible and near-infrared emitting devices are used widely in the electronics industry for indicators and displays, and several diode types, including GaAs and GaAlAs, are commercially available as lasers and laser arrays. Thus, the study of recombination radiation in single crystals is linked to the technological progress of LEDs.

The most common recombination transitions are illustrated in Fig. 5.1; there are two systems of symbols which authors use to label these luminescences and these are given in the Table I and are used to label the

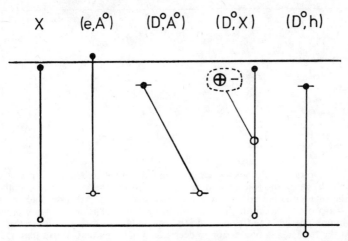

Fig. 5.1. Schematic diagram of principal electron–hole decay processes in semi-conductors. The recombination transitions are: free exciton recombination, X; free-electron to neutral acceptor, (e, A⁰); donor acceptor, (D⁰, A⁰); bound exciton decay at a neutral donor, (D⁰, X); and donor-to-free hole, (D⁰, h). (These processes correspond to E → V(i), C → A, D → A, E → V(ii), and D → V referred to in Chapter 1.)

TABLE I

Symbols in Common Use for Labelling Recombination Processes

Free exciton	X	$+-$
Donor	D^0	$\oplus-$
Acceptor	A^0	$\ominus+$
Exciton bound at neutral donor	D^0, X	$\oplus = +$
Exciton bound at ionized donor	D^+, X	$\oplus - +$
Exciton bound at neutral acceptor	A^0, X	$\ominus -$
Exciton bound at ionized acceptor	A^-, X	$\ominus + -$
Exciton bound at isoelectronic trap		$\odot \pm$
Free electron to acceptor	e, A^0	
Donor to free hole	D^0, h	

transitions in Fig. 5.1. The composite emission in Fig. 5.2 shows schematically the relative positions and the characteristics of the possible transitions in ZnSe. The weak broad band at highest energy is the free exciton, the intense sharp lines are the bound excitons, the weak group of sharp lines are the close donor–acceptor pairs, followed by the broad and overlapping series of free electron-to-bound hole and distant donor–acceptor bands. Finally, there is, at the lowest energy, one example of a donor-to-deep acceptor transition; in this case it is the self-activated emission, a transition from a donor to an acceptor formed from a V_{Zn}–Cl pair. In the last part of this section the various recombination processes are discussed in more detail

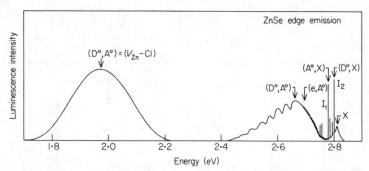

Fig. 5.2. A composite luminescence spectrum of ZnSe excited by ultraviolet light. At the highest energy the free exciton recombination is observed and at slightly lower energies the sharp line bound exciton emissions can be seen. At lower energies again a series of sharp donor–acceptor lines merge into a series of broad bands separated by the LO phonon energy. The free-electron-to-acceptor transitions are also in this region. Finally, at the lowest energy a donor-to-deep-acceptor emission is illustrated.

but it should be pointed out that the purpose of this chapter is to illustrate, by examples from recombination radiation studies, the application of magneto-optical techniques to the study of luminescence and not to present a complete review of the subject. However, an attempt has been made to include references of the most recent papers in this field so that this article supplements the reviews by Reynolds *et al.* (1965), Gershenzon (1966), Dean (1966), Garlick (1967), Halsted (1967), Landsberg (1967), Varshni (1967), Williams (1968), Dean (1969, 1973), and Bergh and Dean (1972).

In the present review, the principles and techniques of Zeeman spectroscopy and optically detected magnetic resonance are discussed in Section 5.2 followed by a description of the application of these techniques to studies of donor–acceptor recombination in II–VI compounds in Section 5.3. Finally, in Section 5.4, these same techniques are considered with reference to bound exciton luminescence in III–V compounds and, very briefly, to triplet excitons in insulators.

5.1.2. Donor–acceptor Recombination

The emission of donor–acceptor pairs has been investigated in both III–V and II–VI compound crystals and reviewed in detail by Dean (1973). Several distinct types of spectra are observed; for example, the sharp line spectra of GaP are characteristic of close donor–acceptor pairs where both the donor and acceptor levels are shallow (Thomas *et al.*, 1964a). Broad band emissions with longitudinal optical (LO) phonon replicas such as the CdS green luminescence are characteristic of distant donor–acceptor recombinations where both centres are weakly bound. It is worth noting that because the levels are shallow these emissions are not observed at room temperature.

Finally, several broad emission bands which are very intense at room temperature have been identified as donor-to-deep acceptor transitions. The emission energies for these transitions are generally much less than the band gap energy; for example, the self-activated emissions in II–VI compounds are discussed in Section 5.3.4 and reviewed by Curie and Prener (1967).

One of the most important characteristics of the donor–acceptor recombination radiation is the dependence of the energy of recombination on the distance between the donors and acceptors. For pairs, separated by r, the energy of emission is given by Thomas *et al.* (1964a)

$$E(r) = E_g - (E_a + E_d) + (e^2/4\pi\varepsilon_0\, \varepsilon r) \qquad (5.1)$$

where E_a and E_d are the acceptor and donor binding energies, E_g is the band-gap energy and ε is the low frequency dielectric constant. When r is small, the donors and acceptors will occupy pairs of lattice sites with specific crystallographic directions and, since only certain values of r will therefore be allowed, the recombination energies given by Eq. (5.1) will be discrete. Thus, the close pairs will give rise to sharp line spectra which can be related to the definite donor–acceptor pairs. However, as r increases the discrete lines will merge together to form a continuum and so the recombination radiation from distant donor–acceptor pairs is in the form of broad emission bands where the highest energy band has a peak energy corresponding to $r \to \infty$ in Eq. (5.1). Both these types of spectra are shown in the upper curve of Fig. 5.3 which is from the work of Merz *et al.* (1973). We note that, in order to observe the sharp pair lines, the excitation intensity must be high;

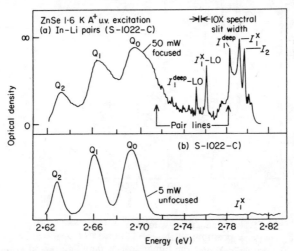

Fig. 5.3. Photoluminescence of ZnSe at $1 \cdot 6$ K showing the donor–acceptor recombination bands Q_0, Q_1, ..., at low excitation power and the additional sharp pair lines at higher power. (After Merz *et al.*, 1973.)

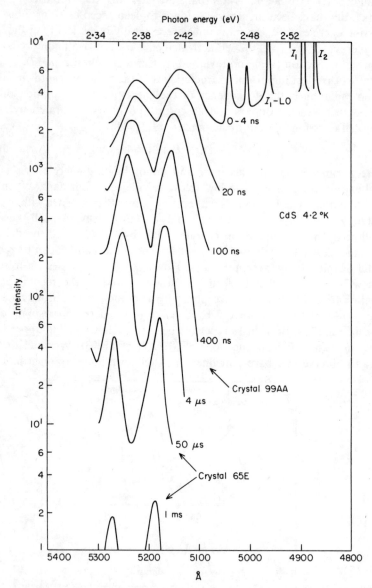

Fig. 5.4. Time-resolved emission spectra for the donor–acceptor recombination bands in CdS. The peak positions of the bands move to longer wavelength as the delay between the excitation pulse and the luminescence sampling period is increased. After Thomas *et al.*, 1967.)

at low power (see the lower curve in Fig. 5.3) only the distant pair bands labelled Q_0, Q_1, and Q_2 are observed. The series limit of the sharp line spectra is $Q_0(E_{r\to\infty})$ and the energy separation between the distant pair bands is the LO phonon energy. The reason for requiring high excitation intensities in order to observe the sharp donor–acceptor lines is that the recombination rate depends on the pair separation. In fact, the radiative transition probability is (Thomas *et al.*, 1965)

$$W(r) = W_0 \exp\left(-2r/a_0\right) \tag{5.2}$$

where a_0 is the donor Bohr radius and W_0 is the transition probability as $r \to 0$ and will be the same for all pairs. Thus, close pairs recombine more rapidly than distant pairs, explaining the observed power dependence of the sharp line spectra.

The dependence of the recombination rate on the donor–acceptor pair separation is very important since it is this property that has enabled investigators to determine which emission bands are donor–acceptor in nature. The experiments involve observing the luminescence at varying time delays after a pulse of excitation radiation; they are called time-resolved spectroscopy measurements and were first carried out on donor–acceptor bands by Colbow (1965) and Thomas *et al.* (1964b, 1967). The results for CdS from the work of Thomas *et al.* (1967) are shown in Fig. 5.4 and it is observed that as the delay after excitation is increased, the two donor–acceptor peaks move to lower energies. The more distant pairs which have slower recombination rates as well as lower recombination energies are preferentially sampled for long delays and hence the emission peaks shift to lower energies.

The emission bands of donor-to-deep acceptor recombinations also show time-resolved spectra. For example, Era *et al.* (1969) studied the time-resolved spectroscopy of self-activated and copper emission bands in ZnS, and Iida (1968) and Bryant and Manning (1974) investigated the self-activated emission in ZnSe. These donor-to-deep acceptor bands do not show the sharp line spectra observed from the close, shallow donor–acceptor pairs and, in general, they are broad featureless bands without phonon structure. However, since these emissions are usually very intense and because they can be observed at room temperature, there is considerable interest in developing light-emitting diodes which show these emissions.

The self-activated emission has been observed in most of the II–VI compounds and has been studied in detail in ZnS and ZnSe. Figure 5.5 illustrates the excitation and emission process for the self-activated emission in ZnS. On excitation with ultraviolet an electron is removed from the self-activated centre, a (V_{Zn}–donor) complex, which becomes a deep acceptor state called the A-centre. Recombination of the electron on the

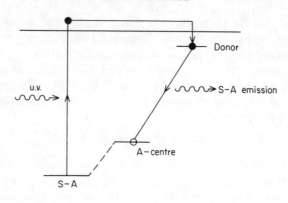

Fig. 5.5. Excitation and emission cycle for the blue self-activated emission in ZnS. Ultraviolet excitation removes an electron from the $(V_{Zn}-Cl)'$ self-activated centre giving an electron in the conduction band and leaving an A-centre, $(V_{Zn}-Cl)$, acceptor. The electron recombines with the hole on the acceptor by donor–acceptor recombination.

donor with the hole on the acceptor gives the self-activated emission. An investigation of this emission by optically detected magnetic resonance is discussed in Section 5.3.4 along with more recent results on the ZnSe self-activated emission. Other recent investigations of the self-activated emission have been reported by Livingstone *et al.* (1973), Jones and Woods (1974), Özsan and Woods (1975), Bouley *et al.* (1975) for ZnSe and by Susa *et al.* (1975) for CdS.

5.1.3. Excitons and Bound Excitons

The excitonic state of a crystal is an excited state of the whole crystal and so the electron–hole pair is mobile and can move throughout the lattice. In semiconductors such as the III–V and II–VI compound crystals the electron–hole pairs are loosely bound and are called Wannier excitons since they are described by crystal lattice states. However, in insulators such as the alkali-halides and alkaline earth fluorides the excitons are tightly bound electron–hole pairs. These excitons exhibit more local atomic character and are called Frenkel excitons; the excitation is still an excited state of the crystal and can exist anywhere throughout the crystal. For a discussion of excitons see Dexter and Knox (1965) or Segall and Marple (1967).

Semiconductors. The sharp line emissions due to bound exciton recombinations, referred to in the first section, are well-known features of the luminescence from II–VI and III–V compound crystals. Although the initial studies of these lines by Zeeman spectroscopy were carried out in the early sixties and a comprehensive literature exists describing the various recombination processes, there is still considerable interest in the study of these lines, particularly in the II–VI compounds. The early investigations were concerned with the characterization of the various recombination processes and the association of each emission line with a particular bound exciton decay process. Zeeman spectroscopy proved to be a valuable way to make these identifications because the sharpness of the lines allowed the Zeeman components to be identified (see Section 5.2.1). When more controlled crystal growth was possible the emphasis of the investigations centred on the association of specific emission lines with known impurities. Thus the spectroscopic study developed into an analytical method for determining the purity of the semiconducting crystals, particularly the III–V compounds. For example, see the work on donors and acceptors in GaP by Dean et al. (1970, 1971). This interest continues and it is hoped that the same analytical approach to the II–VI compounds will be equally successful. In particular, CdTe is receiving considerable attention (Heisinger et al., 1975; Taguchi et al., 1975).

Further impetus to the understanding of exciton emission in II–VI crystals has come from the reports of excitonic molecules and Bose–Einstein condensation in these materials. Since the exciton lifetimes in the II–VI crystals are very short it has been necessary to use pulsed electron beams and pulsed lasers to produce excitons at sufficiently high density such that excitonic molecules and Bose condensation can be observed. In order to study excitonic molecules and exciton–exciton collisions most laboratories use a nitrogen laser giving 10 ns wide pulses and peak excitation intensities of several MW/cm^2 on the sample surface. Although two photon absorption studies (Gale and Mysyrowicz, 1974) are important for studying the high density exciton properties, in general it has been the new luminescence bands which occur with high excitation intensities that have been investigated. Kuroda et al. (1973), Benoît à la Guillaume (1976), and Hanamura (1976) have reviewed the most recent results of high intensity excitation effects in semiconductors and Saito and Shionoya (1974) and Saito et al. (1976) have published comprehensive studies of CdS, CdSe, and ZnSe. The luminescence of a CdSe crystal under increasing excitation intensities up to 2 MW/cm^2, as reported by Kuroda et al. (1973), is shown in Fig. 5.6. The emission band M has been assigned to excitonic molecule decay and the P and P_M bands to decay processes involving inelastic collisions between excitons and between molecules, respectively. However, at this stage the interpretation of these emission bands has not yet been settled. The Tokyo group has also reported Bose condensation of excitons in CdSe (Kuroda et al., 1973) though this is the subject of some controversy (Johnston and Shaklee, 1974). Müller

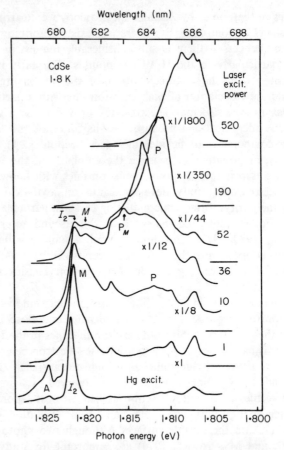

Fig. 5.6. Photoluminescence of CdSe under high intensity excitation from a pulsed nitrogen laser. The maximum excitation shown at the top of the figure corresponds to 2 MW/cm² and can be compared with a spectrum taken with a mercury lamp where the high density exciton bands are not observed. (After Kuroda *et al.*, 1973.)

et al. (1976) have reported excitonic molecule emission in CdS and have suggested that the Q-line, which is observed under excitation intensities of 50MW/cm², is due to a radiating electron–hole plasma. Excitonic molecule emission has also been reported in ZnO by several authors (Hvam, 1974; Miyamoto and Shionoya, 1976; Segawa and Namba, 1976).

Insulators. When alkali halide crystals are excited with X-rays two emission bands are observed due to the radiative decay of self-trapped excitons. These excitons are localized on a pair of adjacent halogen ions and the formation of the self-trapped exciton can be considered as arising from the trapping of an electron by a V_k centre (a self-trapped hole centre). For a discussion of these defect centres see the review by Kabler (1972). Interest in the self-trapped exciton has been stimulated by the detailed information obtained from optically detected magnetic resonance measurements carried out on the low energy emission which is due to transitions from the triplet state of the self-trapped exciton. These resonance data are the only reported results for inorganic crystals since magnetic resonance of excitons in semiconductors has not yet been observed. The first optically detected magnetic resonance measurements were carried out by Marrone et al. (1973) and Wasiela et al. (1973) on KBr and CsBr and a summary of these results can be found in Section 5.4.3. Further studies of the self-trapped exciton in alkali halides have been reported by Wasiela and Duran (1974) and Call et al. (1975a). Wasiela and Duran (1974) have also reported a study of the self-trapped exciton in SrCl$_2$ which has the fluorite structure. Recently, optically detected magnetic resonance of excitons has been reported in many materials such as the alkaline earth fluorides (Call et al., 1975b), KMgF$_3$ (Hayes et al., 1975), and AgCl (Hayes and Owen, 1975; Murayama et al., 1976).

5.2. MAGNETO-OPTICAL SPECTROSCOPY

5.2.1. Magnetic Field Interactions in Solids

The general principle behind the use of magnetic fields to study luminescence is well known since if an atomic model has been proposed to explain an emission band or line, the effect of a magnetic field on the luminescence can be predicted and compared with the magneto-optical measurements in order to confirm that correctness of the model. This approach, which is an example of what can be called perturbation spectroscopy, is also used to investigate atomic systems by stress and electric fields. Unfortunately, experimental limitations may prevent unambiguous analysis of magnetic field effects but before discussing this point we should examine in a general way the effect of a magnetic field on an atomic system.

Luminescence occurs when an atomic system relaxes radiatively from an excited energy state to a lower, possibly the ground, atomic state. Now each atomic state has an allowed value of angular momentum characterized by the quantum number J and associated with each of the states is a magnetic moment, μ, which is proportional to the angular momentum. When the magnetic field is present there is an energy of interaction between the field and the magnetic moment and this is written as

$$\mathcal{H}_z = -\mu \cdot \mathbf{B} \tag{5.3}$$

where $\mathcal{H}_z$ is the Zeeman interaction Hamiltonian. If the system is described by an angular momentum $\mathbf{J}$ then the magnetic moment is

$$\mu = -g\mu_B \mathbf{J} \tag{5.4}$$

and

$$\mathcal{H}_z = g\mu_B \mathbf{B}.\mathbf{J} \qquad (5.5)$$

where, for an atomic system and Russell–Saunders coupling,

$$g = g_{\text{Lande}} = \tfrac{3}{2} + \frac{S(S+1)-L(L+1)}{2J(J+1)} \qquad (5.6)$$

with $\mathbf{J} = \mathbf{L}+\mathbf{S}$ and where $S(S+1)$, $L(L+1)$, and $J(J+1)$ are the eigenvalues of $\mathbf{S}^2$, $\mathbf{L}^2$, and $\mathbf{J}^2$, respectively. If we choose the magnetic field in the z-direction then $B = B_z$ and the interaction energy from Eq. (5.5) is

$$\mathcal{H}_z = g\mu_B B_z J_z \qquad (5.7)$$

where J_z is the component of angular momentum in the z-direction. Now, the allowed values of J_z are given by the quantum number M_J which has the values $+J, J-1, ..., -J$. Thus, we can write the allowed Zeeman energies as

$$\langle \mathcal{H}_z \rangle = E_z = g\mu_B B_z M_J \qquad (5.8)$$

where $\langle \mathcal{H}_z \rangle$ are the average or allowed values of the Zeeman operator. It is clear then that an energy state labelled by J will split into $2J+1$ Zeeman levels labelled by M_J. Since both the initial and final atomic states can split in the presence of a magnetic field, the resulting luminescence spectrum can be complicated. For example, we can consider the emission due to electric dipole transitions between an excited p-state and a ground s-state. The energy levels with and without a magnetic field are shown in Fig. 5.7. The ground state is $^2S_{\frac{1}{2}}$ ($J = \frac{1}{2}$) and splits into two Zeeman levels for $M_J = \pm\frac{1}{2}$ while the p-state is split by spin–orbit coupling into two states $^2P_{\frac{3}{2}}$ ($J = \frac{3}{2}$) and $^2P_{\frac{1}{2}}$ ($J = \frac{1}{2}$) which are then split by the magnetic field into a quartet, $M_J = \pm\frac{3}{2}, \pm\frac{1}{2}$, and a doublet, $M_J = \pm\frac{1}{2}$. The energies can be calculated using Eqs (5.6) and (5.8). The allowed transitions are also shown in Fig. 5.7 and are of two types; there are σ transitions for $\Delta M_J = \pm 1$ which give luminescence which is circularly polarized when observed parallel to the magnetic field or linearly polarized perpendicular to B if observed normal to B. There are also π transitions for $\Delta M_J = 0$ which give radiation linearly polarized parallel to the magnetic field when observed normal to it. In the present example, two emission lines are observed before the application of a magnetic field but, with a field present, a total of 10 lines can be observed; six transitions are σ-polarized and four are π-polarized. All ten lines can be seen when observing perpendicular to B but only the six σ lines are observed along B. Although this simple model describes the well-known yellow emission from sodium atoms, it is important in the present discussion because it provides an introduction to the description of electronic transitions in solids where conduction band states have s-like character and valence band states are p-like. At this stage of the discussion this model allows us to emphasize

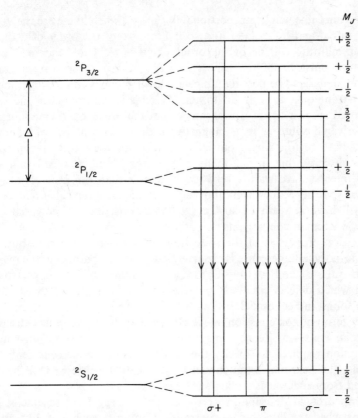

Fig. 5.7. Energy level diagram showing the allowed transitions between a p-state and an s-state in the presence of a magnetic field. The transitions are grouped according to the polarizations, σ_+, π, and σ_-. The quantity Δ is the spin–orbit splitting.

the two effects which the magnetic field has on the emission. Firstly, the emission spectrum splits into Zeeman components and, secondly, these components have specific polarization properties with respect to the magnetic field direction.

The experimental limitations of Zeeman spectroscopy can now be defined more clearly since the observation of magnetic field splittings of sharp emission lines will depend on the natural or phonon width of the emission lines, the spectrometer resolution, the maximum magnetic field strength, and the g-factor. All of these factors must be considered for each problem. However, it is possible to place an approximate general limit to the Zeeman splitting which can be obtained since practical magnetic fields are rarely more than 15 Teslas (150 kgauss) and so for a free electron ($g = 2$) the splitting of the energy levels is of the order of $15\ cm^{-1}$. The maximum splitting of the luminescence line will be proportional to $(g_1 + g_2)$ where these are the g-values of the initial and final states and so if both are of the order of 2 then the Zeeman splitting will be of the order of 30 cm^{-1}. Thus, the spectrometer resolution must be of the order of a few wavenumbers for satisfactory Zeeman analysis to be possible. In general, when one is considering magneto-optical techniques it is useful to classify emissions into sharp line spectra and broad spectra where it is assumed that the broad band emissions show no resolvable splitting in a magnetic field. There will be, of course, an intermediate region where partial resolution of the Zeeman components can be observed, but in fact many of the broad luminescence bands which are of interest have widths very much greater than the general maximum Zeeman splitting value given above. For example, a typical broad band in ZnS will have a width of 4000 cm^{-1} (0·5 eV) at the half power. In these cases the Zeeman components can still be "resolved" by making use of the polarization properties of the transitions which were discussed above. For example, if the luminescence is observed in a direction parallel to the magnetic field then the σ^+ and σ^- transitions can be distinguished by a spectroscopic system which detects circularly polarized light. Examples of this technique can be found in Section 5.3.3. Finally, it should be noted that in some cases a sharp zero-phonon emission line is observed at the high energy extremity of the broad emission. Zeeman measurements on this line may provide the information required to specify the nature of the luminescent centre. For example, the near-neighbour model for the Cd–O centre in GaP has been deduced from such measurements (Henry et al., 1968).

The theoretical description of the atomic energy level scheme is more complex when the luminescent centre is an impurity such as a 3d transition metal ion (e.g. Mn^{2+}, Co^{2+}, Ni^{2+}) in a crystal lattice. The crystal electric field of the ions surrounding the impurity ion can split the energy levels and produce anisotropic magnetic properties. For details of the group theoretical methods for dealing with such problems, the books by Griffith (1961), Judd (1963), and Abragam and Bleaney (1970) should be consulted. Similar theoretical complexities are found when one considers luminescent processes involving free or bound electrons and holes in semiconductors. In general, the g-values of the free carriers, donors, and acceptors will all be determined by the band structure of the material. For a discussion of the calculation of g-values from first principles consult Yafet (1963) and Callaway (1974).

The discussion in the previous paragraph implies that the experimentalist is faced with a formidable calculation in order to gain any valuable information from magneto-optical measurements. However, although the expression (see Eq. 5.6) for the atomic Lande g-factor generally does not apply to the case of luminescent centres in solids, the concept of the g-factor is still useful as an experimental parameter which can be used to interpret and compare results. The formalism used is called the spin Hamiltonian method and a detailed discussion can be found in Abragam and Bleaney (1970) or in other books dealing with the analysis of magnetic resonance spectra. The method employs an effective Hamiltonian which has the appropriate symmetry properties for the problem and contains an effective spin, S_{eff}, which correctly describes the observed degeneracy of the atomic state. The crystal field effects can be included in a very general form, if necessary. The simplest spin Hamiltonian can be written as

$$\mathcal{H} = \mu_B \mathbf{B} \cdot \mathbf{g} \cdot \mathbf{S}_{eff} \qquad (5.9)$$

where the g-factor is now a 3×3 matrix and is a set of parameters which contains the orbital contribution to the magnetic moment. If the g-value is isotropic then we have an equation which is analogous to Eq. (5.5) and is

$$\mathcal{H} = g\mu_B \mathbf{B} \cdot \mathbf{S}_{eff} \qquad (5.10)$$

For example, if the degeneracy of a state in a magnetic field is observed to be 2 then $S_{eff} = \frac{1}{2}$ and if the degeneracy is 4, then $S_{eff} = \frac{3}{2}$, but in both cases the g-parameter need not be 2 since the effective spin need not be the true spin. The values of the conduction electron g-values in semiconductors (see Tables II and III) illustrate this point since in all cases the $S_{eff} = \frac{1}{2}$ but the g-values range from 0·4 to 50. Usually, the magnetic interaction can be described by three g-parameters, g_x, g_y, and g_z. If there is axial symmetry, $g_{\parallel} = g_z$ and $g_{\perp} = g_x = g_y$. Finally, it must be noted that in the end a first principles calculation must be made for each problem but the spin Hamiltonian method is useful as the meeting place for the theorist and the experimentalist.

5.2.2. Zeeman and Magnetic Polarization Spectroscopy

In this section the basic techniques of magneto-optical spectroscopy are briefly reviewed. The usual requirement is for the sample to be at a low temperature and in a high magnetic field with good optical access so that the sample can be excited with laser radiation and the luminescence can be collected for analysis with a high resolution spectrometer or spectrograph. The most important difference between experimental systems is the type of magnet used and so it is worth while briefly discussing the merits of the different magnet systems. For many years the most commonly used system was the conventional iron core magnet with tapered pole pieces which can give fields of 3–4 T (30–40 kgauss). A conventional helium cryostat with windows can be used and the most convenient way to observe the luminescence is in a direct perpendicular to the magnetic field. Observations of the luminescence emitted parallel to the field can be carried out by using mirrors or by boring a hole in a pole piece and directing the light through a quartz light pipe. However, in both these cases caution is needed if polarization measurements are required. Magnetic field measurements can be made with room temperature Hall effect and NMR probes. The major disadvantage of an iron core magnet system is the low maximum field strength determined by the core saturation properties. A derivative of this type of magnet is the air-core solenoid of Bitter or Oxford design. These are copper magnets which dissipate several megawatts of power into water cooling systems; magnetic fields of greater than 15 T (150 kgauss) can be attained. Although, these magnets are generally part of a major high field laboratory installation, the systems are usually convenient to use. Optical access may be restricted due

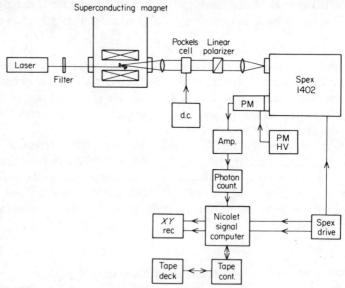

Fig. 5.8. Diagram of a magneto-optical spectrometer which is suitable for magneto-emission studies such as high resolution Zeeman spectroscopy and high and low resolution magnetic-induced polarization studies.

to the size of the magnet and the most convenient direction for observing the luminescence is parallel to the field. In general, it is necessary to operate the measuring equipment by remote control since access to the magnet room is usually restricted when the magnet is operating. Finally, the most common systems in use are the superconducting magnets. Fields in excess of 15 T ([1]50 kgauss) are available and the magnets are small and convenient to use. Optical access is generally very good and split-coil magnets make possible convenient measurements of the luminescence both parallel and perpendicular to the magnetic field. Field measurements are made with a low temperature Hall effect probe or an NMR probe.

The superconducting magnet system used in this laboratory is shown in Fig. 5.8. Both the sample and the magnet are immersed in liquid helium which is overpumped to below the He^4 λ-point (2·2 K) for optical clarity. The samples are excited by radiation from an argon ion laser or a CW tunable dye laser and the luminescence is analysed with a high resolution double monochromator. The light output of the spectrometer is detected with a cooled photomultiplier and a photon-counting system. The spectra are recorded directly onto chart or stored digitally in a signal computer with magnetic tape facilities.

The analysis of circularly polarized light can be made either by ac or dc methods. If the polarization is small in magnitude then the ac or modulation method can be used. A basic analyser consists of a $\lambda/4$ retarder and a linear polarizer and phase shifts of $\pm \lambda/4$ can be achieved at audio and rf frequencies by using an electro-optical modulator such as a Pockels' cell or stress modulator. The output of a phase-sensitive detector is proportional to $(I_{\sigma+} - I_{\sigma-})$ when the detector reference is at the modulator driver frequency. These techniques have been discussed recently by Cavenett and Sowersby (1975). DC measurements can be made with standard $\lambda/4$ plates, Fresnel rhombs, stressed quartz plates, and Pockels' cells with dc voltages.

5.2.3. Optically Detected Magnetic Resonance

It is first important to discuss conventional magnetic resonance which is commonly described as EPR or ESR. Details of the technique and its applications can be found in the many excellent books on the subject; for example, Abragam and Bleaney (1970). In Section 5.2.1 we have seen that the effect of a magnetic field is to split the atomic energy levels into several Zeeman levels with splittings proportional to the g-factor. If the sample is excited with a radio-frequency field, of frequency v, which is oscillating at right angles to the applied dc magnetic field then transitions can be induced between the Zeeman levels when

$$hv = g\mu_B B \qquad (5.11)$$

Thus power is absorbed from the radio-frequency field by the sample at resonance. In general, the sample is placed in the radio-frequency magnetic field of a microwave cavity and the power absorbed is detected by a sensitive microwave bridge circuit. The most common microwave spectrometers operate at 9 GHz (X-band), though both 23 GHz (K-band) and 35 GHz (Q-band) systems are used. For a description of magnetic resonance spectrometers see Wertz and Bolton (1972). Since very precise measurements of g-factors, crystal field effects, and hyperfine interactions can be obtained by using this technique, the ground states of most magnetic ions and defect centres have been characterized in considerable detail. In the case of luminescent systems where a transition between an excited state and the ground state is involved, it would be hoped that conventional EPR measurements of the excited state would be possible. However, at room or lower temperatures the excited states are not occupied and, even when the crystals are excited by radiation in order to populate these states, only very few systems have sufficiently long lifetimes in the excited state for conventional EPR to be observed. For systems with short lifetimes in the excited state the technique of optically detected magnetic resonance can generally be used. Basically

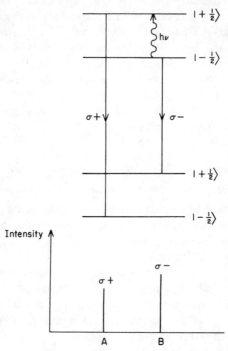

Fig. 5.9. Optically detected resonance is shown in the emitting state by the transition labelled $h\nu$. The luminescence components σ^+ and σ^- shown in the lower part of the diagram change at resonance with σ^+ increasing and σ^- decreasing.

the technique involves monitoring the intensity of the total luminescence or one of the polarized components and detecting changes in the emission intensity when resonance occurs in the emitting (or excited state). In order to record an optically detected magnetic resonance spectrum the magnetic field is swept (as in conventional EPR) and the resulting trace is a record of the change in luminescence intensity versus magnetic field. Unlike conventional EPR, both positive and negative signals can be observed.

The technique of monitoring the luminescence to detect magnetic resonance has been well known in atomic spectroscopy for many years and has been reviewed by Series (1970). Geschwind et al. (1967) were the first to apply this method to solids in an investigation of the Cr^{3+} emission in Al_2O_3. In this classic paper and in a review paper (Geschwind, 1972) three basic techniques for detecting magnetic resonance optically are described, (a) the high resolution technique; (b) detection of circularly polarized radiation method; and (c) the self-absorption method. These techniques have also been reviewed briefly by Davies (1976) and will be discussed below, but it is worth noting that several other techniques have been used. In particular, Mollenauer and Pan (1972) have shown that by monitoring the dichroism in absorption it is possible to observe the excited state resonance in systems with spin memory. Also, in the author's own laboratory studies of electron–hole recombination radiation have shown that in many systems changes in the total luminescence can be observed without self-absorption (see Section 5.3.4 on ZnS). We now discuss the various methods of optically detected magnetic resonance in more detail.

(a) *High resolution technique.* This method is applicable when the emission is sharp and can be resolved into Zeeman components at the magnetic field corresponding to resonance. The ruby R_1 line was the first emission to be investigated by this technique (Geschwind et al., 1967). Consider the transitions shown in Fig. 5.9 for a doublet-to-doublet luminescence. The intensity of one component, A say, is monitored by use of a high resolution spectrometer. At resonance, the transition within the excited state shown in the diagram increases the A line intensity and decreases B, assuming a Boltzmann distribution in the excited state before resonance. Figure 5.10 shows the energy level scheme and the Zeeman components for Tm^{2+} in SrF_2 as measured by Hayes and Smith (1971). The optically detected magnetic resonance from this emission is shown in the same figure.

(b) *Detection of Circularly Polarized Radiation.* The example discussed in the previous section and shown in Fig. 5.9 has two emission components, A and B, which are circularly polarized when observed along the magnetic field direction. Therefore it is possible to observe either A or B luminescence by using a circular polarization analyzer instead of a monochromator. In this example, the A luminescence, σ^+, increases at resonance in the excited state and the B luminescence, σ^-, decreases. This is the same result as obtained by the high resolution method but the obvious advantage of the polarization technique is that the emission Zeeman components need not be resolved and so the method can be used with broad bands. Several examples of the use of this technique are given in Section 5.3.4.

(c) *Self-absorption method.* Magnetic resonance can be detected optically by the self-absorption technique when the emission and absorption energies

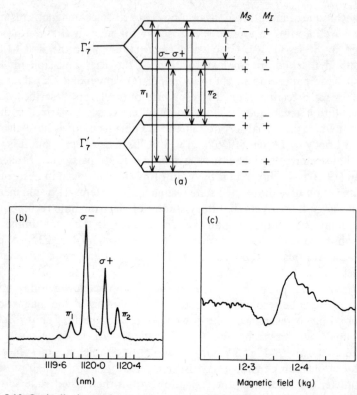

Fig. 5.10. Optically detected magnetic resonance in $SrF_2 : Tm^{2+}$. The Zeeman transitions in (a) give rise to the emission shown in (b). If the intensity of the σ^- component of the luminescence is monitored then the magnetic resonance shown in (c) is observed corresponding to the dashed transition in the Γ_7' state. (After Hayes and Smith, 1971.)

are the same. In this case the total luminescence is observed; no spectral or polarization resolution of the emission components is necessary. Referring to our example in Fig. 5.9, the A component will increase in intensity at resonance and the B component will decrease so without self-absorption no change in the total luminescence will be observed. However, atoms which have not been excited by the radiation will be in their ground states and so will be able to absorb the luminescence from the excited atoms, but we note that absorption A will be greater than B because of the Boltzmann distribution. At resonance more luminescence is emitted at the A energy and since this energy has the higher absorption coefficient the total light intensity decreases.

(d) *Spin-dependent Recombination.* In all of the above cases we have been concerned with transitions between an occupied excited state and an empty ground state; the atom is in one or the other state at any instant. However, when one considers the detection of magnetic resonance via recombination radiation, two particles, the electron and the hole, take part in the luminescence and both the electron and hole populations will determine the transition rates in a magnetic field. For example, Fig. 5.11(a) shows the allowed transitions for distant D–A recombinations in CdS. It is clear that the populations of both the donor and acceptor states determine the strengths of the σ^+ and σ^- radiation and at low temperatures $n_3 > n_4$ for the donor states and $n_2 > n_1$ for the acceptor states. Thus, for example, when microwave resonance occurs between the donor states σ^- light increases and σ^+ light decreases. For full details see Section 5.3.4.

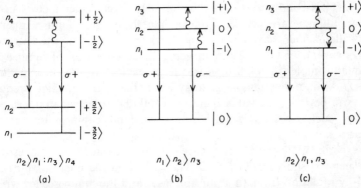

Fig. 5.11. Spin-dependent electron–hole recombination processes. (a) Energy levels and allowed transitions for donor–acceptor recombination in a magnetic field. Resonance in the donor state increases σ^- and decreases σ^+ radiation. (b), (c) Triplet state energy levels for closely associated electron–hole pairs (excitons) and the allowed optical transition. The populations can be such that changes in σ^+ and σ^- are of opposite sign (b) or the same sign (c).

Figures 5.11(b) and (c) show examples of magnetic resonance between triplet exciton states. The spins of the electron and the hole are strongly coupled when the two particles are bound on the same impurity or on a close donor–acceptor pair and the allowed recombination transitions are as shown, $|+1\rangle \rightarrow |0\rangle$ and $|-1\rangle \rightarrow |0\rangle$ for observation along the field. If the three triplet states $|+1\rangle$, $|0\rangle$, and $|-1\rangle$ are populated according to a Boltzmann distribution then at resonance σ^+ increases and σ^- decreases as shown in Fig. 5.11(b). However, if the recombination rates are fast compared with the spin lattice relaxation time, thermalization will not occur and the population of $|0\rangle$ will be greater than the other two states. Magnetic resonance will then increase both σ^+ and σ^-. This is observed in the study of triplet excitons as discussed in Section 5.4.3.

Experimental Techniques. An optically detected magnetic resonance spectrometer with a superconducting magnet as used in the author's laboratory will be described. However, at first we should note that the magnitude of the resonance signal will depend on the population difference between the Zeeman levels and since this is determined, in general, by the Boltzmann distribution the experiments are carried out at low temperatures. In fact, it is important to immerse the sample in overpumped liquid helium when high intensity laser excitation and high power microwave oscillators are used. Such a system is restricted to a very small temperature range ($\sim 1 \cdot 6$–$2 \cdot 0$ K) but the temperature of the sample is known accurately. It would also be expected that the maximum benefit of the Boltzmann distribution could be utilized if the microwave frequency and hence the magnetic field were higher than 9·5 GHz and 0·33 T (3300 gauss), the approximate X-band spectrometer figures. However, other factors become important at higher frequencies. For example, the cutting of optical apertures in the walls of microwave cavities has much more effect on the Q-factor at 35 GHz than at 9·5 GHz. Thus, in general, the collection of luminescence from small samples in small high frequency cavities is less efficient than at 9·5 GHz. Also, if the sample is lossy, then the effect on high frequency cavities will be more than at X-band. As would be expected, an intermediate frequency such as 23 GHz (K-band) is a good compromise for most experiments.

The experimental arrangement of the optically detected magnetic resonance spectrometer is shown in Fig. 5.12. The sample in the microwave cavity is excited by radiation from an argon ion laser and the luminescence which is emitted parallel to the magnetic field is analyzed by a circular polarizer. The intensity is monitored by a photomultiplier. Microwave radiation from a 1·5 W klystron is chopped at low frequency either by applying a square wave voltage to the klystron reflector or by using a diode microwave modulator. When the magnetic field is swept, changes in the luminescence intensity at resonance are detected by a phase-sensitive detector and recorded on chart. The technique is very sensitive since no ac signal is present at the photomultiplier unless the microwaves interact with the sample to give an ac change in the luminescence. There are two important questions that have to be answered when a resonance signal has been detected. These are (a) What is the sign of the luminescence change: is there an increase or decrease in light with the microwaves on? (b) Is the resonance due to all or only part of the luminescence being monitored? The first question is most easily answered by using a multi-channel signal averager. On one half of the memory the wave form of the chopped microwaves is displayed as shown in the upper trace of Fig. 5.13. The ac resonance signal is then taken directly from the photomultiplier to the second half of the memory and the wave form is displayed by repetitively scanning and recording the wave form N times giving a signal-to-noise improvement of $\sqrt{N}$. The two signals can then

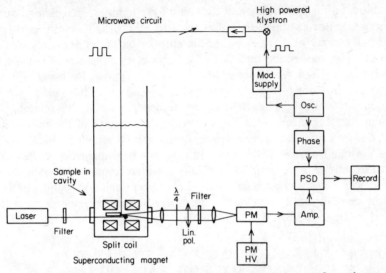

Fig. 5.12. Optical magnetic resonance spectrometer. Luminescence from the sample in the microwave cavity is monitored with a photomultiplier and changes in the luminescent intensity are recorded at resonance. This system uses a superconducting magnet and either σ^+ or σ^- radiation is monitored in a direction parallel to the magnetic field.

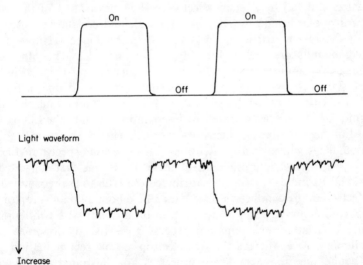

Fig. 5.13. A comparison of the microwave and light waveforms enables the sign of the change in light intensity to be determined in an optically detected magnetic resonance experiment. In the case illustrated the emission increases when the microwaves are switched on.

be compared on the display and the sign of the luminescence change can be determined. For example, the lower trace in Fig. 5.13 shows an increase in the luminescence at resonance (photomultiplier output is negative). We can now consider the second question posed above which concerns the spectral dependence of the resonance. This can be determined by using a mono-chromator in front of the photomultiplier instead of a filter (see Fig. 5.12). The magnetic field is adjusted to resonance and then, by scanning the monochromator through the luminescence, the strength of the resonance is recorded as a function of wavelength. In this way a resonance due to a weak emission band present under more intense but non-magnetic bands can be identified and if more than one optically detected resonance is observed the resonances can be associated with the correct emission bands or lines. Examples of optically detected magnetic resonance measurements will be found in Sections 5.3.4, 5.4.2, and 5.4.3.

5.3. MAGNETO-OPTICAL STUDIES OF DONOR–ACCEPTOR RADIATION

5.3.1. Introduction to CdS and ZnSe

Historically, the CdS recombination radiation has been discussed in terms of the blue and the green emissions. The blue emission consists of a series of sharp lines due to the recombination of bound excitons and has been investigated in detail by Zeeman spectroscopy. A description of this work, which was primarily due to Thomas, Hopfield, and collaborators, can be found in the review article by Halsted (1967). Analogous centres in the III–V compounds are reviewed briefly in Section 5.4.1. The other edge emission component is the green luminescence which is illustrated in Fig. 5.14 for a crystal at 4·2 K. Two spectra are shown; the solid line is for low intensity excitation and the dashed line is for high excitation intensity. Each spectrum shows two series of bands with separation equal to the LO phonon energy. For high excitation intensities the high energy spectrum (labelled HES) dominates while for low intensities the low energy spectrum (LES) is stronger. The correct interpretation of these bands was first suggested by Pedrotti and Reynolds (1960), who attributed the HES to the recombination of an electron in the conduction band with a hole bound on an acceptor and the LES to a recombination of an electron on a donor with a hole on an acceptor. This latter recombination process gives rise to donor–acceptor (D–A) luminescence whereas the free electron-to-hole recombination gives free-to-bound luminescence. These processes are illustrated in Fig. 5.15. The change in intensities of the HES and LES of Fig. 5.14 can now be understood. At low excitation levels most of the free electrons are captured at donors before recombination takes place. These electrons then recombine

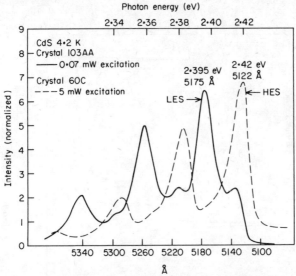

Fig. 5.14. The CdS edge emission showing the presence of two series of emission bands. The donor–acceptor transitions are stronger at low excitation intensities (solid line) and the free-to-bound bands predominate at higher excitation intensities. (After Thomas *et al.*, 1967.)

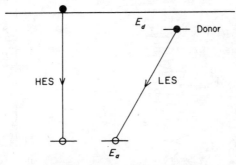

Fig. 5.15. Schematic illustration of the transitions which give rise to the free-to-bound emissions (HES) and the donor–acceptor bands (LES).

with holes on acceptors. However, this process can be saturated so that on increasing the excitation intensity the HES emission increases more rapidly than the LES emission and soon dominates the spectrum. A similar change in the relative intensities is observed between 4 K and 77 K. At low temperatures more electrons are bound at the donors and so, although the HES emission is larger at 77 K, the LES is more intense at 2 K.

We have already noted in Section 5.1.2 that under conditions of high intensity excitation a series of sharp lines is observed on the high energy

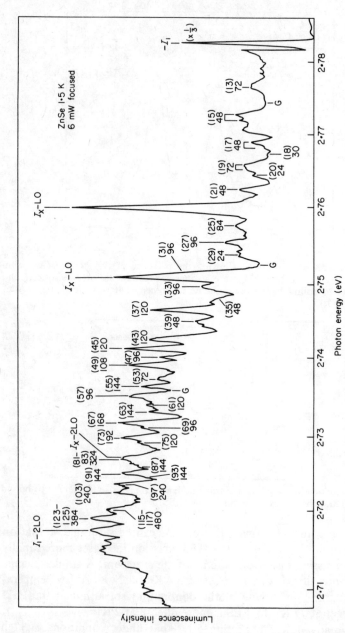

Fig. 5.16. Photoluminescence of ZnSe observed under high excitation intensities and showing the sharp emission lines which are due to electron–hole recombinations via donor–acceptor pairs. The labels give the shell number (in parentheses) and the total number of pairs with the same lattice spacing. (After Dean and Merz, 1969.)

tail of the first D–A band. Zeeman measurements on these lines in CdS were carried out by Henry *et al.* (1969) who were able to confirm the near D–A pair nature of these transitions. Details of these investigations can be found in the next section. The broad D–A and the close pair emissions for ZnSe as measured by Merz *et al.* (1973) have been illustrated in Fig. 5.3. Only at high excitation intensities are the pair lines clearly observed and these sharp lines are shown in more detail in Fig. 5.16 which is from Dean and Merz (1969). The series nature of these sharp lines is well illustrated if the photon energy is plotted against the D–A separation. This has been carried out for ZnSe by Dean and Merz (1969) and is shown in Fig. 5.17. The

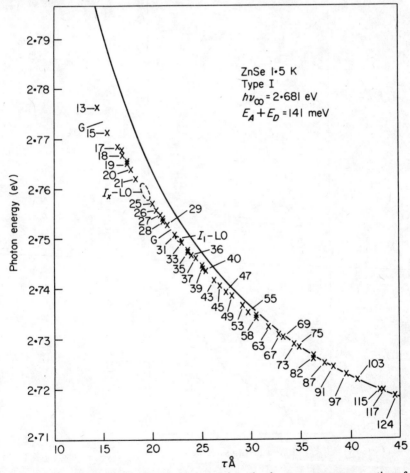

Fig. 5.17. Plot of sharp line peak energies against the donor–acceptor separations for ZnSe. The labels on the points are the shell numbers and the solid line is a fit of the expression $E(r) = E_g - (E_a + E_d) + (e^2/4\pi\varepsilon_0\varepsilon r)$. (After Dean and Merz, 1969.)

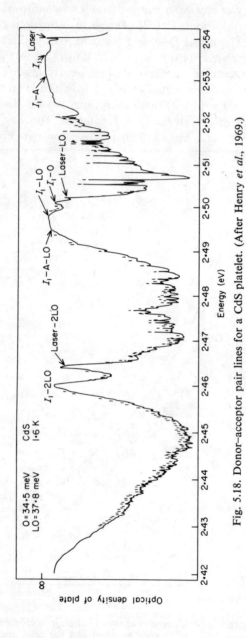

Fig. 5.18. Donor–acceptor pair lines for a CdS platelet. (After Henry *et al.*, 1969.)

labels are the shell numbers (see Dean, 1973) and the solid line is the fit of the distant pair expression given in Eq. (5.1). The fit is very good for separations greater than 30 Å but cannot be satisfactory for close pairs since the Coulomb term in Eq. (5.1) diverges. The close pairs must be treated as bound excitons as discussed by Jefferson et al. (1975). The energy limit for the D–A pair series corresponds to the first of the broad emission bands confirming the interpretation from the time-resolved spectroscopy studies that these bands are the result of distant D–A pair recombinations.

In the following sections we briefly review the magneto-optical measurements carried out on II–VI crystals, starting with Zeeman studies of CdS and followed by magnetic-field-induced polarization studies of ZnSe. Optically detected magnetic resonance measurements on both the CdS and ZnSe D–A bands and then similar measurements on the self-activated emissions in ZnS and ZnSe are then discussed.

5.3.2. Zeeman Spectroscopy of Sharp Donor–Acceptor Lines

Using Zeeman spectroscopy, Henry et al. (1969) studied the sharp emission lines from D–A recombinations in CdS. The experiments showed conclusively the close pair nature of these lines and similar results have also been reported by Reynolds and Collins (1969).

The CdS samples used by Henry et al. (1969) were strain-free platelets of CdS grown from the vapour using argon transport. Excitation radiation at 488 nm from an argon ion laser was directed along the crystal c-axis and the luminescence was observed in a direction perpendicular to the magnetic field with a 2 m spectrograph. Although large portions of the pair spectrum were obscured by LO phonon replicas of the I_1 line and Raman scattering at LO and 2LO, approximately 170 resolvable lines were observed at 1·6 K. The observed pair spectrum is shown in Fig. 5.18 for 20 mW of focused excitation radiation. Complicated magnetic splittings are expected for C_{3v} symmetry because of the large number of equivalent pairs with different orientations with respect to the magnetic field direction. However, each pair splits into only four lines except for the case where B is parallel to the c-axis, when, in general, only two lines are observed. Typical Zeeman splittings are shown in Fig. 5.19 for B both perpendicular and parallel to the c-axis. The circles are the experimental points and both the solid and dashed lines are the calculated splittings which we discuss below.

The analysis of the Zeeman spectra is based on a consideration of the energies of donors and acceptors which interact via an exchange interaction. Thus, the Hamiltonian is

$$\mathcal{H} = g_e \mu_B \mathbf{B}.\mathbf{S} + g_h \mu_B \mathbf{B}.\mathbf{J} + a_J \mathbf{J}.\mathbf{S} \tag{5.12}$$

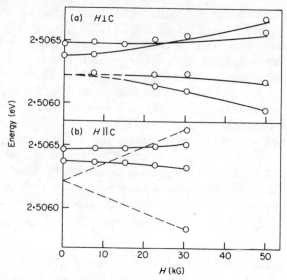

Fig. 5.19. Zeeman splitting of a typical pair line in CdS for (a) $B_\perp$ c-axis and (b) $B_\parallel$ c-axis. (After Henry et al., 1969.)

where the first two terms are the interactions of the isolated donors and acceptors with the magnetic field and the third term describes the j–j coupling or exchange energy. The donor state will have the near spherical symmetry of the conduction band so that the donor g-factor, g_e, will be isotropic with $S = \frac{1}{2}$ (the small anisotropy of the g-factor can be measured by magnetic resonance; see Table II). The acceptor state is derived from the spin–orbit split-off valence band, $J = \frac{3}{2}$. For zincblende symmetry, T_d, this state would be fourfold degenerate, but in the case of wurtzite symmetry of CdS it is convenient to consider a strain splitting of the valence band and the acceptor levels. This strain perturbation splits the fourfold degenerate acceptor state into two Kramers doublets $\left|\pm \frac{3}{2}\right\rangle$ and $\left|\pm \frac{1}{2}\right\rangle$ which are mixed by the electrostatic interaction existing between the donor and the acceptor. Thus the ground states of the acceptor are

$$|\phi_{\pm \frac{3}{2}}\rangle = \left|\pm \tfrac{3}{2}\right\rangle - \frac{\left|\pm \tfrac{1}{2}\right\rangle \left\langle \pm \tfrac{1}{2}\right| V \left|\pm \tfrac{3}{2}\right\rangle + \left|\mp \tfrac{1}{2}\right\rangle \left\langle \mp \tfrac{1}{2}\right| V \left|\pm \tfrac{3}{2}\right\rangle}{\Delta} \qquad (5.13)$$

where the strain valence band splitting is Δ and the Coulomb interaction due to the presence of the donor is V. Now the Hamiltonian given in Eq. (5.12) can be written in terms of the shift operators, $B^\pm$, $J^\pm$, and $S^\pm$ and so

$$\mathcal{H} = g_e \mu_B B S_z + \tfrac{1}{2} g_e \mu_B (B^+ S^- + B^- S^+) + g_h \mu_B B J_z$$
$$+ \tfrac{1}{2} g_h \mu_B (B^+ J^- + B^- J^+) + a_J (J_z S_z + \tfrac{1}{2} J^+ S^- + \tfrac{1}{2} J^- S^+) \quad (5.14)$$

TABLE II

Magnetic Resonance Results on Donor and Conduction Electrons
in the II–VI Compounds

	Doping	Temperature	g	$g_{\parallel}$	$g_{\perp}$	Excitation	Reference
ZnO	Zn	77		1·956	1·955	UV	1
	Ga, Al	77	1·96			UV	2
	Ga	4·2		1·9576	1·9561	—	3
ZnS	Cl, Br	77	1·883			UV	4
(cubic)	I	77	1·88225			UV	1
	Cl	77	1·8835			UV	1
	Al–Zn	77	1·8849			UV	1
	I	1·8	1·886			UV	5
ZnS	Cl	77	1·888			UV	4
(hex)	Cl	77		1·8933	1·8860	UV	1
ZnSe		77	1·14				
		2	1·125			UV	6
ZnTe		1·9	0 401			488 nm	13
CdTe		1·7	1·59			800 nm	7
CdS	Cl	4·2		1·79	1·78		8
	Ga	77		1·792	1·775	550 nm	9
		1·5		1·783	1·764	524 nm	10
	Ga	4·2	1·78				11
		2·0		1·789	1·769	488 nm	12
CdSe	Ga	4·2	0·68				11

1. Müller and Schneider (1963).   2. Kokes (1962).   3. Schultz (1975).   4. Kasai and Omoto
(1962).   5. James *et al.* (1975).   6. Dunstan *et al.* (1977a).   7. Cavenett *et al.* (1978).   8. Lambe
and Kikuchi (1958).   9. Dieleman (1963).   10. Morigaki (1964).   11. Piper (1967).
12. Brunwin *et al.* (1976).   13. Killoran *et al.* (1978).

The splittings of the pair lines are given by the eigenvalues of $\mathscr{H}$ within the
D–A product states $|\phi_{\pm\frac{3}{2}}, \pm\frac{1}{2}\rangle$ where the second label, $\pm\frac{1}{2}$, refers to the
electron states on the donor. The detailed calculation makes use of three
parameters which involve the j–j coupling interaction factor, a_J; these are

$$A = -a_J\langle\phi_{\frac{3}{2}}|J_z|\phi_{\frac{3}{2}}\rangle, \quad C = a_J\langle\phi_{\frac{3}{2}}|J^-|\phi_{\frac{3}{2}}\rangle$$
and
$$D = a_J\langle\phi_{-\frac{3}{2}}|J^-|\phi_{\frac{3}{2}}\rangle \tag{5.15}$$

For B parallel to the c-axis and writing $g_{h\parallel} = -2g_h A/a_J$ the energies are

$$E = \tfrac{1}{2}\{A \pm [(g_{h\parallel}-g_e)^2(\mu_B B)^2 + |D|^2]^{\frac{1}{2}}\} \tag{5.16}$$

for the allowed transitions, and

$$E = \tfrac{1}{2}[-A \pm (g_{h\parallel}+g_e)\mu_B B] \tag{5.17}$$

for the forbidden transitions. Here g_e and $g_{h\parallel}$ are the g-values of the isolated donor and acceptor and the values measured by Thomas and Hopfield (1962) of $g_e = 1.76$ and $g_{h\parallel} = 2.76$ are assumed. For, B perpendicular to the c-axis both transitions become allowed and the calculated splittings are

$$E = \frac{1}{2}\left[[A^2 + (g_e\mu_B B)^2]^{\frac{1}{2}} \pm |D|\left(1 + \frac{[A - (2\alpha/A)(g_e\mu_B B)^2]}{[A^2 + (g_e\mu_B B)^2]^{\frac{1}{2}}}\right)\right] \quad (5.18)$$

$$E = -\frac{1}{2}\left[[A^2 + (g_e\mu_B B)^2]^{\frac{1}{2}} \pm |D|\left(1 - \frac{[A - (2\alpha/A)(g_e\mu_B B)^2]}{[A^2 + (g_e\mu_B B)^2]^{\frac{1}{2}}}\right)\right] \quad (5.19)$$

where we note that $\alpha = g_{h\parallel}/2g_e = 0.78$. The magnetic splitting is therefore described by the j–j coupling constant, A, and the crystal field splitting, $|D|$. These parameters were determined from the zero field splitting and, for a common value of $A = 0.20$ meV, crystal field splitting of 0–0.25 meV were observed. The excellent agreement between the observed and calculated Zeeman splittings confirms the D–A nature of these lines. Another important consequence of this study was the correlation of the pair spectrum with the distant pair peak at 5176 Å implying the D–A nature of this emission band (see Section 5.3.4).

More recently Henry *et al.* (1970) reported a double donor–neutral acceptor pair spectrum. The double donor consists of two electrons bound at centres such as sulphur vacancies or cadmium interstitials. The observed pair line spectra were more intense than the lines discussed above and no

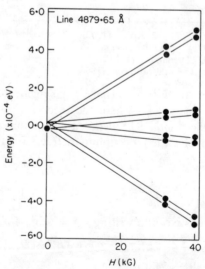

Fig. 5.20. Zeeman splitting of a donor-to-double acceptor line in CdS for $B_\parallel$ c-axis. (After Reynolds *et al.*, 1975.)

distant pair peak was found. Measurement of the Zeeman splittings confirmed the assignment. Reynolds *et al.* (1975) have reported double D–A pair lines in CdS. The double acceptor was tentatively identified as a cadmium vacancy and these authors investigated the Zeeman spectra of three lines at 4872·80 Å, 4877·01 Å, and 4879·65 Å. The Zeeman results for the latter line is shown in Fig. 5.20 where it is seen that there is an eightfold splitting of the line which results from the donor mixing the two hole states on the double acceptor. The final state of the complex after electron–hole recombination is a single hole on the acceptor.

5.3.3. *Magnetic-induced Polarization of Donor–Acceptor Bands*

In the previous sections the highest energy broad edge emission bands in CdS and ZnSe were shown to be the series limit of the sharp D–A recombination lines. Thus these bands and the LO phonon replicas are due to the recombination of electrons and holes via distant D–A pairs where the separations are of the order of 100 Å. Because these emission bands are ~0·02 eV wide it has not been possible to carry out Zeeman measurements to confirm this interpretation. However, in Section 5.2.1 we noted that the optical transitions have specific polarization properties in a magnetic field and so it is possible to measure the donor and acceptor g-values by observing the magnetic-field-induced polarization of the emission.

Cavenett and Hagston (1975) have studied the polarization of the ZnSe D–A emission bands, shown in Fig. 5.21a, in magnetic fields up to 6·6 T (66 kgauss). The evidence from the sharp D–A lines suggesting that the pairs responsible for the broad bands are distant pairs implies that the exchange energy term in Eq. (5.12) can be neglected. The Hamiltonian describing the energies of the isolated donors and acceptors in the magnetic field is then

$$\mathcal{H} = g_e \mu_B \mathbf{B}.\mathbf{S} + g_h \mu_B \mathbf{B}.\mathbf{J}. \qquad (5.20)$$

ZnSe has zincblende crystal structure which has T_d symmetry. The donor is described by $S = \frac{1}{2}$ and the acceptor by $J = \frac{3}{2}$; both g-factors are isotropic and g_h is assumed to be positive so that the electronic state $|J = \frac{3}{2}, M_J = \frac{3}{2}\rangle \equiv |\frac{3}{2}, \frac{3}{2}\rangle$ has the highest acceptor energy as shown in Fig. 5.22. The allowed transitions for electron–hole recombinations are given in the figure with the relative probabilities and polarizations. It is important to note that the observed luminescence comes from a set of such D–A pairs, each having a different separation, energy (see Eq. 5.1) and recombination rate. In the absence of a magnetic field the radiative transition probability for a D–A pair separated by a distance r is (see Eq. 5.2)

$$W(r) = W_0 \exp(-2r/a_0)$$

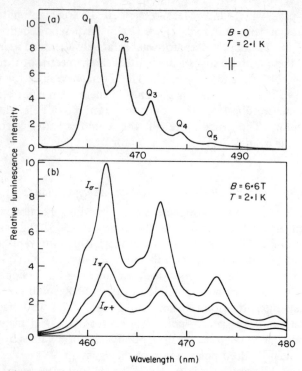

Fig. 5.21. (a) ZnSe edge emission without magnetic field showing the Q series of peaks which correspond to the notation of Merz *et al.* (1969). (b) Magnetic-induced polarization showing $I_{\sigma-}$ and $I_{\sigma+}$ measured parallel to the magnetic field and I_π measured perpendicular to the field. (After Cavenett and Hagston, 1975.)

The recombination rate is modified by the magnetic field since the spin selection rules only allow certain transitions to take place and so we let W_0 go to $\tilde{W}_0$ in a magnetic field. The function $\tilde{W}_0$ can be calculated by noting that in Fig. 5.22 the transition probabilities depend on the populations of both the donor and acceptor states. Assuming a Boltzmann distribution of electrons within the $|+\tfrac{1}{2}\rangle$ and $|-\tfrac{1}{2}\rangle$ states and of holes in the four acceptor states, the total probability of transition $|+\tfrac{1}{2}\rangle \rightarrow |\tfrac{3}{2}, \tfrac{3}{2}\rangle$, for example, is proportional to the product of $\exp(-\tfrac{1}{2}g_e x)$ and $\exp(\tfrac{3}{2}g_h x)$, where $x = \mu_B B/kT$. For this transition we can write (Hagston and Cavenett, 1978),

$$P_1 = C \exp \tfrac{1}{2}x(3g_h - g_e) \tag{5.21}$$

where C is the constant of proportion. Similar expression can be found for $P_2, \dots, P_6$. The total relative probability of any transition occurring is the sum of these terms P_1 to P_6 and this must be equal to unity. Thus, by gathering

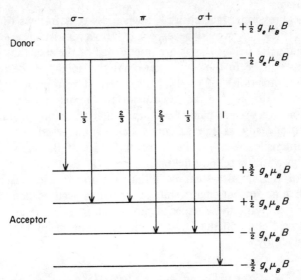

Fig. 5.22. Energy levels and allowed transitions for donor–acceptor recombination in cubic ZnSe. The polarizations and the relative strengths are given for each transition.

the pairs of terms together, we have

$$1 = 2C[\cosh \tfrac{1}{2}x(3g_\text{h}-g_\text{e}) + \cosh \tfrac{1}{2}x(g_\text{h}-g_\text{e})$$
$$+ \cosh \tfrac{1}{2}x(g_\text{e}+g_\text{h}) + \cosh \tfrac{1}{2}x(3g_\text{h}+g_\text{e})] \quad (5.22)$$
$$= 2CD \quad (5.23)$$

and substituting for C in Eq. (5.21) the relative probabilities are ($P_1 \ldots$ etc. are labelled from left to right in Fig. 5.22)

$$P_1 = (1/2D)\exp \tfrac{1}{2}x(3g_\text{h}-g_\text{e}), \quad P_4 = (1/2D)\exp -\tfrac{1}{2}x(g_\text{h}-g_\text{e})$$
$$P_2 = (1/2D)\exp \tfrac{1}{2}x(g_\text{h}+g_\text{e}), \quad P_5 = (1/2D)\exp -\tfrac{1}{2}x(g_\text{h}+g_\text{e}) \quad (5.24)$$
$$P_3 = (1/2D)\exp \tfrac{1}{2}x(g_\text{h}-g_\text{e}), \quad P_6 = (1/2D)\exp -\tfrac{1}{2}x(3g_\text{h}-g_\text{e})$$

with D defined by Eqs (5.22) and (5.23). Now the total average strength of an emission at energy E was determined by W_0 before the application of a magnetic field. With the field present, we have, using the relative strengths shown in Fig. 5.22,

$$\tilde{W}_0 = W_0[P_1 \times 1 + P_2 \times \tfrac{1}{3} + P_3 \times \tfrac{2}{3} + P_4 \times \tfrac{2}{3} + P_5 \times \tfrac{1}{3} + P_6 \times 1]$$
$$= W_0(2/D)\,[\cosh \tfrac{1}{2}x(3g_\text{h}-g_\text{e}) + \tfrac{2}{3}\cosh \tfrac{1}{2}x(g_\text{h}-g_\text{e}) + \tfrac{1}{3}\cosh \tfrac{1}{2}x(g_\text{e}+g_\text{h})] \quad (5.25)$$

that is,

$$\tilde{W}_0 = W_0 g(x) \quad (5.26)$$

From this expression, since $\cosh z \to 1$ when $z \to 0$, it follows that as $B \to 0$, $x \to 0$ and $g(x) \to 1$ as would be expected. However, when $x \to \infty$, $g(x) \to 0$ for large magnetic fields; this is a consequence of the fact that in this case the electrons will be in the $|-\frac{1}{2}\rangle$ state and the holes will be in the $|\frac{3}{2}, \frac{3}{2}\rangle$ state but the transition $|-\frac{1}{2}\rangle \leftrightarrow |\frac{3}{2}, \frac{3}{2}\rangle$ has zero transition probability. Thus the effect of increasing the magnetic field is to reduce the transition probability out of every D–A pair. This means that the more distant pairs can saturate more easily as the magnetic field is increased with a constant excitation intensity, resulting in an overall shift of the emission band to higher energy with increasing field. However, Cavenett and Hagston (1975) show that if the D–A system is initially saturated with laser excitation radiation then no further energy shift of the bands occurs when the magnetic field is applied. With these experimental conditions the relative intensities of the polarized components $I_{\sigma+}$, $I_{\sigma-}$, and I_{π} can be calculated and the peak ratios are given by

$$\left.\begin{array}{l} \dfrac{I_{\sigma-}}{I_{\sigma+}} = \dfrac{\exp\left[\frac{1}{2}x(3g_h - g_e)\right] + \frac{1}{3}\exp\left[\frac{1}{2}x(g_e + g_h)\right]}{\frac{1}{3}\exp\left[-\frac{1}{2}x(g_h + g_e)\right] + \exp\left[\frac{1}{2}x(g_e - 3g_h)\right]} \\[3ex] \dfrac{I_{\pi}}{I_{\sigma+}} = \dfrac{\frac{2}{3}\{\exp\left[\frac{1}{2}x(g_h - g_e)\right] + \exp\left[\frac{1}{2}x(g_e - g_h)\right]\}}{\frac{1}{3}\exp\left[-\frac{1}{2}x(g_h + g_e)\right] + \exp\left[\frac{1}{2}x(g_e - 3g_h)\right]} \end{array}\right\} \qquad (5.27)$$

The magnetic-field-induced polarization of the ZnSe D–A luminescence was measured by the methods discussed in Section 5.2.2 and the intensities of $I_{\sigma+}$, $I_{\sigma-}$, and I_{π} at 2·1 K and 6·6 T (66 kgauss) are shown in Fig. 5.21(b). The centre two peaks are less affected by the free-to-bound transitions and the ratios of $I_{\sigma-}^{\text{peak}}/I_{\sigma+}^{\text{peak}}$ and $I_{\pi}^{\text{peak}}/I_{\sigma+}^{\text{peak}}$ for these two bands are the same. Analysis of these ratios using Eqs (5.27) gives a donor g-value of $g_e = 1·22 \pm 0·06$ and an acceptor g-value of $g_h = 0·40 \pm 0·01$. These results can be compared with previously obtained measurements of the donor g-value; the acceptor g-value had not been reported before. Schneider et al. (1968) measured g_e using conventional magnetic resonance and obtained a value of 1·14. Also, Fleury and Scott (1971) using spin–flip Raman scattering, obtained a value of $g_e = 1·18 \pm 0·03$. Dunstan et al. (1977) have also measured the donor g-value using optically detected resonance—see Section 5.3.4. Again the value of 1·115 is in close agreement with the value obtained by the polarization measurements demonstrating that this technique can yield reliable estimates of donor and acceptor g-factors. Finally, it is worth noting that, although the optically detected magnetic resonance experiments give more precise g-values than the polarization method, the latter technique enables both the donor and acceptor g-values to be measured.

5.3.4. *Optically Detected Magnetic Resonance in II–VI Crystals*

In this section we illustrate how the technique of optically detected magnetic resonance can be used to study broad emissions such as the CdS and ZnSe D–A recombination bands which have just been discussed. In particular we illustrate how the technique can be used to relate optical and conventional microwave resonance information. For example, it is known from optical studies that a portion of the green CdS edge emission is due to D–A recombination (the LES) and since the donor resonance has been observed by conventional EPR one would expect to be able to detect this resonance optically and show that it is observed only via the D–A emission bands. This is exactly what can be achieved. It is worth noting that the detection of the donor resonance by optical methods can be carried out on

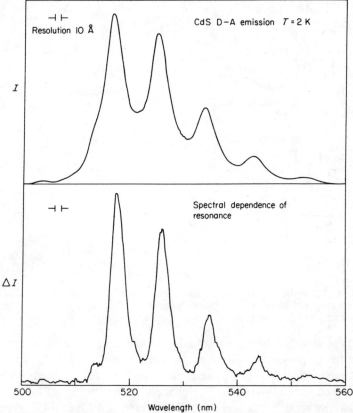

Fig. 5.23. The upper curve shows the green CdS edge emission with both the free-to-bound and donor–acceptor bands. The lower curve is the spectral dependence of the optically detected resonance shown in Fig. 5.24. The resolution is the same for both spectra.

undoped materials and since the resonance measurement is directly linked to optical data, donor resonance can be distinguished from free-electron resonance; this is not generally possible by conventional techniques because of the closeness of the two g-values.

CdS. The principles and techniques of optically detected resonance have been discussed in Section 5.2.3. Brunwin *et al.* (1976) used the experimental arrangement shown in Fig. 5.12 and investigated the green luminescence shown in Fig. 5.23. The microwaves at 9·5 GHz were chopped at 1 KHz and resonance-induced changes in the intensity of the circularly polarized components of the luminescence were observed. A polished (0001) face of a

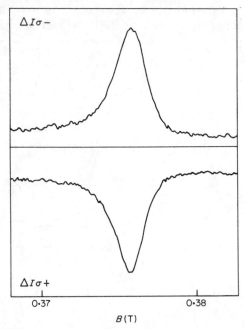

Fig. 5.24. The optically detected magnetic resonance of donors in CdS. The upper curve shows the increase of the $I_{\sigma-}$ luminescence at resonance and the lower curve shows the corresponding decrease of the $I_{\sigma+}$ luminescence.

CdS crystal was illuminated by radiation at 488 nm from an argon ion laser and the luminescence was observed from the same face. The crystal c-axis, the magnetic field direction, and the direction of observation of the luminescence were all parallel. The changes in light intensity are shown in Fig. 5.24; the upper curve represents an increase of 0·4% of the $I_{\sigma-}$ light and the lower curve shows the corresponding decrease in the $I_{\sigma+}$ radiation. A value of $g_{\parallel} = 1·789 \pm 0·02$ was measured which was found to agree with the measurements on CdS donors using conventional magnetic resonance techniques (Lambe and Kikuchi, 1958; Dielman, 1963; Morigaki, 1964). In order to confirm that the optically detected resonance was due to donors and not to conduction electrons the spectral dependence of the resonance was measured by placing a monochromator in front of the photomultiplier as described in Section 5.2.3. This dependence is shown in Fig. 5.23 and can be compared with the total green emission measured at the same temperature and with the same resolution shown in the upper part of the figure. It can clearly be seen that the resonance is due to the D–A luminescence (the LES bands) and not the free-to-bound transitions (HES) which are at slightly higher energy.

The observation of the donor resonance can be explained in terms of the D–A recombination model discussed in Section 5.3.2. Since the luminescence is due to distant D–A pairs the exchange term of the Hamiltonian in Eq. (5.12) can be neglected and only the separate electron and hole energies need to be considered as was the case for the D–A bands in ZnSe discussed in the previous section. For the CdS wurtzite crystal structure with B parallel to the c-axis the energy states for the donor and the acceptor are shown in Fig. 5.25. The donor state has $S = \frac{1}{2}$ and g values, $g_{\parallel}$ and $g_{\perp}$, which are very close in value for the nearly spherical conduction band. The acceptor state

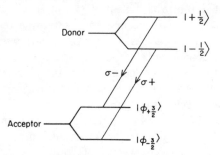

Fig. 5.25. Donor–acceptor recombination model for $B_{\parallel}$ c-axis. The allowed optical transitions are shown.

comes from the $J = \frac{3}{2}$ valence band and, as shown in Section 5.3.2, the doublet which is occupied by holes at 2 K is labelled $|\phi_{\pm\frac{3}{2}}\rangle$. The allowed electron–hole recombination transitions are also shown in the figure and the transition rates for each of $I_{\sigma+}$ and $I_{\sigma-}$ will depend on the product of the populations of the donor and acceptor states involved. If it is assumed that initially there is a Boltzmann distribution of electrons in the donor states, then saturation with microwaves will equalize the populations of the two donor states and $I_{\sigma-}$ will increase and $I_{\sigma+}$ will decrease. These assumptions predict maximum changes of $\Delta I_{\sigma+}/I_{\sigma+} = -11\cdot6\%$ and $\Delta I_{\sigma-}/I_{\sigma-} = +10\cdot3\%$ which are considerably larger than the observed $\pm0\cdot4\%$. However, the resonance signal showed no sign of saturation with the microwave power available (~ 1 W). The acceptor resonance was not observed in these experiments but this was not unexpected since magnetic resonance transitions between the acceptor levels is only weakly allowed for B parallel to the c-axis (Thomas and Hopfield, 1968).

ZnS. D–A recombination luminescence has also been investigated in ZnS by James *et al.* (1975a, b). This work is important because for the first time the technique of optically detected magnetic resonance was used to confirm the D–A nature of the luminescent transitions and used to identify the acceptor. The particular luminescence investigated is called the self-activated emission and is a broad blue band peaking at 2·6 eV. It was so named because the emission had been observed many times in nominally pure ZnS crystals. Prener and Williams (1956) proposed that the self-activated centre consisted of a zinc vacancy paired with a substitutional halogen ion on an adjacent sulphur site. The coactivators need not be halogen ions but can be Al or Ga which are also donors in ZnS. The self-activated centre is negatively charged with respect to the lattice and can be written as $(V_{Zn}–Cl)'$. The centre is diamagnetic and so conventional magnetic resonance is not observed in crystals containing these centres. However, if the crystals are illuminated with ultraviolet light, the electron is removed from the self-activated centre and neutral $(V_{Zn}–Cl)$ centres are produced. Kasai and Otomo (1962) were the first to observe the resonance of these centres which they named A-centres. Details of the structure of the A-centre were investigated by Dischler *et al.* (1964) and Schneider *et al.* (1965) using conventional magnetic resonance methods and the association between the A-centre and the self-activated centre was confirmed by Watkins (1973), who showed that the wavelength dependence for polarized excitation of the A-centre resonance is the same as the polarized luminescence measurements of Koda and Shionoya (1964) for the self-activated emission. At the time of these investigations two models were in favour. The polarized emission results were explained on a localized model where the transitions giving rise to the self-activated emission were considered to take place within the self-activated centre and between the 4s orbital of the Cl ion associated with the Zn vacancy and a hole on one of the surrounding S ions. However, the time-resolved measurements of Era *et al.* (1969) implied that the self-activated emission was D–A in nature with the transitions taking place from the donor state of an isolated chlorine ion to the acceptor state of a $(V_{Zn}–Cl)$ centre.

Optically detected magnetic resonance measurements were carried out by James *et al.* (1975a, b) on vapour-grown ZnS crystals. The resonance results are shown in Fig. 5.26. The sharp resonance at high field is isotropic and has $g = 1·886 \pm 0·003$. This corresponds to the donor or conduction band resonance observed by Müller and Schneider (1963) using conventional

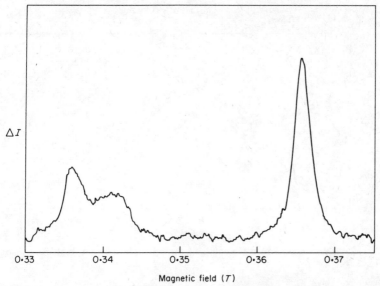

ΔI

Magnetic field (T)

Fig. 5.26. Change, ΔI, of the total luminescence induced by microwaves at 9·65 GHz in a ZnS sample at 2 K as a function of the magnetic field along the c-axis.

magnetic resonance techniques and ultraviolet irradiated ZnS samples. The low field resonance lines in the figure correspond to A-centre resonances with g-values between 2·02 and 2·05. The spectral dependences of the donor and the acceptor resonances were found to be the same and each corresponded to the self-activated emission (see Fig. 5.27). Thus, both the donor and the A-centre acceptor are involved in the self-activated emission.

The observed resonances can be explained using the D–A model shown in Fig. 5.28. In a magnetic field the donor and acceptor levels split as shown. The probability of radiative emission for the allowed transitions shown in the diagram is proportional to $(p_1 q_1 + p_2 q_2)$. If the spin lattice relaxation time is short compared with the optical decay time both donor and acceptor states will have a Boltzmann distribution and saturation of either resonance with microwaves gives 1·3%. This compares with the measured value of 0·3%.

The excitation and emission cycle for the self-activated emission has been shown in Fig. 5.5. The ultraviolet radiation ionizes the self-activated centre producing an electron in the conduction band and a neutral A-centre acceptor. The electron is captured by a donor centre and the self-activated emission occurs when the electron and the hole recombine.

ZnSe. Optically detected magnetic resonance measurements have been carried out on the blue ZnSe D–A emission (Dunstan *et al.*, 1977). The energy levels for distant D–A pairs in a magnetic field have been given in

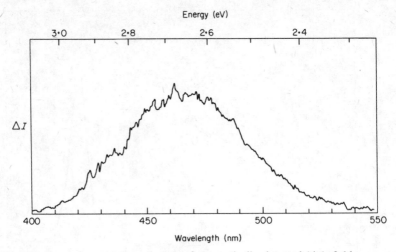

Fig. 5.27. Dependence on wavelength of the optically detected high field resonance of Fig. 5.26.

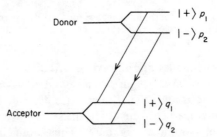

Fig. 5.28. Donor–electron and acceptor–hole spin levels. Allowed optical transitions are shown.

Fig. 5.22 and the recombination process is discussed in Section 5.3.3. The observed magnetic resonance of the donors at $g = 1 \cdot 115 \pm 0 \cdot 010$ is shown in Fig. 5.29. The expected maximum changes in the intensity of $I_{\sigma+}$ and $I_{\sigma-}$ for observations along the magnetic field can be calculated using the transition probabilities shown on the energy level diagram. Assuming a Boltzmann distribution before resonance and microwave saturation at resonance, the calculation predicts $\Delta I_{\sigma+}/I_{\sigma+} = -5 \cdot 0$ and $\Delta I_{\sigma-}/I_{\sigma-} = +6 \cdot 3$. The observed resonance was $\pm 0 \cdot 1 \%$ without microwave saturation. No acceptor resonance was observed.

A deep acceptor due to Zn vacancies has been observed in ZnSe by optically detected magnetic resonance (Dunstan *et al.*, 1977). The purpose of the investigation was to attempt to identify the emission bands in the 600 nm region since some were believed to be donor–acceptor in nature but the acceptors had not been identified. The emission of iodine transported

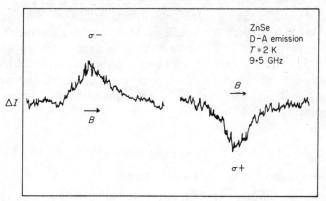

Fig. 5.29. Optically detected donor resonance in ZnSe showing the increase of σ^- radiation and the decrease of σ^+. (After Dunstan *et al.*, 1977.)

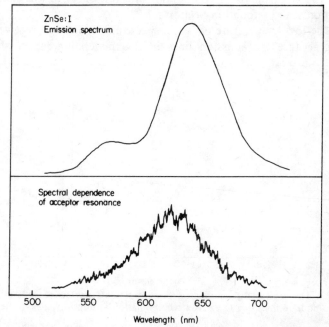

Fig. 5.30. Optically detected resonance of the V^- acceptor in ZnSe: (a) the emission spectrum of ZnSe : I,Cu, (b) the spectral dependence of the V^- resonance shows which part of the emission is associated with the donor-to-V^- acceptor recombination. (After Dunstan *et al.* 1977.)

ZnSe : Cu is shown in the upper part of Fig. 5.30. The optically detected magnetic resonance spectrum is similar to ZnS with both donor and acceptor resonances observed. The measured donor g-value of $1 \cdot 135 \pm 0 \cdot 010$ corresponds to that reported by Schneider *et al.* (1968). The acceptor resonance was fitted to the V⁻ centre g-values, $g_\parallel = 1\cdot9548$ and $g_\perp = 2\cdot2085$, reported by Watkins (1974). The spectral dependence of the acceptor resonance is shown in the lower part of Fig. 5.30 and shows that the donor-to-V⁻ recombination peaks at 625 nm (632 nm after correction for detection response). Figure 5.30 clearly illustrates how important a spectral dependence measurement of the resonance signal can be when the principal emission bands do not contribute to the resonance.

5.4. BOUND EXCITON RECOMBINATION RADIATION

5.4.1. Zeeman Spectroscopy of Bound Excitons

In this section the examples of spectroscopy of bound exciton luminescence are taken from studies of III–V compound crystals. Similar examples can be found in the work on II–VI compounds and references to the more recent investigations can be found in Section 5.1.3.

GaP. The low temperature emission spectrum of GaP has received much attention because the emission lines in this material are very sharp. In

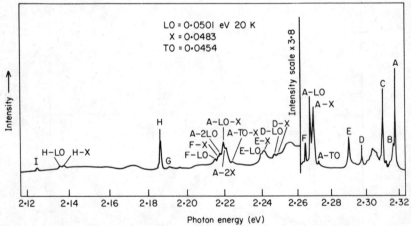

Fig. 5.31. The high energy emission spectrum of GaP crystals at 20 K showing the sharp zero-phonon lines due to exciton recombination at impurity sites. (After Thomas *et al.*, 1963.)

particular, the exciton emission lines have been characterized in considerable detail. Figure 5.31 shows the emission from an undoped solution-grown GaP crystal which was studied by Thomas *et al.* (1963). The three lines labelled A, B, and C were investigated by Zeeman spectroscopy and a summary of this early work can be found in the review by Gershenzon (1966). The A and B lines were shown to be related and were attributed to the recombination of excitons at ionized donors. However, in a later paper

Thomas and Hopfield (1966) showed that these lines are due to excitons bound at nitrogen isoelectronic traps. In fact, it is not possible to distinguish between the two possibilities by using Zeeman spectroscopy since the energy states are the same for both cases, but by doping the GaP before growth with GaN, the nitrogen emission spectrum was observed as shown in Fig. 5.32.

Let us consider the Zeeman effect of an exciton bound at an isoelectronic trap. The crystal structure of GaP has T_d symmetry and the electrons with $S = \frac{1}{2}$ have an isotropic g-factor. The holes can be described by angular

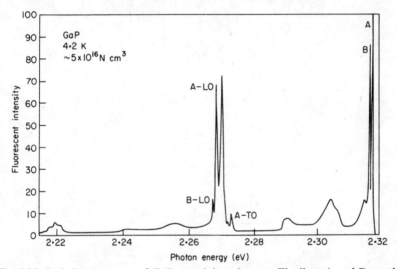

Fig. 5.32. Emission spectrum of GaP containing nitrogen. The lines A and B are the zero-phonon emissions due to the decay of excitons bound to the nitrogen isoelectronic traps. (After Thomas and Hopfield, 1966.)

momentum $J = \frac{3}{2}$. Since the isoelectronic trap has no particle to contribute to the exciton complex we are concerned with the energy states of an electron and a hole bound together on the trap. The theory has been outlined by Yafet and Thomas (1963). They show that the angular momenta of the electron and the hole combine to form states $J = 1$ and $J = 2$ which are split apart by the electron–hole jj coupling. The $J = 2$ level is split further by the cubic crystal field and the Zeeman Hamiltonian is

$$\mathscr{H}_z = g_e \mu_B \mathbf{B}.\mathbf{S} + K \mu_B \mathbf{B}.\mathbf{J} + L(J_x^3 B_x + J_y^3 B_y + J_z^3 B_z) \qquad (5.28)$$

where g_e is the electron g-factor and K and L are the spin Hamiltonian g-factors for the hole. Making the assumptions that the cubic field splitting is negligible and that the g-values are isotropic with $g_h = g_{h\frac{3}{2}} = g_{h\frac{1}{2}}$, Thomas et al. (1963) proposed the energy level diagram shown in Fig. 5.33(a). In the

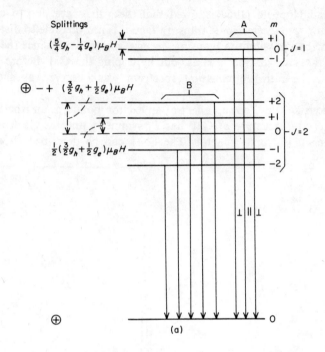

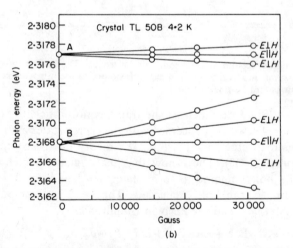

Fig. 5.33. (a) Energy level scheme and observed transitions for excitons trapped at isoelectronic traps. The A and B lines shown in Fig. 5.32 are due to transitions from the $J = 1$ and $J = 2$ states respectively. (b) The Zeeman splitting of the A and B lines showing the observed polarizations. (After Thomas *et al.*, 1963.)

presence of a magnetic field the $J = 1$ state splits into three states and the A emission is observed when transitions take place from these states to the ground state. For observations of the luminescence perpendicular to the magnetic field there are two components polarized perpendicular to the field (the σ-components) and one parallel to the field (the π-component). The B emission line results from forbidden transitions between the $J = 2$ quintet and the ground state and the observed polarizations are shown on the diagram. At high temperatures the allowed A line is much stronger but at low temperatures (~ 2 K) the B line is dominant showing thermalization throughout the energy levels. The splitting of the A and B lines in a magnetic field is shown in Fig. 5.33(b) and the observed g-values were found to be $g_e = 2 \cdot 02 \pm 0 \cdot 12$ and $g_h = 0 \cdot 99 \pm 0 \cdot 06$ for the simplified model of Thomas et al. (1963). A detailed analysis by Yafet and Thomas (1963) yielded the Hamiltonian parameters as $g_e = 2 \cdot 02 \pm 0 \cdot 12$, $K = 0 \cdot 77 \pm 0 \cdot 16$, and $L = 0 \cdot 11 \pm 0 \cdot 07$.

We now turn our attention to the analysis of the C line which is due to the recombination of excitons bound to neutral donors. Three particles, two electrons and a hole, have to be considered for the analysis of the Zeeman spectra. In fact the electrons are in a singlet or paired state required by the Pauli principle and so the exciton states in a magnetic field are determined by the hole states. The energy level scheme proposed by Thomas et al. (1963) is shown in Fig. 5.34(a) where it can be seen that the ground state is an isolated donor with $S = \frac{1}{2}$ and an isotropic g-factor, g_e. The polarizations for the allowed transitions for observation perpendicular to the magnetic field are shown in the figure and observation of the strengths of the components as a function of temperature showed that the excited exciton state was thermalized. Although six lines are predicted by the energy level diagram only four were observed and so it was assumed that accidental coincidences occurred at two lines. The splittings of the C line in a magnetic field are shown in Fig. 5.34(b). The centre two lines were unpolarized with each corresponding to the superposition of two lines, one with the electric vector $E \perp B$ and the other $E \parallel B$. The parameters used to describe the splitting when B is parallel to [111] were found by Thomas et al. (1963) to be $g_e = 1 \cdot 89 \pm 0 \cdot 1$, $g_{h\frac{1}{2}} = 1 \cdot 12 \pm 0 \cdot 1$, and $g_{h\frac{3}{2}} = 0 \cdot 93 \pm 0 \cdot 04$. It was observed that when the crystal was rotated the positions of all the lines changed and in order to describe these anisotropic effects the general Hamiltonian (Eq. 5.28) must be used.

The above description of excitons bound to neutral donors is only applicable to phosphorus site donors such as the group VI elements S, Se, and Te since it was recognized by Morgan (1968) that only these donors can be described by s-like ground states. Because of the band structure symmetry at the Ga sites, donors at these sites have p-like ground states

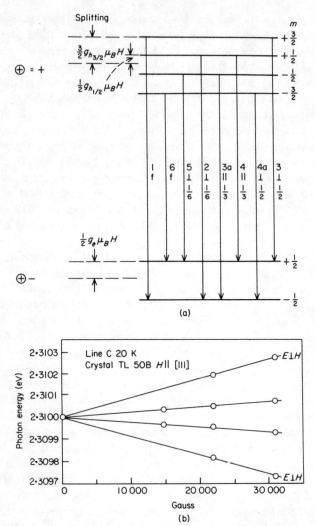

Fig. 5.34. (a) Energy level scheme for the decay of an exciton bound at a neutral donor. The initial state is that of an unpaired hole and the final state is that of a single donor electron. The C line emission in Fig. 5.31 is an example of such a process. (b) The Zeeman splitting of the C line showing that the outer lines are polarized $E \perp B$ but the inner lines are not polarized. (After Thomas et al., 1963.)

which are split by spin–orbit interaction into $p_{\frac{3}{2}}$ and $p_{\frac{1}{2}}$ states. The group IV donors such as C, Si, Ge, and Sn on the Ga sites have been investigated in considerable detail by optical (Dean et al., 1968; Morgan et al., 1969; Vink et al., 1972) and magneto-optical measurements (Dean et al., 1970, 1974) and one consequence of the orbital triplet character of the ground state of these donors is that the zero-phonon donor–acceptor transitions are only weakly allowed. This was verified by Dean *et al.* (1968) and Morgan *et al.* (1969) for Si and C donors. The recombination radiation from excitons bound at Sn donors on Ga sites has been investigated by Dean *et al.* (1970). The observed transitions are shown in Fig. 5.35 where the orbital character of the donor state can be seen. The ground state of the donor was shown to be $p_{\frac{3}{2}}$ by magneto-optical measurements on the zero-phonon line Sn_1^0 (see Fig. 5.35). The Zeeman splittings for this line with B parallel to [110] and [111] are shown in Fig. 5.36. The energy level diagram is similar to Fig. 5.34(a) for p-site donors but in this case the transitions are from weakly bound hole states ($J = \frac{3}{2}$) to the $p_{\frac{3}{2}}$ donor state with a g-factor equal to $-\frac{1}{3}g_e$. The results were satisfactorily fitted to the Hamiltonian given in Eq. (5.28) with the donor g-factor, g_e, changed to the above value and the parameters were found to be $K = 0.67 \pm 0.02$, $L = 0.17 \pm 0.02$, and

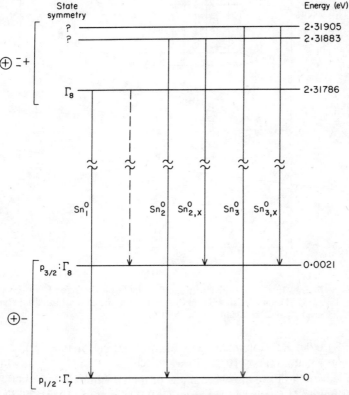

Fig. 5.35. Energies and allowed transitions for the decay of excitons bound to neutral Sn donors on Ga sites in GaP. The donor has p-like symmetry which is illustrated by the spin–orbit split final state. (After Dean *et al.*, 1970.)

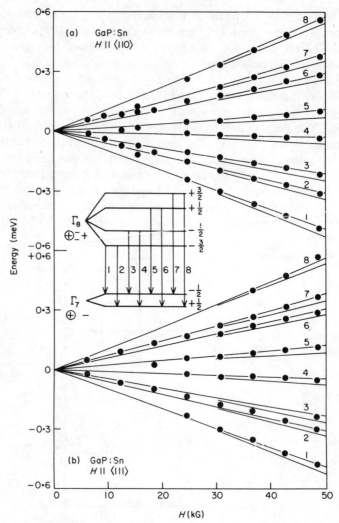

Fig. 5.36. Energy level diagram and magnetic field splittings for the transition Sn_1^0 shown in Fig. 5.35. The initial state is that of an unpaired hole and the final state is the spin doublet level Γ_7. (After Dean *et al.*, 1970.)

$g_e = 1.95 \pm 0.06$. For a comparison with magnetic resonance results of Title (1967) and Mehran *et al.* (1972a, b) see Table III.

Detailed studies have also been made of the recombination radiation from excitons bound at neutral acceptors in GaP. Dean *et al.* (1972) and Dean and Ilegems (1971) have reported the optical properties of Ga-site acceptors such as Be, Mg, Zn, and Cd and, in particular, have investigated the Zeeman

TABLE III

Magnetic Resonance Results on Donor and Conduction Electrons
in the III–V Compounds

	Doping	Temperature	g	Reference
GaP	Te	77	1·9978	1
	Sn	4·2	1·995	2
	Te	4·2	1·9935	2
	Ge	4·2	2·000	3
GaAs		1·7	0·51	4
GaSb		1·7	9·3	5
InP		1·7	1·26	6
InAs		2 K	15·28	7 (SFR)†
InSb		4·2	48·8 to 50·7	8
		1·4	−43·4 to −51·3	9

† SFR = Spin flip Raman.
1. Title (1967).   2. Mehran *et al.* (1972a).   3. Mehran *et al.* (1972b).   4. Lampel (1974).   5. Hermann and Lampel (1971).   6. Weisbuch and Hermann (1975).   7. Eng *et al.* (1974).   8. Bemski (1960).   9. Isaacson (1968).

spectra of zero-phonon bound exciton transitions. There are two $J = \frac{3}{2}$ holes which have to be considered when an exciton is bound to a neutral acceptor and the only states allowed by the Pauli principle are $J = 0$ and $J = 2$. For a Ga-site bound exciton the electron has p-like symmetry, as discussed above, so that from the combined electron and hole states a total of 12 zero-phonon lines were predicted for zero magnetic field (Dean *et al.*, 1971). The splittings of the lowest energy Cd zero-phonon line is shown in Fig. 5.37 along with the energy level diagram showing the transitions from the lowest spin doublet of the bound exciton states to the final state which is that of a weakly bound hole, $J = \frac{3}{2}$. The solid lines in the figure were calculated using the Hamiltonian given in Eq. (5.28) with g_e replaced by $g_{\text{effective}}$ and, using the values $g_{\text{eff}} = -0·32 \pm 0·04$, $K = 0·99 \pm 0·04$, and $L = 0·07 \pm 0·02$, a satisfactory fit of the data was obtained.

In conclusion, we note that, although the examples above show clearly how important Zeeman studies have been to the investigation of recombination radiation in GaP, an equally important part of the research programme has been the preparation of well-characterized crystals. Thus the most important experiments have involved the combination of magneto-optical spectroscopy and samples prepared under carefully controlled conditions. Reports of Zeeman measurements on other III–V compounds include studies of donors and acceptors in GaAs and InP (White *et al.*, 1972, 1974a; White, 1973) and isoelectronic traps in InP (White *et al.*, 1974b).

GaSb. Not all of the III–V emission spectra can be characterized to the same extent as GaP because in materials such as GaAs and GaSb, for example, many of the emission lines are too broad for Zeeman analysis to be possible. However, Benôit à la Guillaume and Lavallard (1973) have measured the bound electron and hole *g*-factors in GaSb by magneto-absorption and magneto-emission experiments where the polarization of the transition was investigated. The experiments were carried out on the emission

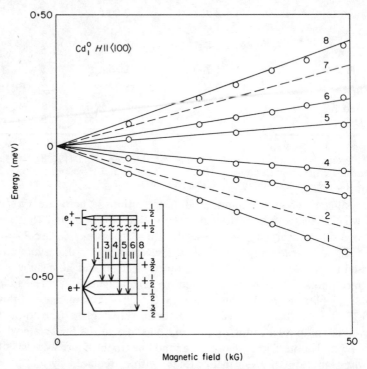

Fig. 5.37. Energy level scheme and magnetic field splittings of an exciton decaying at a cadmium neutral acceptor site. The initial state is that of an unpaired electron and the final state is the hole state of the neutral acceptor. (After Dean *et al.*, 1971.)

of excitons bound at neutral acceptors which are described by the energy level diagram shown in Fig. 5.37 with the exciton level having $J = \frac{1}{2}$ ($j = 0$ for the two holes and $j = \frac{1}{2}$ for the electron). The separation between the $I_{\sigma+}$ and $I_{\sigma-}$ emissions was measured as a function of magnetic field up to 5 T (50 kgauss) and a value of $g_e = 8.65 \pm 0.5$ was obtained which is in agreement with the optically detected magnetic resonance value of $|g_e| = 9.3 \pm 0.3$ measured by Hermann and Lampel (1971) (see next section). The splitting of the free exciton emission band also gave a value,

$g_e = -8\cdot8 \pm 1\cdot0$, which was in agreement with the above results and the magneto-absorption measurement on the bound exciton band gave a hole g-factor of $K = 0\cdot30 \pm 0\cdot05$ (see Hamiltonian 5.28).

5.4.2. Optically Detected Magnetic Resonance via Bound Exciton Luminescence

One of the basic assumptions made in all of the discussions so far is that the distribution of electrons and holes corresponds to thermal equilibrium. Although this is generally true, even for fast-decaying excitons in a magnetic field, it is possible to change the populations of the Zeeman levels from the equilibrium values by exciting with circularly polarized radiation. This process is called optical pumping and has been investigated extensively in the study of gases (Series, 1970) and radiation damage centres in solids (Mollenauer and Pan, 1972). Lampel (1968) was the first to observe the effects of optical pumping in semiconductors when he investigated dynamic polarization of silicon nuclei enhanced by polarized conduction electrons which were produced by pumping with circularly polarized light.

Parsons (1969) showed that a more general method for monitoring the spin polarization of the photo-created electrons involved the observation of the degree of polarization of the luminescence produced by circularly polarized pump light in zero magnetic field. He showed that in p-type GaSb the degree of polarization which is defined by

$$\rho = \frac{L_F(\sigma^+) - L_F(\sigma^-)}{L_F(\sigma^+) + L_F(\sigma^-)} \tag{5.29}$$

was equal to $\rho = 0\cdot21$ for σ^+ polarized excitation light and zero for unpolarized pump light ($L_F(\sigma^\pm)$ is the intensity of the luminescence with polarization $\sigma^\pm$). The results showed that for circularly polarized pump light there was a steady-state electronic polarization of 42%. Since these first investigations there have been many studies of polarized conduction electrons in III–V and II–VI crystals and a summary of the results can be found in the review papers by Zakharchenya (1972) and Lampel (1974). If a longitudinal magnetic field is present, transitions between Zeeman levels can be induced and magnetic resonance detected by the change in the intensity of the polarized emission components. In this section we will describe several of these optically detected magnetic resonance experiments where exciton recombination radiation was investigated and electron g-values were observed.

GaSb. The principles behind the detection of conduction electrons by optical techniques have been given by Hermann and Lampel (1971). The steady-state electronic polarization is changed by a radio-frequency magnetic field which induces transitions between the two conduction electron spin levels. Since the polarization of the recombination light depends on the spin polarization of the conduction electrons, the strengths of the components of polarized radiation change at resonance. The experimental arrangement is shown in Fig. 5.38(a) and this can be compared with the equivalent microwave system shown in Fig. 5.12. Hermann and Lampel used radio-frequency

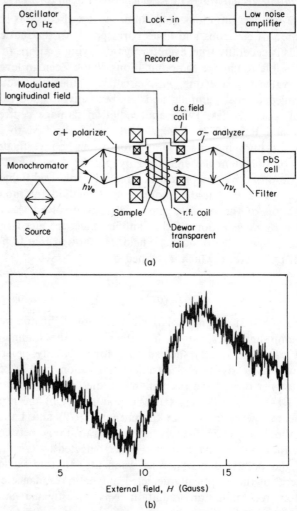

(a)

(b)

External field, H (Gauss)

Fig. 5.38. (a) Experimental arrangement for the optical detection of magnetic resonance of conduction electrons in GaSb. Polarized luminescence is monitored by a PbS cell and, with radiation of frequency $\nu = 151 \cdot 6$ MHz applied to the sample, the resonance signal shown in (b) is observed. The derivative is observed by modulating the magnetic field at 70 Hz and detecting light changes at this frequency. (After Hermann and Lampel, 1971.)

magnetic fields at 150 MHz and the small dc magnetic field required for resonance at this frequency was provided by Helmholtz coils. A second set of coils was used to modulate the dc magnetic field at 70 Hz so that the derivative of the resonance signal was recorded. Circularly polarized exciting light was focused onto one face of the sample and the I_1 luminescence, which is due to free electron-to-acceptor recombination, was monitored with a PbS cell. The signal observed is shown in Fig. 5.38(b) and the g-value is $9 \cdot 3 \pm 0 \cdot 3$. This value agrees with the value measured by Benôit à la Guillaume and Lavallard (1973) and Bimberg and Ruhle (1974) using magneto-optical techniques, but differs significantly from the theoretical estimates.

InP. The optical detection of conduction electron magnetic resonance has also been observed in InP by Weisbuch *et al.* (1974) and Weisbuch and Hermann (1975). Optical pumping techniques were used and the resonance was detected by monitoring the intensity of the circularly polarized emission from several transitions of excitons bound at donors and acceptors. The InP emission and polarization spectra are shown in Fig. 5.39(a) and the apparatus used for the experiment was similar to that shown in Fig. 5.38(a). However, in order to improve the signal-to-noise ratio a pulsed GaAs laser diode excitation source was used and the luminescent intensity was monitored with a boxcar detector. Using 40 W of radio-frequency at 151 MHz the resonance signal shown in Fig. 5.39(b) was obtained with a g-value of $1 \cdot 26 \pm 0 \cdot 05$. The authors noted that the resonance did not depend on which of the lines shown in Fig. 5.39(a) was monitored and so, as in the case of GaSb, it was concluded that the observed resonance was that of conduction electrons. In fact, the measured g-value of $1 \cdot 26$ agreed well with the calculated value for free electrons, $g = 1 \cdot 20$, by Lawaetz (1971), and was also close to the value of $g = 1 \cdot 15 \pm 0 \cdot 05$ found by Zeeman measurements of bound excitons in InP by White *et al.* (1974b).

GaP. Recently, optically detected magnetic resonance has been observed in GaP with nitrogen isoelectronic traps (Cavenett *et al.*, 1977). The emission spectrum of excitons bound at these centres is shown in Fig. 5.32 and the apparatus which was used for the investigation is shown in Fig. 5.12. Circularly polarized radiation was monitored and both σ^+ and σ^- increased at resonance. The observed g-value was $1 \cdot 99$ which corresponds to the donor or free-electron value (see Table III). The energy level scheme for the recombination of excitons bound at nitrogen isoelectronic traps is shown in Fig. 5.33(a) and for observation of the luminescence in a direction parallel to **B**, the transitions $|\pm 1\rangle \rightarrow |0\rangle$ are circularly polarized. The g-values of the bound excitons can be predicted from the expressions given in the figure and using the magneto-optical data from Thomas and Hopfield (1966); resonance in the $J = 1$ state, which would be observed via the A emission line, should occur at $g = 0.75$ and in the $J = 2$ state (B line) at $g = 1.25$. Neither of these values corresponds to the observed result. In order to understand the results a spectral dependence of the resonance was measured. This is shown in the lower part of Fig. 5.40 and can be compared with the emission spectrum in the upper part of the figure which shows the A and B lines with the phonon side-band. Clearly the resonance is observed from both lines

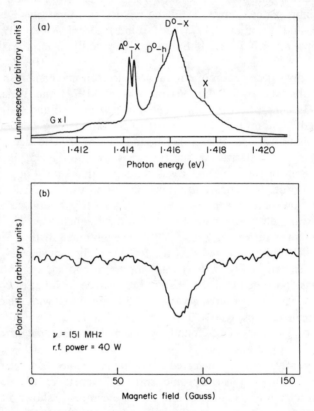

Fig. 5.39. Optically detected resonance of conduction electrons in InP. (a) Photoluminescence of an n-type InP crystal at 1·7 K showing the various bound exciton decay emissions. (b) Resonance line observed with radiation, $\nu = 151$ MHz. (After Weisbuch and Hermann, 1975.)

and the broad band. Since the bound exciton complexes are formed by the capture of electrons at the isoelectronic traps which then trap holes, the resonance has been interpreted as that of the electron bound to the isoelectronic trap. The intensity of the luminescence changes at resonance because the formation of the exciton is spin dependent.

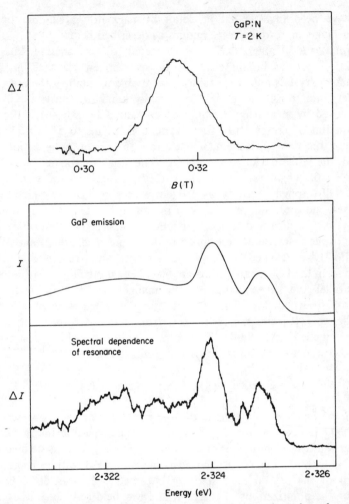

Fig. 5.40. (a) Optically detected resonance in GaP from excitons bound at nitrogen isoelectronic traps. The emission is shown in the upper part of (b) and the spectral dependence of the resonance (lower curve) shows that all of the emission is associated with the resonance. (After Cavenett *et al.*, 1977.)

5.4.3. *Optically Detected Magnetic Resonance of Self-trapped Excitons*

The general character of excitons in insulators has been outlined in Section 5.1.3 where it was noted that, in a material such as KCl, a hole which is self-trapped at a chlorine molecular ion site can trap an electron to give what is known as a self-trapped exciton. Recombination radiation from these bound excitons gives rise to singlet and triplet emissions and the latter have been investigated in detail by magneto-optical and optically detected magnetic resonance techniques as described below.

Marrone and Kabler (1971) investigated the magnetic-field-induced polarization of the luminescence from self-trapped excitons in several alkali halide crystals and, for example, were able to confirm the assignment of triplet state transitions to the low energy band (545 nm) in KBr. This was achieved by measuring the induced circular polarization (see Eq. 5.29) as a function of temperature for magnetic fields up to 5·7 T (57 kgauss) and fitting the data to a calculated expression for the polarization which involved the g-factor as a parameter. The analysis yielded a value of $g = 1·95 \pm 0·05$ which was consistent with the assignment of a triplet state for the self-trapped excitons associated with the 545 nm emission. The results also indicated the possibility of investigating the self-trapped exciton triplet state by optically detected magnetic resonance. The first observations of these triplet resonances were made by Marrone et al. (1973) and Wasiela et al. (1973) who excited the luminescence with an X-ray tube which was shielded from the field of an electro-magnet. The emitted light was monitored both parallel and perpendicular to the magnetic field and, for KBr, the results are shown in Fig. 5.41. The spectra illustrated were obtained with B parallel to [110] and the resonances from excitons trapped at different sites are labelled accordingly. The spin Hamiltonian which describes the observed resonance spectra is (Marrone et al., 1973).

$$\mathcal{H} = \mu_B \mathbf{B}.g.\mathbf{S} + D(S_z^2 - \tfrac{2}{3}) + E(S_x^2 - S_y^2) + \sum_{i=1}^{2}(a I_i.\mathbf{S} + b I_{iz} S_z) \quad (5.30)$$

where $S = 1$ has been substituted and the parameters D and E describe the crystal field interaction for D_{2h} symmetry. The hyperfine interaction is a sum over the two bromine nuclei and the Br–Br axis, [110], is defined as the z-direction with [001] and [1$\bar{1}$0] chosen as the x- and y-directions respectively. The sign of the crystal field splitting parameter, D, was determined by observing the polarizations of the high and low field resonance lines and D was found to be positive for KBr and CsBr. The triplet state is not in thermal equilibrium and the population distribution corresponds to that shown in Fig. 5.11(c) where the $m_s = 0$ state is more highly populated than both the $m_s = \pm 1$ states because radiative transitions from the $m_s = 0$ to the ground state are forbidden and the recombination rate from the $m_s = \pm 1$ states is faster than the spin lattice relaxation rate. Thus microwave saturation of the allowed transitions produces an increase in both the σ^+ and σ^- radiation.

The hyperfine lines observed in the spectra are due to the interaction of the excitons with the nuclear spins ($I = \tfrac{3}{2}$) of the pairs of bromine ions. For most orientations of the magnetic field the hyperfine pattern was not resolved and Wasiela et al. (1973) reported only the isotropic hyperfine coupling

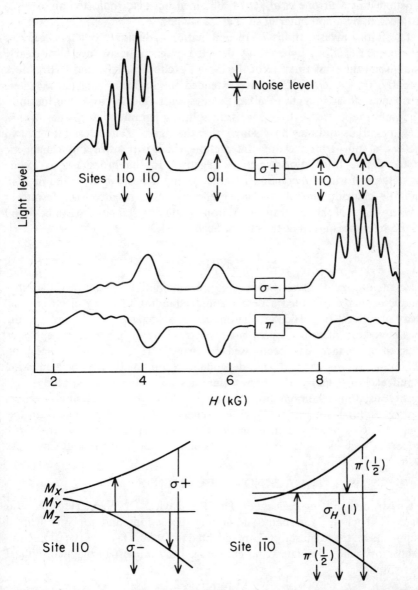

Fig. 5.41. Optically detected resonance and energy level schemes for self-trapped excitons in KBr. The three resonance curves were obtained by monitoring the intensity of the polarized emission components which are indicated. The triplet energy level scheme shows both the allowed optical and magnetic resonance transitions. (After Wasiela *et al.*, 1973.)

constant, but Marrone *et al.* (1973) obtained both isotropic and anisotropic parameters by a computer fit of the line shapes.

The initial investigations of recombination radiation from the decay of self-trapped excitons by optically detected magnetic resonance have clearly established the triplet nature of the bound excitons in KBr and CsBr. More recently, Call *et al.* (1975a) have extended the optically detected magnetic resonance measurements to alkali halides with chlorine and fluorine ions. They have reported that the zero field splitting parameter, D, is negative in these materials and they have shown that the sign of D is determined by two factors, a spin–orbit exchange interaction which is positive and a magnetic dipole–dipole interaction which is negative. In the bromide and iodide crystals the former interaction is larger giving a positive D, while in NaF and the chloride crystals the latter effect gives a negative D. Results of investigations of the self-trapped exciton in other crystal lattices can be found in the references given at the end of Section 5.1.3.

5.5. CONCLUSION

Magneto-optical investigations of recombination radiation in semi-conductors have provided a detailed understanding of many electron–hole decay processes. In particular, the impurities associated with bound exciton lines have been identified by Zeeman spectroscopy. More recently, optically detected resonance has been used to investigate broad donor–acceptor recombination bands and spin-dependent recombination associated with bound excitons. Currently these techniques are being used to study high density exciton recombination radiation, the nature of bound exciton complexes, and nuclear polarization effects. Very much linked to the luminescence studies are investigations of other spin-dependent transport properties such as photoconductivity (Lepine, 1972; Brunwin and Cavenett, 1976).

ACKNOWLEDGEMENTS

I wish to thank my colleagues, Dr J. J. Davies, Dr W. E. Hagston, and Dr J. E. Nicholls, for many stimulating discussions and for very helpful comments on the manuscript. I am indebted to Dr W. Hayes of the Clarendon Laboratory who stimulated and encouraged my interest in magneto-optical spectroscopy.

APPENDIX

Considerable progress in the application of optically detected magnetic resonance (ODMR) to the study of recombination processes in semiconductors has been made since the completion of this article and so this appendix briefly reports the most recent developments.

ZnS

The original work of James *et al.* (1975, 1976) has been extended to a study of cubic ZnS as well as hexagonal material. A detailed analysis of the mechanism giving rise to the optically detected resonance has also been carried out and, whereas in James *et al.* (1975, 1976) thermalized donors and acceptors were considered, the more recent investigations suggest that the donors and acceptors should be considered as pairs which have not thermalized. Then the relative spin populations in the exited state of the donor–acceptor pair are determined by the recombination rates from the spin levels. Details of these processes can be found in Davies *et al.* (1978) and Nicholls *et al.* (1978a).

The experimental arrangement for optically detected magnetic resonance is shown in Fig. 5.12 and it can be seen that an argon ion laser is used to excite the luminescence. However, if the time dependences of the ODMR signals are required, a pulsed excitation source and pulsed microwaves must be used. Dawson *et al.* (1978a) have developed a pulsed ODMR system which has been used to investigate time-dependent recombination processes in ZnS and CdS. The pulsed system involves a two-channel sample and hold unit with a digital phase-sensitive detector used with a pulsed nitrogen laser and either a magnetron or a klystron plus travelling wave amplifier. The luminescence can be sampled with gatewidths variable down to 35 ns. The microwaves are pulsed at half the sampling rate so that the detector output is proportional to the microwave-induced changes in the luminescence. In ZnS, separate donor and acceptor resonances are observed from distant pairs when millisecond delay times are used, whereas, for the shorter-lived pairs, a new resonance at $g = 2 \cdot 1$ corresponding to close-pair or excitonic recombination is observed. These time-resolved ODMR measurements are reported by Dawson and Cavenett (1978).

ZnO

Donor–acceptor pair recombination in lithium-doped ZnO has been investigated by ODMR by Block *et al.* (1978) and Cox *et al.* (1978). A yellow luminescence centred at 600 nm was shown to be due to the recombination between shallow donors and lithium acceptors. An electron resonance was observed at $g = 1 \cdot 957 \pm 0 \cdot 001$ and two lithium acceptor resonance were observed at $g = 2 \cdot 0018 \pm 0 \cdot 001$ and $g = 2 \cdot 0182 \pm 0 \cdot 001$. The spin-dependent recombination processes were explained by considering unthermalized pairs as in the ZnS case discussed above.

ZnSe

The analogue of the ZnS self-activated centre in ZnSe has recently been studied by ODMR (Dunstan *et al.*, 1978b, Nicholls *et al.*, 1978b). Melt-grown ZnSe crystals doped with chlorine showed an acceptor resonance with many resolved components at 23 GHz. A fit of the spectrum for the [111] direction using the g-values of Holton et al. (1966) confirmed that the acceptor was the A-centre. A spectral dependence of the A-centre resonance showed that the 620 nm emission (after correction) is the self-activated emission in ZnSe.

CdS

Optically detected electron resonances have been observed in CdS by monitoring the exciton emissions. In particular the emissions at 486·5 nm and 488·8 nm, labelled I_{2A} and I_{1A}, have been investigated. The resonances are only observed as changes in the polarized emission components and for I_{2A}, $\sigma-$ decreases and $\sigma+$ increases. In the case of the I_{1A} emission it is found that $\sigma-$ increases and $\sigma+$ decreases at resonance. The I_{2A} line is due to the recombination of an exciton at a neutral donor and so the electron resonance arises because of a spin-dependent exciton binding process. At low temperatures, the donor electrons are predominantly in the $|-\frac{1}{2}\rangle$ state, but since only the exciton $(-\frac{3}{2}, \frac{1}{2})$ will be bound for this state, the $\sigma-$ emission is greater than the $\sigma+$ emission. At resonance of the donor, more $(\frac{3}{2}, -\frac{1}{2})$ excitons can be bound and the $\sigma+$ emission increases. With fewer $(-\frac{3}{2}, \frac{1}{2})$ excitons bound the $\sigma-$ emission decreases. This process is discussed in more detail by Dawson et al. (1978b) and Dunstan et al. (1978a). These papers also discuss the observation of the electron resonance on the I_{1A} line which is due to recombination of excitons at neutral acceptors. In this exciton complex the holes pair leaving a single electron which is loosely bound. This electron is partially thermalized and at resonance the emission changes are similar to that discussed in Section 5.3.4 for the donor resonance. In order to confirm that the electron in the exciton complex was being observed, optical pumping of the ODMR signals with circularly polarized excitation light was carried out on both the donor resonance, as observed from the donor–acceptor pair emission, and the I_{1A} resonance. The behaviour of the two resonances was very different showing that, as expected, the donor electron was more thermalized than the electron in the exciton complex.

CdTe

The observation of conduction electron resonance was observed in CdTe by Cavenett et al. (1978) using the optical pumping technique described in Section 5.4.2. The maximum polarization memory was observed on the (A^0, X) emission at 780 nm but the resonance, which has $g_e = -1·59 \pm 0·02$, was observed on several emission lines. The negative sign of the resonance was determined by observing a shift of the electron resonance position when the Cd and Te nuclei were polarized by fixed circularly polarized excitation light (see Weisbuch, 1976). Unlike the materials discussed above, the g-value of the electrons in CdTe had not been determined with certainty; a value of $-1\cdot1\pm0\cdot1$ was reported from combined measurements of the nuclear relaxation time and the Knight shift (Look and Moore, 1972), $-0\cdot74\pm0\cdot03$ was obtained by spin–flip Raman measurements (Walker et $al.$, 1972) and $1\cdot6803\pm0\cdot0005$ was obtained from CdTe doped with donors using conventional EPR (Alekeenko and Veringer, 1964). The spin-dependent process giving rise to the resonance has not been determined with certainty and is still the subject of further investigations.

ZnTe

The value of the electron g-value in ZnTe had also been uncertain because· of very different reported values. Hollis et $al.$ (1973) measured $g_e = 0\cdot57$ using spin–flip Raman measurements but in later magneto-optical measurements the value was revised to $1\cdot74$ (Scott and Hollis, 1976). Magneto-reflectance studies by Venghaus et $al.$ (1977) on the free exciton in ZnTe have given a value of $g_e = -0\cdot57$ and Dean et $al.$ (1978) have obtained a value of $-0\cdot38\pm0\cdot05$ by investigating the Zeeman splitting of excitons bound to neutral acceptors. More recently, Killoran et $al.$ (1978) have investigated ZnTe using ODMR and by observing both (A^0, X) and (D^0, A^0) recombination in the 520–550 nm region, an electron resonance at $0\cdot401\pm0\cdot004$ was measured. Spectral dependence measurements have not determined which of the two processes is responsible for the resonance but (D^0, A^0) is more likely. The spin dependent recombination process for T_d symmetry has been discussed for ZnSe in Section 5.3.4.

GaSe

The section on self-trapped excitons (5.4.3) was included because at the time of preparation of this article there were no examples of exciton resonances in semiconductors. However, recently triplet exciton resonances have been observed in the layered semiconductor GaSe (Dawson et $al.$, 1978c; Morigaki et $al.$, 1978a). GaSe has axial symmetry and the ODMR measurements were carried out with $B // c$-axis. Two resonances were observed from the bound exciton emission region, a low-field increase in $\sigma+$ emission and a high-field increase in $\sigma-$ emission. These resonances are characteristic of an unthermalized triplet exciton where the populations of the $|1\rangle$ and $|-1\rangle$ states drop below the $|0\rangle$ state because of the larger decay rate from the $|\pm1\rangle$ states. Analysis of the results gave $g_{\parallel} = 1\cdot85$ and a crystal field splitting of $D = +0\cdot11$ cm^{-1}. Electron and free-hole resonances have also been observed as changes in the circularly polarized emission components. The electron g-value is $1\cdot13$ and the hole has $g_{\parallel} = 1\cdot72$. These resonances are observed on both free and bound exciton emissions and are interpreted as due to spin memory in the formation of free excitons. Although GaSe shows a strong free exciton emission at 588 nm, so far no free exciton ODMR signal has been observed. Magnetic circular dichroism measurements have also been carried out as a function of magnetic field strength. A larger signal is observed at 1·5 kG corresponding to the crossing of the $|0\rangle$ and $|-1\rangle$ bound exciton levels. This value is consistent with the observed g-value and zero field splitting obtained from the ODMR experiment.

a-Si

The investigation of amorphous semiconductors by conventional magnetic resonance has recently become very important (for example, see review by Bishop *et al.*, 1977). The detection of spin resonance by spin-dependent photoconductivity measurements has also been carried out in the case of a-Si by Solomon *et al.* (1977). Since the nature of the recombination centres and mechanisms were not clearly understood, an investigation of ODMR of a-Si was undertaken by Morigaki *et al.* (1978c). Independent studies were also carried out at the same time by Lampel *et al.* (1978) and Biegelson *et al.* (1978). Morigaki *et al.* (1978c) observed three resonances from the total emission band, two narrow lines with $g = 2·006 \pm 0·001$ (D_2) and $g = 2·018 \pm 0·002$ (D_1) which were decreases of the emission at resonance and a broad resonance at $g = 1·999 \pm 0·010$ which was an increase in light. Spectral dependence measurements showed that the acceptor resonance came principally from an emission at 950 nm while the D_1 and D_2 resonances extended over the whole emission. Time-resolved emission measurements showed the existence of two bands, one at 950 nm and the other at 870 nm. The latter decayed rapidly with a time constant of the order of 10 ns, whereas the longer wavelength decay was approximately 2 µs. The 950 nm band showed a shift of the emission peak position with decay and is consistent with a donor–acceptor emission where the centres involved are the D_1 and acceptor centres. The high energy emission is either excitonic-like or a recombination between an electron on a D_1 centre and a hole in the valence band edge. The D_2 centre is believed to be a non-radiative recombination centre such as a dangling bond. More details of the time-resolved measurements of both the emission and the ODMR signals can be found in Morigaki *et al.* (1978b).

Conclusion

The work described in this appendix illustrates many of the current fields where ODMR is contributing to the study of recombination processes in semiconductors. More details of these investigations can be found in the reviews by Cavenett (1977, 1978a, b). Undoubtedly, the interest in the application of ODMR to amorphous materials will grow since the nature of the defects in these materials is, in most cases, unknown. A similar situation exists in the case of deep traps in semiconductors. It is currently believed that many of these defects are transition-metal ions which, by changing charge states, act as recombination centres which reduce the efficiency of the visible luminescence process. Near infrared emissions have been associated with these impurities but, so far, ODMR experiments in this spectral region have not been successful. Nevertheless, this work is continuing.

Finally, it is intersting to note that there is considerable interest in the observation of ODMR signals from free excitons in semiconductors. There are many materials where such an observation may be possible and it is quite likely that the first observation will be made before this article is published, such is the activity in the ODMR field.

REFERENCES

Abragam, A. and Bleaney, B. (1970). "Electron Paramagnetic Resonance of Transition Ions." Clarendon Press, Oxford.
Alekeenko, M. V. and Veringer, A. I. (1974). *Soviet Phys. Semiconductors*, **8**, 143–145.
Bemski, G. (1960). *Phys. Rev. Lett.* **4**, 62–64.
Benoît à la Guillaume, C. (1976). "Proc. Int. Conf. Luminescence, Tokyo, 1975." *J. Luminescence*, **12/13**, 57–66.
Benoît à la Guillaume, C. and Lavallard, P. (1973). *Phys. Stat. Sol.* **59b**, 545–549.
Bergh, A. and Dean, P. J. (1972). *Proc. IEEE*, **60**, 156–223.
Biegelsen, D. K., Knights, J. C., Street, R. A., Tsang, C., and White, R. M. (1978). *Phil. Mag.* B **37**, 477–488.
Bimberg, D. and Ruhle, W. (1974). "Proc. 12th Int. Conf. on the Physics of Semiconductors", (M. H. Pilkuhn, ed.) pp. 561–565. Teubner, Stuttgart.
Bishop, S. G., Strom, U., and Taylor, P. C. (1977). *Phys. Rev.* **B15**, 2278–2294.
Block, D., Cox, T. R., Herve, A., Picard, R., Santier, C., and Helbig, R. (1978). "Proc. 3rd Colloque Ampère, Dublin," (To be published in *J. Semiconductors and Insulators.*)
Bouley, J. C., Blanconnier, P., Hermann, A., Ged, Ph., Henoc, P., and Noblanc, J. P. (1975). *J. Appl. Phys.* **46**, 3549–3555.
Brunwin, R. F. and Cavenett, B. C. (1976). "Proc. Int. Conf. Semiconductors, Rome" (F. G. Fumi, ed.) 1016–1019.
Brunwin, R. F., Cavenett, B. C., Davies, J. J., and Nicholls, J. E. (1976). *S.S. Com.* **18**, 1283–1285.
Bryant, F. J. and Manning, P. S. (1974). *J. Phys. Chem. Solids*, **35**, 97–101.
Call, P. J., Hayes, W. and Kabler, M. N. (1975a). *J. Phys. C*, **8**, L60–L62.
Call, P. J., Hayes, W., Huzimura, R., and Kabler, M. N. (1975b). *J. Phys. C*, **8**, L56–L59.
Callaway, J. (1974). "Quantum Theory of the Solid State", Part B. Academic Press, New York and London.
Cavenett, B. C. (1977). "Proc. Symposium on Microwave Diagnostics of Semiconductors, Porvoo" (Paananen, R., ed.), pp. 29–52. Helsingfors, Helsinki.
Cavenett, B. C. (1978a). "Proc. Luminescence Conference, Paris." (To be published in *J. Luminescence.*)

Cavenett, B. C. (1978b). "Proc. Applic. of High Magnetic Fields in Semiconductor Physics." (Unpublished.)

Cavenett, B. C. and Hagston, W. E. (1975). *Solid State Commun.* **16**, 1235–1238.

Cavenett, B. C. and Sowersby, G. (1975). *J. Phys. E*, **8**, 365–368.

Cavenett, B. C., Brunwin, R. F., and Nicholls, J. E. (1977). *Solid State Commun.* **23**, 71–74.

Cavenett, B. C., Herman, C., Lampel, G., Nakamura, A., Paget, D., and Weisbuch, C. (1978). "Proc. 3rd Colloque Ampère, Dublin." (To be published in *J. Semiconductors and Insulators.*)

Colbow, K. (1965). *Phys. Rev.* **139**, A274–749.

Cox, R. T., Block, D., Herve, A., Picard, R., Santier, C., and Helbig, R. (1978). *Solid State Commun.* **25**, 77–80.

Curie, D. and Prener, J. S. (1967). *In* "Physics and Chemistry of II–VI Compounds", (M. Aven and J. S. Prener, eds), Chapter 9. North Holland, Amsterdam.

Davies, J. J. (1976). *Contemporary Physics*, **17**, 275–294.

Davies, J. J., Nicholls, J. E., and Cavenett, B. C. (1978). "Proc. 3rd Colloque Ampère, Dublin." (To be published in *J. Semiconductors and Insulators.*)

Dawson, P. and Cavenett, B. C. (1978). "Proc. Luminescence Conference, Paris." (To be published in *J. Luminescence.*)

Dawson, P., Cavenett, B. C., and Sowersby, G. (1978a). "Proc. Conf. on Recombination in Semiconductors, Southampton." (To be published in *J. Solid State Electronics.*)

Dawson, P., Dunstan, D. J., and Cavenett, B. C. (1978b). "Proc. 3rd Colloque Ampere, Dublin." (To be published in *J. Semiconductors and Insulators.*)

Dawson, P., Morigaki, K., and Cavenett, B. C. (1978c). "Proc. XIV Semiconductor Conf., Edinburgh." (To be published.)

Dean, P. J. (1966). "Luminescence of Inorganic Solids" (P. Goldberg, ed.), pp. 119–203. Academic Press, New York and London.

Dean, P. J. (1969). "Applied Solid State Science" (R. Wolfe and C. J. Kriessman, eds), Vol. VI, pp. 1–151. Academic Press, New York and London.

Dean, P. J. (1973). "Progress in Solid State Chemistry" (J. O. McCaldin and G. Somorjai, eds), Vol. 8, pp. 1–126. Pergamon Press, Oxford.

Dean, P. J. and Ilegems, M. (1971). *J. Luminescence*, **4**, 201–230.

Dean, P. J. and Merz, J. L. (1969). *Phys. Rev.* **178**, 1310–1318.

Dean, P. J., Frosch, C. J., and Henry, C. H. (1968). *J. Appl. Phys.* **39**, 5631–5646.

Dean, P. J., Faulkner, R. A., and Kimura, S. (1970). *Phys. Rev.* **B2**, 4062–4076.

Dean, P. J., Faulkner, R. A., Kimura, S., and Ilegems, M. (1971). *Phys. Rev.* **B4**, 1926–1944.

Dean, P. J., Schairer, W., Lorentz, M., and Morgan, T. N. (1974). *J. Luminescence*, **9**, 343–379.

Dean, P. J., Venghaus, H., Pfister, J. C., Schaub, B., and Manne, J. (1978). *J. Luminescence*, **16**, 363–394.

Dexter, D. L. and Knox, R. S. (1965). "Excitons." Interscience, New York.

Dielman, J. (1963). "Proc. Colloque Ampère, Eindhoven," pp. 409–413. North Holland, Amsterdam.

Dischler, B., Rauber, A., and Schneider, J. (1964). *Phys. Stat. Sol.* **6**, 507–510.

Dunstan, D. J., Nicholls, J. E., Cavenett, B. C., Davies, J. J., and Reddy, K. V. (1977). *Solid State Commun.* **24**, 677–680.

Dunstan, D. J., Cavenett, B. C., Dawson, P., and Nicholls, J. E. (1978a). *J. Phys. C.* (To be published.)

Dunstan, D. J., Nicholls, J. E., Cavenett, B. C., Davies, J. J., and Reddy, K. V. (1978b). *J. Phys. C.* (To be published.)

Eng, R. S., Mooradian, A., and Fetterman, H. R. (1974). *Appl. Phys. Letters*, **25**, 453–454.

Era, K., Shionoya, S., and Washizawa, Y. (1969). *J. Phys. Chem. Solids*, **29**, 1827–1841.

Fleury, P. A. and Scott, J. F. (1971). *Phys. Rev.* **B3**, 1970–1985.

Gale, G. M. and Mysyrowicz, A. (1974). "Proc. 12th Int. Conf. on the Physics of Semiconductors" (M. H. Pilkuhn, ed.), pp. 133–141. Teubner, Stuttgart.

Garlick, G. F. J. (1967). *Rept. on Prog. Physics*, **30**, 491–560.

Gershenzon, M. (1966). *In* "Semiconductors and Semimetals", (R. K. Willardson and A. C. Beer, eds), Vol. 2, Chapter 13. Academic Press, New York and London.

Geschwind, S. (1972). "Electron Paramagnetic Resonance" (S. Geschwind, ed.), Chapter 5. Plenum Press, New York.

Geschwind, S., Devlin, G. E., Cohen, R. L., and Chinn, S. R. (1967). *Phys. Rev.* **137**, 1087–1100.

Griffith, J. S. (1961). "The Theory of Transition-metal Ions." Cambridge University Press.

Hagston, W. E. and Cavenett, B. C. (1978). *J. Phys. C.* (To be published.)

Halsted, R. E. (1967). "The Physics and Chemistry of II–VI Compounds" (M. Aven and J. S. Prener, eds), Chapter 8. North Holland, Amsterdam.

Hanamura, E. (1976). "Proc. Int. Conf. Luminescence, Tokyo, 1975." *J. Luminescence*, **12/13**, 119–129.

Hayes, W. and Owen, I. B. (1975). *J. Phys. C*, **9**, L69–L71.

Hayes, W. and Smith, P. H. S. (1971). *J. Phys. C*, **4**, 840–851.

Hayes, W., Owen, I. B., and Pilipenko, G. I. (1975). *J. Phys. C*, **8**, L407–L409.

Heisinger, P., Suga, S., Willmann, F., and Dreybodt, W. (1975). *Phys. Stat. Sol.* **67**, 641–652.

Henry, C. H., Dean, P. J., and Cuthbert, J. D. (1968). *Phys. Rev.* **166**, 754–756.

Henry, C. H., Faulkner, R. A., and Nassau, K. (1969). *Phys. Rev.* **183**, 798–806.

Henry, C. H., Nassau, K., and Shiever, J. W. (1970). *Phys. Rev. Lett.* **24**, 820–822.

Hermann, C. and Lampel, G. (1971). *Phys. Rev. Lett.* **27**, 373–376.

Hollis, R., Ryan, J. F., Toms, D. J., and Scott, J. F. (1973). *Phys. Rev. Lett.* **31**, 1004–1007.

Holton, W. C., de Wit, M., and Estle, T. L. (1966). "International Symposium on Luminescence", (N. Riehl and H. Kallmann, eds), pp. 454–459. Verlag Karl Thiemig, Munich.

Hvam, J. M. (1974). *Phys. Stat. Sol.* (b) **63**, 511–517.

Iida, S. (1968). *J. Phys. Soc. Japan*, **25**, 177–184.

Isaacson, R. A. (1968). *Phys. Rev.* **169**, 312–314.

James, J. R., Nicholls, J. E., Cavenett, B. C., Davies, J. J., and Dunstan, D. J. (1975). *Solid State Commun.* **17**, 969–972.

James, J. R., Cavenett, B. C., Nicholls, J. E., Davies, J. J., and Dunstan, D. J. (1976). "Proc. Int. Conf. Luminescence, Tokyo, 1975." *J. Luminescence*, **12/13**, 447–452.

Jefferson, J. H., Hagston, W. E., and Sutherland, H. H. (1975). *J. Phys. C*, **8**, 3457.

Johnston, W. D. and Shaklee, K. L. (1974). *Solid State Commun.* **15**, 73–75.

Jones, G. and Woods, J. (1974). *Luminescence*, **9**, 389–405.

Judd, B. R. (1963). "Operator Techniques in Atomic Spectroscopy." McGraw-Hill, New York.

Kabler, M. (1972). In "Point Defects in Solids" (J. H. Crawford, Jnr and L. M. Slifkin, eds), Vol. 6, Chapter 6. Plenum Press, New York.

Kasai, P. H. and Otomo, Y. (1962). J. Chem. Phys. 37, 1263–1275.

Killoran, N., Cavenett, B. C., and Dean, P. J. (1978). Solid State Commun. (To be published.)

Koda, T. and Shionoya, S. (1964). Phys. Rev. 136, A541–555.

Kokes, R. J. (1962). J. Phys. Chem. 66, 99–103.

Kuroda, H., Shionoya, S., Saito, H., and Hanamura, E. (1963). J. Phys. Soc. Japan, 35, 534–542.

Lambe, J. and Kikuchi, C. (1958). J. Phys. Chem. Solids, 9, 492–494.

Lampel, G. (1968). Phys. Rev. Lett. 20, 491–493.

Lampel, G. (1974). "Proc. XII Int. Conf. on the Physics of Semiconductors", pp. 743–750. (M. H. Pilkuhn, ed.). Teubner, Stuttgart.

Lampel, G., Rosso, M., and Solomon, I. (1978). Private communication.

Landsberg, P. T. (1967). Solid State Electronics, 10, 513–537.

Lawaetz, P. (1971). Phys. Rev. B4, 3460–3467.

Lepine, D. J. (1972). Phys. Rev. 6, 436–441.

Livingston, A. W., Turvey, K., and Allen, J. W. (1973). Solid State Electronics, 16, 351–356.

Look, D. C. and Moore, D. L. (1972). Phys. Rev. 5B, 3406–3412.

Marrone, M. J. and Kabler, M. N. (1971). Phys. Rev. Lett. 27, 1283–1285.

Marrone, M. J., Patten, F. W., and Kabler, M. N. (1973). Phys. Rev. Lett. 31, 467–471.

Mehran, F., Morgan, T. N., Title, R. S., and Blum, S. E. (1972a). Solid State Commun. 11, 661–662.

Mehran, F., Morgan, T. N., Title, R. S., and Blum, S. E. (1972b). Phys. Rev. B10, 3917–3926.

Merz, J. L., Nassau, K., and Shiever, J. W. (1973). Phys. Rev. B8, 1444–1452.

Miyamoto, S. and Shionoya, S. (1976). "Proc. Int. Conf. Luminescence, Tokyo, 1975." J. Luminescence, 12/13, 563–567.

Mollenauer, L. F. and Pan, S. (1972). Phys. Rev. B6, 772–787.

Morgan, T. N. (1968). Phys. Rev. Lett. 21, 819–823.

Morgan, T. N., Plaskett, T. S., and Pettit, G. D. (1969). Phys. Rev. 180, 845–851.

Morigaki, K. (1964). J. Phys. Soc. Japan, 19, 1253.

Morigaki, K., Dawson, P., and Cavenett, B. C. (1978a). Solid State Commun. (To be published.)

Morigaki, K., Dawson, P., Cavenett, B. C., Dunstan, D. J., Nitta, S., and Shimakawa, K. (1978b). "Proc. XIV Semiconductor Conf. Edinburgh," (To be published.)

Morigaki, K., Dunstan, D. J., Cavenett, B. C., Dawson, P., Nicholls, J. E., Nitta, S., and Shimakawa, K. (1978c). Solid State Commun. 26, 981–985.

Müller, G. O., Rosler, M., Weber, H. H., and Zimmerman, R., Jacobson, M. A., Michailov, G. V., Razbirin, B. S., and Ural'tsev, I. N. (1976). "Proc. Int. Conf. Luminescence, Tokyo, 1975". J. Luminescence, 12/13, 557–562.

Müller, K. A. and Schneider, J. (1963). Phys. Lett. 4, 288–291.

Murayama, K., Morigaki, K., Sakuragi, S., and Kanzaki, H. (1976). "Proc. Int. Conf. Luminescence, Tokyo, 1975." J. Luminescence, 12/13, 309–314.

Nicholls, J. E., Davies, J. J., Cavenett, B. C., James, J. R., and Dunstan, D. J. (1978a). *J. Phys. C*. (To be published.)

Nicholls, J. E., Dunstan, D. J., and Davies, J. J. (1978b). "Proc. 3rd Colloque Ampère, Dublin." (To be published in *J. Semiconductors and Insulators*.)

Oӡsan, M. E. and Woods, J. (1975). *Solid State Electronics*, 18, 519–527.

Parsons, R. R. (1969). *Phys. Rev. Lett.* 23, 1152–1154.

Pedrotti, L. S. and Reynolds, D. C. (1960). *Phys. Rev.* 120, 1664–1669.

Piper, W. W. (1967). "II–VI Semiconducting Compounds" (D. G. Thomas, ed.), pp. 839–849. Benjamin, New York.

Prener, J. S. and Williams, F. E. (1956). *J. Chem. Phys.* 25, 361.

Reynolds, D. C. and Collins, T. C. (1969). *Phys. Rev.* 188, 1267–1271.

Reynolds, D. C., Litton, C. W., and Collins, T. C. (1965). *Phys. Stat. Sol.* 12, 3–55.

Reynolds, D. C., Litton, C. W., and Collins, T. C. (1975). *Solid State Commun.* 17, 15–18.

Saito, H., Kuroiwa, A., Kuribayashi, S., Aogaki, Y., and Shionoya, S. (1976). "Proc. Int. Conf. Luminescence, Tokyo, 1975." *J. Luminescence*, 12/13, 575–580.

Saito, H. and Shionoya, S. (1974). *J. Phys. Soc. Japan*, 37, 423–430.

Schneider, J., Rauber, A., Dischler, B., Estle, T. L., and Holton, W. C. (1965). *J. Chem. Phys.* 42, 1839–1841.

Schneider, J., Dischler, B., and Rauber, A. (1968). *J. Phys. Chem. Solids*, 29, 451–462.

Schulz, M. (1975). *Phys. Stat. Solidi* 27, K5-8.

Scott, J. F. and Hollis, R. L. (1976). *Solid State Commun.* 20, 1125–1128.

Segall, B. and Marple, D. T. F. (1967). *In* "Physics and Chemistry of II–VI Compounds" (M. Aven and J. S. Prener, eds), Chapter 7. North Holland, Amsterdam.

Segawa, Y. and Namba, S. (1976). "Proc. Int. Conf. Luminescence, Tokyo, 1975." *J. Luminescence*, 12/13, 569–573.

Series, G. W. (1970). "Quantum Optics" (S. M. Kay and A. Maitland, eds), pp. 395–482. Academic Press, London and New York.

Solomon, I., Biegelsen, D., and Knights, J. C. (1977). *Sol. State Commun.* 22, 505–508.

Susa, N., Watanabe, H., and Wada, M. (1975). *Jap. J. Appl. Phys.* 14, 1733–1737.

Taguchi, T., Shirafuji, J., and Inuishi, Y. (1975). *Phys. Stat. Sol.* 68, 727–738.

Thomas, D. G. and Hopfield, J. J. (1962). *Phys. Rev.* 128, 2135–2148.

Thomas, D. G. and Hopfield, J. J. (1966). *Phys. Rev.* 150, 680–689.

Thomas, D. G. and Hopfield, J. J. (1968). *Phys. Rev.* 175, 1021–1032.

Thomas, D. G., Gershenzon, M., and Hopfield, J. J. (1963). *Phys. Rev.* 131, 2397–2404.

Thomas, D. G., Gershenzon, M., and Trumbore, F. A. (1964a). *Phys. Rev.* 133, A269–279.

Thomas, D. G., Hopfield, J. J., and Colbow, K. (1964b). "7th Inf. Conf. on the Physics of Semiconductors, 4, Radiative Recombination Symposium", pp. 67–80. Dunod, Paris.

Thomas, D. G., Hopfield, J. J., and Augustyniak, W. M. (1965). *Phys. Rev.* 140, A202–A220.

Thomas, D. G., Dingle, R., and Cuthbert, J. D. (1967). "II–VI Semiconducting Compounds" (D. G. Thomas, ed.), pp. 863–876. Benjamin, New York.

Title, R. S. (1967). *Phys. Rev.* 154, 668–671.

Varshni, Y. P. (1967a). *Phys. Stat. Sol.* 19, 459–514.

Varshni, Y. P. (1967b). *Phys. Stat. Sol.* 20, 9–36.

Venghaus, H., Simmonds, P. E., Lagois, J., Dean, P. J., and Bimberg, D. (1977). *Solid State Commun.* **24**, 5–9.

Vink, A. T., Bosman, A. J., Van der Does de Bye and Peters, R. C. (1972). *J. Luminescence*, **5**, 57–68.

Walker, T. W., Litton, C. W., Reynolds, D. C., Collins, T. C., Wallace, W. A., Gorrell, J. H., and Jungling, K. C. (1972). 'Proc. XI Conf. Semiconductors, Warsaw" (Miasek, M., ed.), pp. 376–381 PWN-Polish Scientific, Warsaw.

Wasiela, A. and Duran, J. (1974). "Proc. 18th Ampère Congress" (P. S. Allen, E. R. Andrew, and C. A. Bates, eds), pp. 217–218. Nottingham University.

Wasiela, A., Ascarelli, G., and Merle d'Aubigné, Y. (1973). *Phys. Rev. Lett.* **31**, 993–996.

Watkins, G. D. (1973). *Solid State Commun.* **12**, 589–592.

Watkins, G. D. (1974). *Phys. Rev. Letters* **34**, 223–225.

Weisbuch, C. (1976). "Proc. Congress Ampère Heidelberg" (Brunner, H., Hausser, K. H., and Schweitzer, D., eds), pp. 293–296.

Weisbuch, C. and Hermann, C. (1975). *Solid State Commun.* **16**, 659–661.

Weisbuch, C. and Lampel, G. (1974). *Solid State Commun.* **14**, 141–144.

Weisbuch, C., Hermann, C. and Fishman (1974). "Proc. 12th Int. Conf. on the Physics of Semiconductors" (M. H. Pilkuhn, ed.), pp. 761–765. Teubner, Stuttgart.

Wertz, J. E. and Bolton, J. R. (1972). "Electron Spin Resonance". McGraw-Hill, New York.

White, A. M. (1973). *J. Phys. C*, **6**, 1971–1974.

White, A. M., Dean, P. J., Taylor, L. L., Clarke, R. C., Ashen, D. J., and Mullin, J. B. (1972). *J. Phys. C*, **5**, 1727–1737.

White, A. M., Dean, P. J., and Day, B. (1974a). *J. Phys. C*, **7**, 1400–1411.

White, A. M., Dean, P. J., Fairhurst, K. M., Bardsley, W., and Day, B. (1974b). *J. Phys. C*, **7**, L35–L39.

Williams, F. E. (1968). *Phys. Stat. Sol.* **25**, 493–512.

Yafet, Y. (1963). *Adv. Solid State Physics*, **14** (F. Seitz and D. Turnbull, eds), 1–98. Academic Press, New York and London.

Yafet, Y. and Thomas, D. G. (1963). *Phys. Rev.* **131**, 2405–2408.

Zakharchenya, B. P. (1972). "Proc. 11th Int. Conf. on the Physics of Semiconductors", pp. 1315–1326. Elsevier, Amsterdam.

Subject Index

A

A-centre (*see also* Self-activated emission), 305, 339
Absorption, 23–29, 82, 98, 102–106, 151–154
  coefficient, 25, 151
  extinction coefficient, 103, 151
  filters, 173
  optical density, 104, 151
  spectra, 14, 19, 48, 102, 105, 244, 258
Acceptors, *see* Impurities
Activator ion, 68
ADP (ammonium dihydrogen phosphate), 164, 166
AgCl, 309
Aliphatic molecules, 117, 118
Alkali earth halides, 17
Alkali halides, 17, 22, 41, 42, 308, 356
Al_2O_3: Cr^{3+} (ruby), 9–13, 27, 34–38, 61, 62, 71, 72, 162, 229, 317
Amorphous silicon, 362
Analogue divider, 160
Anthracene, 110, 113, 115, 135, 241, 244, 249, 256, 270, 272, 280, 283, 284, 295
  magnetic field effects on, 263, 265, 277, 278, 280, 283, 284, 288, 289, 290, 291
  9-methyl, 135
  tetracene doped, 266, 284
Arc light sources, 158–160
Aromatic hydrocarbons, 95, 110–117, 241
Assaying, of organic materials, 119
Azulene, 113, 115, 117, 245

B

Band gap (*see also* Semiconductor materials), 4, 13, 76, 77–81
$Be_3Al_2(SiO_3)_6$: Cr^{3+} (emerald), 35, 36
Becquerel phosphoroscope, 222
Benz(a)-anthracene, 293
Benzene, 95, 100, 105, 110, 111, 113, 134
  isomeric state of, 127
2,3-Benzocarbazole, 291, 292
Biacetyl, 117
Biphenyl, 110, 112, 113

C

Birefringent filters, 173
Black body radiation, 189
Born–Oppenheimer approximation, 54, 241
Bound excitons, *see* Excitons
Boxcar detector, 211, 221
Brightness, *see* Luminance
Broadband emission, 60, 302, 325, 331, 335

$Ca_5(PO_4)_3F$: Sb^{3+}, Mn^{2+}, 68
Cathodoluminescence, 2, 156, 171
CdS, 304, 305, 307, 319, 322, 323, 325–331, 335–338, 360
CdS_xSe_{1-x}, 162
CdSe, 307, 308, 329
CdTe, 307, 329, 360
Charge-transfer band, 9
Charge-transfer state, 286, 294
Clar's p, α and β nomenclature, 244
Co^{2+} ion, 19, 67, 229, 312
Collisional quenching, *see* Quenching
Concentration quenching, 70, 130
Conduction band, 17, 77
Configurational coordinate diagram, 55, 56, 61, 106, 125
Coronene, 241
Cr^{3+} ion, 9, 10, 27, 34–39, 52, 61–63, 67, 72, 229
Cryostats, 229
Crystal field, 25, 29, 42–64, 313
CsBr, 309, 358

D

Davydov splitting, 122, 257
Debye equation, 122
Debye–Waller factor, 45
Decay time, *see* Fluorescence lifetime, Phosphoresence lifetime
Defects, 4, 309
Delayed fluorescence, 100, 137, 249, 269, 279
  dye-sensitized, 267, 287
  E- and P-type, 138
  magnetic field effects on, 272–278, 293

369